IET POWER AND ENERGY SERIES 68

Distribution System Analysis and Automation

Other volumes in this series:

Distribution System Analysis and Automation

Juan M. Gers, PhD

*Adjunct Professor, T&D Program,
Gonzaga University, Spokane WA, USA*

*Consultant Engineer,
GERS USA, Weston FL, USA*

The Institution of Engineering and Technology

Published by The Institution of Engineering and Technology, London, United Kingdom

The Institution of Engineering and Technology is registered as a Charity in England & Wales (no. 211014) and Scotland (no. SC038698).

The Institution of Engineering and Technology

Michael Faraday House
Six Hills Way, Stevenage
Herts, SG1 2AY,
United Kingdom

www.theiet.org

British Library Cataloguing in Publication Data
A catalogue record for this product is available from the British Library

ISBN 978-1-84919-659-8 (hardback)
ISBN 978-1-84919-660-4 (PDF)

Typeset in India by MPS Limited
Printed in the UK by CPI Group (UK) Ltd, Croydon

Contents

List of figures

List of tables

Preface

Distribution Systems started to receive great attention in the last years and in particular towards the end of the twentieth century. This was motivated by the great development achieved in software and hardware technology, communications, feeder equipment like capacitors, switches, relays, etc. At the beginning of this century, Distribution Automation was incorporated to the big umbrella of the newer topic coined as Smart Grids that encompasses many other fields not only associated to distribution but also to transmission and generation systems.

The material of this book has been prepared based on discussions on many international meetings, real applications on utility systems and personal research. The topics are treated first presenting the fundamental concepts but also the current state of the art as far as it was practical.

This book has been divided in twelve chapters. In the first chapter, an overview to Smart Grids is presented. In the second, the main functions of Distribution Automation are introduced, some of which are developed in detail in further chapters. The fundamental concepts of distribution system analysis are reviewed in Chapters Three and Four, which cover topics of basic laws of electricity, load flows for network and radial systems, and all that is related to short circuit calculation.

A general revision of concepts of reliability is done in Chapter Five which are then applied in the reconfiguration and restoration of distribution systems, treated in Chapter Six. Undoubtedly this is one of the most important topics of Distribution Automation. Chapter Seven deals with the technique of using voltage regulating devices and reactive power controls to maintain voltage levels within the accepted ranges at all points of the distribution system under all loading conditions.

Basic theory of harmonics and their impact on distribution systems are analyzed in Chapter Eight. Chapter Nine presents modern concepts of protection applied to distribution systems. Chapter 10 deals with communications including an overview to the Standard IEC 61850. The last two Chapters 11 and 12 are devoted to more general aspects not only of Distribution Automation but also to Smart Grids which are the concepts of interoperability for integrating all the components and maturity models to define road maps.

The chapters that involve more practical considerations are illustrated with examples which are thoroughly developed and some with illustrations in MATLAB®.

I certainly hope that the material covered will be useful for students and novel engineers working in Distribution Automation and Smart Grid. It can also be a good reference to engineers working already in these fields where the book can give them important references.

In the preparation of the book several colleagues assisted me with great ideas which certainly were very valuable. First of all I want to thank Prof. K. L. Lo from Strathclyde University in the UK who introduced me to the Distribution Automation and gave me good guidance in the first years working on the field. I want to thank the contributions of Jose Munoz, Luis Aragon and James Ariza on the communications chapter and the help of Andres Perez with topics on synchrophasors. Above all I would like to extend my gratitude to Carlo Viggiano who helped me with the organization of the material and was a great support in the great effort that the book demanded. I am also indebted to the books publishing team at the Institution of Engineering and Technology whose assistance was vital during the edition of the book. Finally I want to thank the patience and understanding of my wife and kids to whom I had to steal many hours of family gathering.

Juan M. Gers
Weston, October 2013

Chapter 1
Smart Grid overview

Smart Grid (SG) is a rather new concept that includes aspects of energy generation, transmission and distribution and aims for a more reliable service, higher efficiency, more security, two-way utility-user communications, and promotion of green energy, among other goals.

When the term "Smart Grid" was first used, some people associated it with remote metering, which was later called AMR (automatic meter reading). The activities of AMR were encompassed within those of a broader field that was eventually called AMI (advanced metering infrastructure). Clearly, the metering system is one of the major elements of the Smart Grid, but certainly is not the only one. Many elements are included in the overall field of Smart Grids. Those pertaining to distribution systems and in particular to the automation of distribution systems (or distribution automation) will be considered in this book.

1.1 Smart Grid for distribution systems

Distribution systems operated for many years autonomously, with only occasional manual setting changes and a rather primitive automation that is known today as local intelligence. Automation in fact was first implemented on generation and transmission systems and gradually became popular also on distribution systems.

A good example of local intelligence is that used in reclosers operating in coordination with sectionalizers. After a local fault, reclosers start a set of reclosing operations, before locking out. Other good example is the operation of capacitor bank switches which rely on local signals, such as voltage level, power factor, or even time.

Recently, in response to the growing demand to improve reliability and efficiency of the power system, more automation was introduced to distribution systems.

The Smart Grid policy requirements as outlined in the Energy Independence and Security Act (EISA) of December 2007 give a better understanding of the benefits and challenges of distribution automation for all of its stakeholders.

The ideas behind Smart Grids evolved from greater attention that distribution systems started to receive in the 1980s. Until then most of the attention was

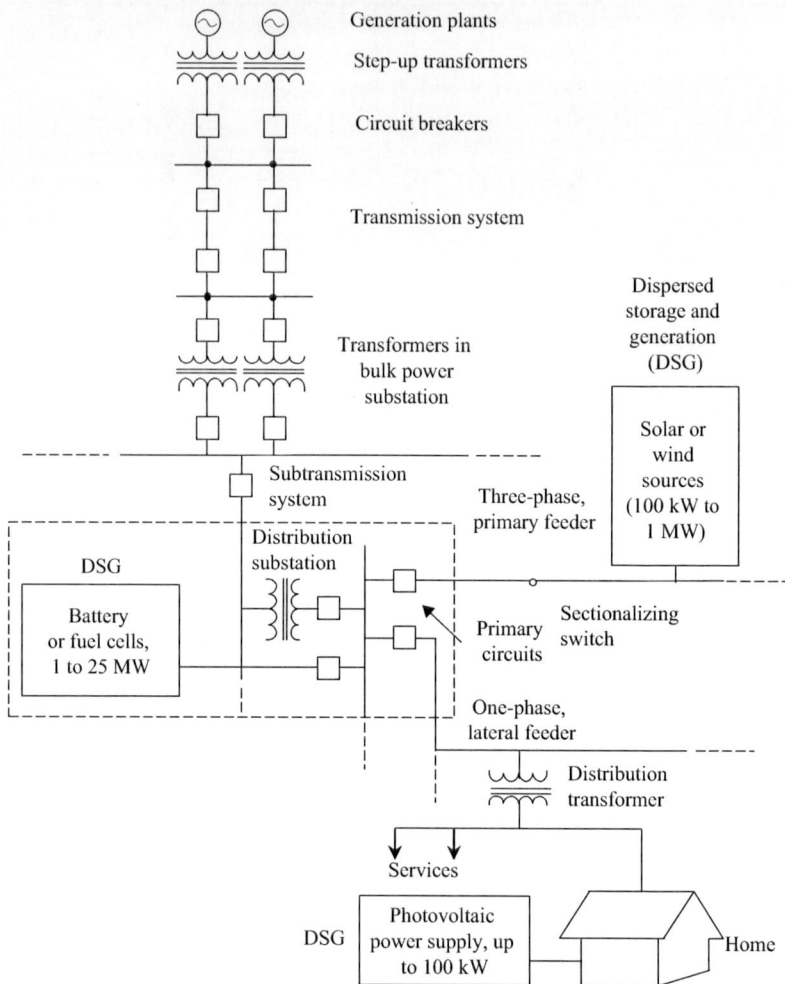

Figure 1.1   Power system as envisaged in 1982 (from "Automated power distribution," published by IEEE Spectrum, April 1982)

given to generation and transmission systems. Figure 1.1, taken from the paper "Automated power distribution" by Arthur C.M. Chen, published by IEEE spectrum in April 1982, shows a representation of a projected distribution system. The paper anticipated immediate detection and isolation of a faulted feeder and a reduction of the amount of time a line crew should spend to locate and fix it. It also referred to the increase of more distributed generation, called by then as dispersed storage and generation (DSG) systems. Finally, the paper made a point on the importance of distribution automation for maintaining a reliable supply and to reduce operating costs. It is really interesting to see how all those provisions is now a reality.

Governments and utilities funding the development and modernization of grids have defined the functions required for Smart Grids. According to the United States Department of Energy's Modern Grid Initiative Report, a modern Smart Grid must satisfy the following requirements:

- motivate consumers to actively participate in operations of the grid,
- be able to heal itself,
- resist attack,
- provide a higher quality power that will save money wasted during outages,
- accommodate all generation and storage options,
- enable electricity markets to flourish,
- run more efficiently,
- enable higher penetration of intermittent power generation sources.

In order to achieve the goals of Smart Grid mentioned above and in particular the improvement in reliability, security, and efficiency, it is essential to have a well-developed digital technology. Among the significant challenges facing development of a Smart Grid are the cost of implementing it, and the new standards that regulatory bodies have to enact. Interoperability standards certainly will allow the operation of highly interconnected systems that include distributed generation plants.

Another big difficulty that the implementation of Smart Grid and distribution automation faces is the huge variety of technologies produced by multiple vendors. Establishing a proper development path is highly recommended to any utility before embarking on a comprehensive project. Maturity models that are discussed in a latter chapter help to establish this plan.

The implementation of the new technologies of Smart Grid will bring about changes that need to be addressed. For example, the connection of small generation plants on distribution feeders will bring the possibility of having short circuit currents in two ways, and therefore the feeders should not be regarded as radial any longer. Problems with reclosing features should be closely examined to avoid out of synchronism closing. Likewise, the connection of charging stations for electrical vehicles (EV) will change the normal operation of the feeders.

Smart Grid has the great advantage of allowing two-way communication, i.e. utility-user and user-utility. This will allow a more friendly and effective relationship between the user and the utility. The latter will be able to monitor and control small appliances of each user. The user will have in turn the great advantage of getting information regarding the consumption level, new rates available, and load management schemes. This of course would require powerful communication systems that need to be flexible and reliable.

1.2   Definitions of Smart Grid

Many definitions have been written to describe Smart Grids. Every utility might have its own definition. Some possible meaningful definitions are the following.

EPRI: The Intelligent Grid

"An intelligent electric power delivery infrastructure (Intelligent Grid) that integrates advances in communications, computing, and electronics to meet society's electric service needs in the future."

Xcel Energy: The Smart Grid

"While details vary greatly, the general definition of a Smart Grid is an intelligent, auto balancing, self-monitoring power grid that accepts any source of fuel (coal, sun, wind) and transforms it into a consumer's end use (heat, light, warm water) with minimal human intervention.

It is a system that will allow society to optimize the use of renewable energy sources and minimize our collective environmental footprint.

It is a grid that has the ability to sense when a part of its system is overloaded and reroute power to reduce that overload and prevent a potential outage situation, a grid that enables real-time communication between the consumer and utility allowing us to optimize a consumer's energy usage based on environmental and/or price preferences."

DOE (Department of Energy) Definition

"An automated, widely distributed energy delivery network, the Smart Grid will be characterized by a two-way flow of electricity and information and will be capable of monitoring everything from power plants to customer preferences to individual appliances. It incorporates into the grid the benefits of distributed computing and communications to deliver real-time information and enable the near-instantaneous balance of supply and demand at the device level.

People are often confused by the terms Smart Grid and smart meters. Are they not the same thing? Not exactly. Metering is just one of hundreds of possible applications that constitute the Smart Grid; a smart meter is a good example of an enabling technology that makes it possible to extract value from two-way communication in support of distributed technologies and consumer participation."

The BC Hydro Definition of Smart Grid

"Smart Grid refers to a modern, intelligent electricity transmission and distribution system that incorporates traditional and advanced power engineering to enhance grid performance and support a wide array of functionality for customers and the economy. In other words: modernization and automation of the current power delivery system."

In summary, Smart Grid refers to a sustainable modernization of the electricity grid, integrating information and communication technologies to intelligently manage and operate generation, transmission, distribution, consumption, or even the electric energy market. The components are shown in Figure 1.2.

The Smart Grid concept comprises many or almost all of the elements of the utility and the relationships among the elements that constitute them. Figure 1.3 includes some of these elements, like smart meters, generation, transmission, substation, and feeders.

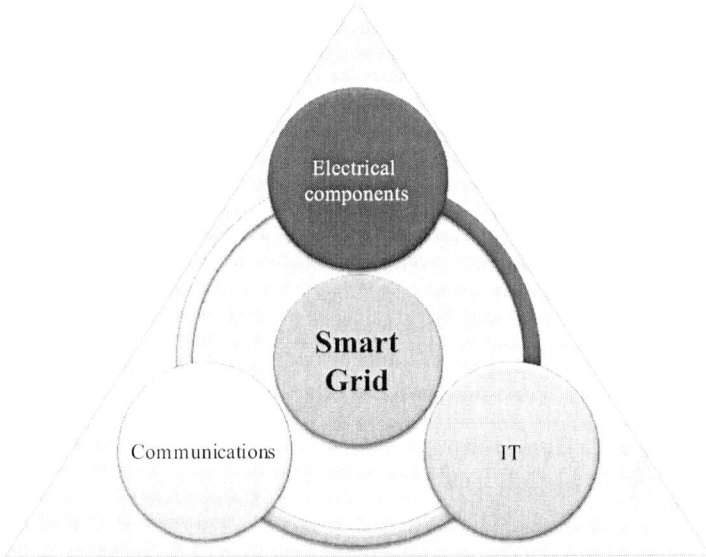

Figure 1.2 Smart Grid concept

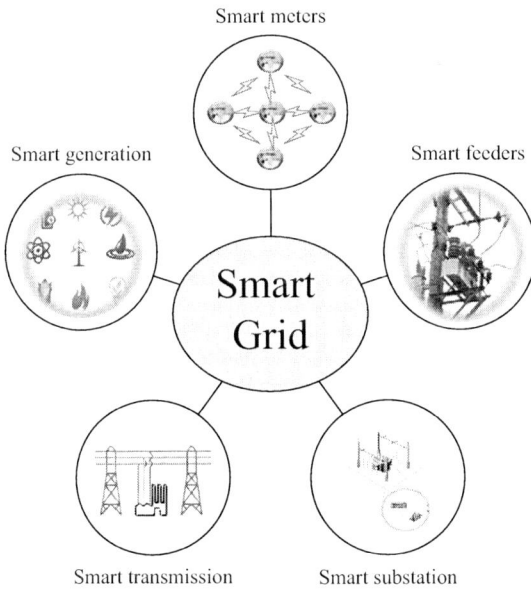

Figure 1.3 Smart Grid components

1.3   Benefits of the Smart Grid on distribution systems

The benefits of implementing a Smart Grid are many. They can be summarized in the following categories.

1.3.1   Enhancing reliability

The Smart Grid dramatically reduces the cost of power disturbances. This can be achieved by means of system reconfiguration using switches placed along the feeders. Communications and control technologies greatly help to isolate faults and allow more rapid restoration of service.

1.3.2   Improving system efficiency

The reduction of losses in electrical systems, both technical and non-technical, is a purpose of every utility in the world. It not only reduces the demanded power but also contributes to the environment. The reduction of system losses results in capital deferral which gives an attractive payback of the investments. Power capacitors, voltage regulators, and proper design criteria are required to achieve this.

1.3.3   Distributed energy resources

Construction of generation plants at the user level is every day more frequent. These sources are referred to as distributed generation or distributed energy resources and are getting much attention from government authorities and environmental institutions as they alleviate the pollution levels that some plants have, in particular those burning coal and oil. Distributed energy resources also contribute to better operating conditions of distribution systems as they are sources connected directly to the users' loads, increasing control of the voltage.

1.3.4   Optimizing asset utilization and efficient operation

Real-time data makes it possible to more effectively utilize assets during both normal and adverse conditions and to reduce the costs of outages. This results in a longer service life of the assets.

1.4   Quality indices

Quality indices, also called reliability indices, measure the performance of a power system. The most important benefit of Smart Grids is that they improve reliability performance of power systems. Given the importance of these indices to enhance a Smart Grid, a proper consideration to their meaning and application is explained in this section.

Several key definitions relating to distribution reliability include:

- *Fault*: an abnormal operating condition of an electrical system which normally develops a short circuit. It can be caused by natural events, rough weather conditions, animal presence, equipment failure, and even vandalism. Faults can be categorized as self-clearing, temporary, and permanent. A self-clearing fault will extinguish itself without any external intervention. A temporary fault will clear if de-energized and then re-energized. A permanent fault lasts until repaired by human intervention.

- *Contingency*: an unexpected event such as a fault or an open circuit. Another term for a contingency is an unscheduled event.
- *Outage*: An outage occurs when a piece of equipment is de-energized. Outages can be either scheduled or unscheduled. Scheduled outages are known in advance (e.g., outages for periodic maintenance). Unscheduled outages result from contingencies. Different utilities have different criteria to define an outage. Some utilities consider outages those interruptions that exceed 1 min. Others use 2 min and other up to 5 min.
- *Open circuit*: a point in a circuit that interrupts load current without causing fault current to flow. An example of an open circuit is the false tripping of a circuit breaker.
- *Momentary interruption*: a momentary interruption occurs when a customer is de-energized for a very short time of period, usually less than a minute. Most momentary interruptions result from reclosing or automated switching. Multiple reclosing operations result in multiple momentary interruptions.
- *Sustained interruption*: a sustained interruption occurs when a customer is de-energized for normally for more than one minute. Most sustained interruptions result from open circuits and faults.

Poor reliability on the part of the electrical utility is penalized based on quantification by reliability indices. Some utilities also pay bonuses to utility personnel, based on outstanding performance. Commercial and industrial customers inquire about reliability indices when locating a new facility. Most regulatory bodies have established targets for reliability indices. If utilities do not fulfill them (have figures higher than those defined), they can be penalized. The most important indices to measure the reliability performance are the following:

1.4.1 *System Average Interruption Duration Index (SAIDI)*

SAIDI is defined as the average interruption duration for customers served specific time period, which normally is a year. It is calculated by summing the customer minutes off for each interruption during a specified time period and dividing the sum by the average number of customers served during the period. The unit is minutes.

The index enables the utility to report the time (normally in minutes) customers would have been out of service if all customers were out at one time. The benchmarking survey SAIDI average for the United States is around 90 min.

$$\text{SAIDI} = \frac{\sum \text{Customer Interruption Durations}}{\text{Total Number of Customers Served}} \text{ min} \qquad (1.1)$$

1.4.2   *System Average Interruption Frequency Index (SAIFI)*

SAIFI is defined as the average number of times that a customer is interrupted during a specific time period, which also is a year. It is calculated by dividing the total number of customers interrupted in that time period by the average number of customers served. (This means outages per customer, not the number of interruptions.)

The resulting unit is "interruptions per customer." The benchmarking survey SAIFI average for the United States is around 1.2 interruptions per customer per year.

$$\text{SAIFI} = \frac{\text{Total Number of Customer Interruptions}}{\text{Total Number of Customers Served}} \tag{1.2}$$

1.4.3   *Customer Average Interruption Duration Index (CAIDI)*

CAIDI is the ratio of the SAIDI over the SAIFI and is defined with the following expression:

$$\text{CAIDI} = \frac{\sum \text{Customer Interruption Durations}}{\text{Total Number of Customer Interruptions}} (\text{min}) \tag{1.3}$$

Example 1.1: Table 1.1 shows an excerpt from one utility's customer information system (CIS) database for feeder 25, which serves 1500 customers with a total load of 3 MW. In this example, Feeder 25 constitutes the "system" for which the indices are calculated. More typically the "system" combines all circuits together in a region or for a whole company.

Table 1.1 Example of outage data

Date	Time	Time on	Total time (min)	Circuit	Event code	Number of customers	Load (kVA)	Interruption type
2/15	14:10:17	14:25:39	15.37	25	51	250	431	S
4/20	17:20:39	17:38:41	18.03	25	306	368	800	S
5/1	06:33:36	07:14:49	41.22	25	468	23	150	S
6/2	23:18:10	23:18:57	0.78	25	522	590	1200	M
6/8	02:39:52	03:55:34	76.63	25	634	87	200	S
9/29	09:29:05	09:30:02	0.95	25	811	1500	3000	M
11/14	17:15:49	17:16:18	0.48	25	963	700	1500	M
12/4	12:16:32	13:01:44	45.2	25	1021	1000	1800	S

S indicates a sustained interruption; M, a momentary interruption.

Applying the previous equations it is possible to find these quality indices.

$$\text{SAIFI} = \frac{250 + 368 + 23 + 87 + 1000}{1500} = 1.152$$

$$\text{SAIDI} = \frac{(15.37 \times 250) + (18.03 \times 368) + (41.22 \times 23) + (76.63 \times 87) + (45.2 \times 1000)}{1500}$$

$$= 42.2 \text{ min}$$

$$\text{CAIDI} = \frac{\text{SAIDI}}{\text{SAIFI}} = \frac{42.2}{1.152} = 36.63 \text{ min}$$

1.4.4   Momentary Average Interruption Frequency Index (MAIFI) and Momentary Average Interruption Event Frequency Index (MAIFI$_E$)

The MAIFI as defined by IEEE Standard 1366-2003 indicates the average frequency of momentary interruptions. Mathematically, this is given in the following equation:

$$\text{MAIFI} = \frac{\sum \text{Total Number of Customer Momentary Interruptions}}{\text{Total Number of Customers Served}} \quad (1.4)$$

The MAIFI$_E$, as defined by the same IEEE Standard, indicates the average frequency of momentary interruption events, not including the events immediately preceding a lockout. Mathematically, this is given by the following equation:

$$\text{MAIFI}_E = \frac{\sum \text{Total Number of Customer Momentary Interruption Events}}{\text{Total Number of Customers Served}}$$

$$(1.5)$$

To calculate these indices, the following equations can be used:

$$\text{MAIFI} = \frac{\sum IM_i N_{\text{mi}}}{N_T} \quad (1.6)$$

$$\text{MAIFI}_E = \frac{\sum IM_E N_{\text{mi}}}{N_T} \quad (1.7)$$

where:

IM_i Number of momentary interruptions
IM_E Number of momentary interruption events
N_{mi} Number of interrupted customers for each momentary interruption event during the reporting period
N_T Total number of customers served for the areas

Example 1.2: To better illustrate the concepts of momentary interruptions, sustained interruptions, and the associated indices, consider Figure 1.4. The figure illustrates a circuit composed of a circuit breaker (B), a recloser (R), and a sectionalizer (S).

For this scenario, 1000 customers would experience a momentary interruption and 500 customers would experience a sustained interruption. Calculations for SAIFI, MAIFI, and MAIFI$_E$ on a feeder basis are shown in the following equations. Notice that the numerator of MAIFI is multiplied by 2 because the recloser took two shots. However, MAIFI$_E$ is multiplied by 1 because it only recognizes that a series of momentary events occurred.

$$\text{SAIFI} = \frac{500}{3000} = 0.167$$

$$\text{MAIFI} = \frac{2 \times 1500}{3000} = 1$$

$$\text{MAIFI}_E = \frac{1 \times 1500}{3000} = 0.5$$

Table 1.2, taken from the paper "Electricity Reliability: Problems, Progress and Policy Solutions" by Galvin Electricity Initiative, February 2011, presents a good comparison of reliability indices from European countries.

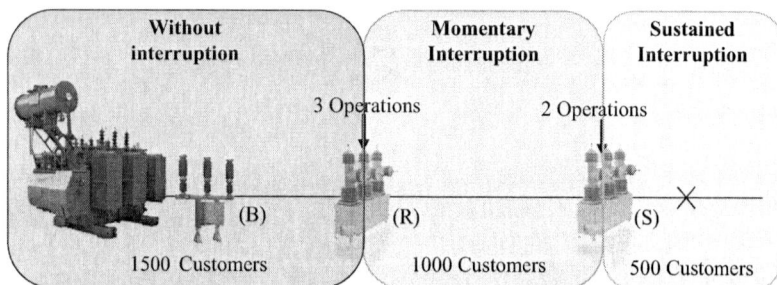

Figure 1.4   Representation of events used in calculating indices

Table 1.2   International comparison of quality indices in 2007

Country	SAIDI	SAIFI
United States	240	1.5
Austria	72	0.9
Denmark	24	0.5
France	62	1.0
Germany	23	0.5
Italy	58	2.2
Netherlands	33	0.3
Spain	104	2.2
UK	90	0.8

Proposed exercise

1. Table 1.3 shows database for feeder 43, which serves 3500 customers with a total load of 5 MW. Complete the table and calculate the quality indices SAIFI, SAIDI, and CAIDI.

Table 1.3   Database for exercise 1.1

Date	Time	Time on	Total time (min)	Number of customers	Load (kVA)	Interruption type
1/23	13:24:28	14:45:43		343	567	
2/28	12:15:46	12:23:58		576	776	
3/11	04:06:09	04:07:02		102	120	
3/11	18:23:24	19:01:01		102	120	
3/18	01:03:43	01:53:24		432	683	
4/10	12:13:56	12:13:59		1500	2480	
4/30	08:20:18	08:20:56		700	1315	
5/02	05:40:11	05:44:23		1400	2269	
6/29	03:56:32	04:38:12		1973	2984	
8/12	17:48:39	17:59:04		2534	3542	
8/13	13:08:02	14:09:01		28	35	
8/13	13:25:36	13:56:26		28	35	
9/22	23:49:27	23:54:24		1769	2647	
10/16	20:09:37	20:44:55		2046	3018	
12/05	19:31:15	19:31:46		378	588	

Chapter 2
Distribution automation functions

Distribution automation started in the 1970s. It allows utilities to implement modern techniques in order to improve the reliability, efficiency, and quality of electric service. Distribution automation is also referred to as feeder automation. It has been defined by the IEEE as follows: "Distribution Automation is a system that enables an electric utility to remotely monitor, coordinate and operate distribution components in a real-time mode from remote locations."

Distribution automation's main applications are categorized into four groups: the first group is fault location and automatic sectionalizing/service restoration, which primarily depends on a switchgear as illustrated in Figure 2.1. The second is Volt/VAR control and optimization, which mainly employs voltage regulators and capacitors. The third is the integration of distributed generation, which requires appropriate protection equipment and a robust SCADA system among other components. Finally, the last group is associated to the advanced asset management, where sensors and metering devices are necessary. Figure 2.2 shows the applications and the benefits rendered by the implementation of the four groups.

Distribution automation, commonly known as DA, has evolved into advanced distribution automation, known as ADA, which incorporates advanced communication schemes, new computer technology, state-of-the-art equipment technologies and high-speed power electronic devices.

2.1 Electrical system automation

Automation of power systems can be handled from control centers connected to generation and transmission systems on one hand and to distribution systems on the other. It could be said that they meet at the HV/LV substations which also feature their own automation. This is shown in Figure 2.3.

Control centers for generation/transmission systems and for distribution systems are handled by using software applications customized for each application. They are referred to as energy management systems (EMS) and distribution management systems (DMS), respectively. These are illustrated in Figure 2.4.

Figure 2.1   Typical installation of a switch on an overhead distribution feeder

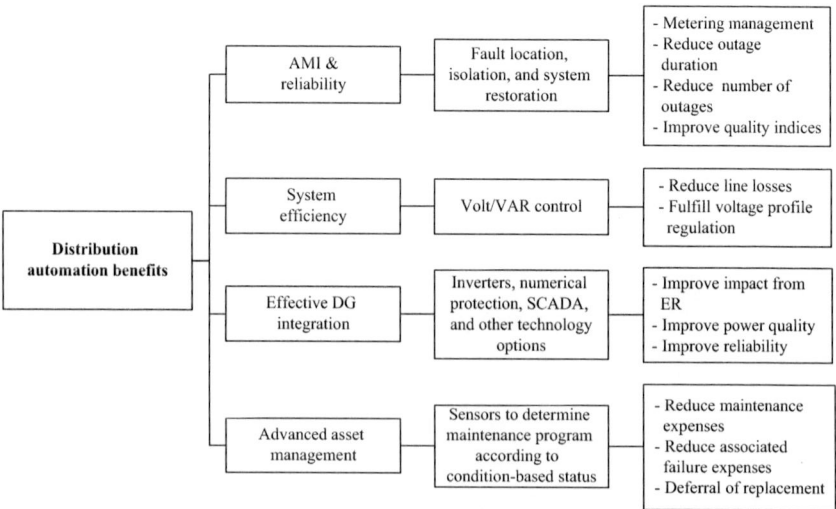

Figure 2.2   Main benefits of distribution automation

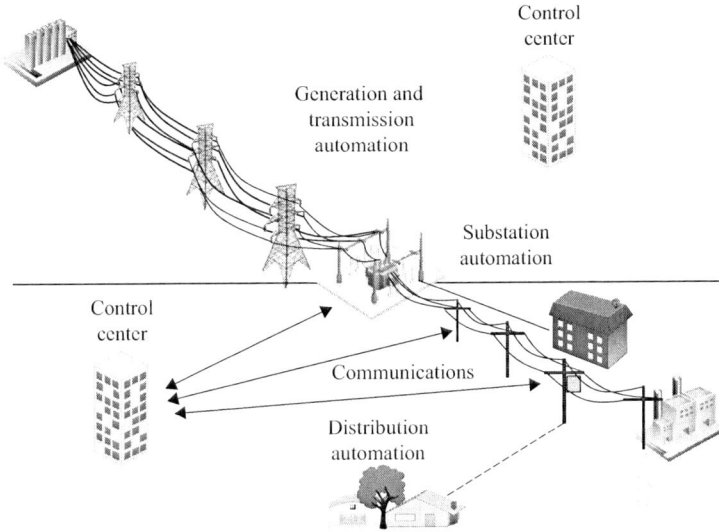

Figure 2.3   Power system automation components

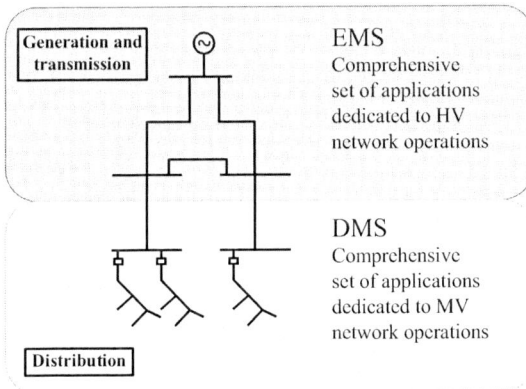

Figure 2.4   Network management EMS/DMS

2.2   EMS functional scope

The EMS is based on two main subsystems, supervisory control and data acquisition (SCADA) and network analysis. Each one incorporates several applications as follows:

The SCADA subsystem incorporates applications like multiple remote terminal unit (RTU) protocols in a non-proprietary environment, load shed, sequence-switching management, and disturbance storage and analysis. The network analysis

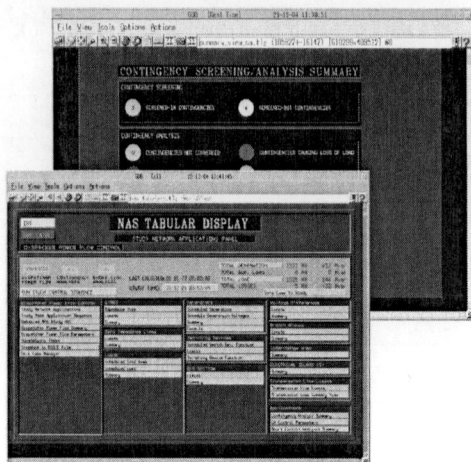

Figure 2.5   Screen shots of a typical EMS

applications offer the possibility of obtaining power flow, state estimation, contingency analysis, short circuit, security enhancement, optimal power flow, and Volt/VAR dispatch information. Figure 2.5 presents a typical screen shot of an EMS.

2.3   DMS functional scope

The SCADA subsystem offers functions like feeder topology, coloring and circuit tracing, and mesh detection. The network analysis offers functions like real-time and power flow (balanced or unbalanced), short circuit, Volt/VAR control and optimal VVC, automatic online feeder reconfiguration, fault location and automatic service restoration, protection system, substation automation, customer load management, advanced metering infrastructure/automated meter reading, demand side management, and load management. Figure 2.6 presents a typical screen shot of a DMS.

2.4   Functionality of DMS

The functionality of DMS can be divided into steady-state performance improvement and dynamic performance improvement.

2.4.1   Steady-state performance improvement

In this section the main functions associated with the analysis of the steady-state performance of the system are considered. These functions are characterized by fulfilling targets previously defined in planning studies.

2.4.1.1   Volt/VAR control

Volt/VAR control deals mainly with the detection and prediction of voltage violations which require close control of the relationship between the voltage and the reactive power equipment. The equipment involved includes mainly

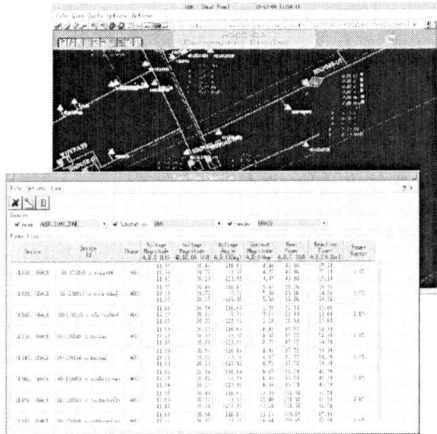

Figure 2.6   Screen shots of a typical DMS

capacitor, static VAR control (SVC) devices, load tap changers and voltage regulator controls.

2.4.1.2   Feeder reconfiguration

Feeder reconfiguration refers to the operation of switches, breakers, or reclosers to reconfigure the topology of feeders and improve the operating condition of the system. It is done under normal conditions in order to reduce losses and increase reliability. It also helps to minimize voltage drop on the feeders. Feeder reconfiguration represents great benefits. However, it is not always possible to achieve all the objectives simultaneously.

2.4.1.3   Demand side management (DSM)

DSM refers to a method for controlling the load at the user's premises, based on an agreement previously established with the utility.

2.4.1.4   Advanced metering infrastructure (AMI)/automatic meter reading (AMR)

AMI/AMR plays an important role as it aids utilities in establishing competitive strategies that conduct load profile gathering in order to characterize the value of individual customers to the utility. It identifies the major customers (aggregated load) and offers load information to customers as a special service.

One of the technologies used by AMI is power line communication – PLC – that uses the wires of the distribution system including the customer's incoming lines. The signal rides on the fundamental power line frequency and modulation occurs at the zero-cross. It does not require repeaters or line conditioning equipment, resulting in lower installation costs. This technology does not affect power quality. Figure 2.7 illustrates a typical PLC system that uses the technology known as TWACS (two-way automatic communication system). This system uses modulation for the outbound signals and modulation for inbound signals as shown in Figure 2.8.

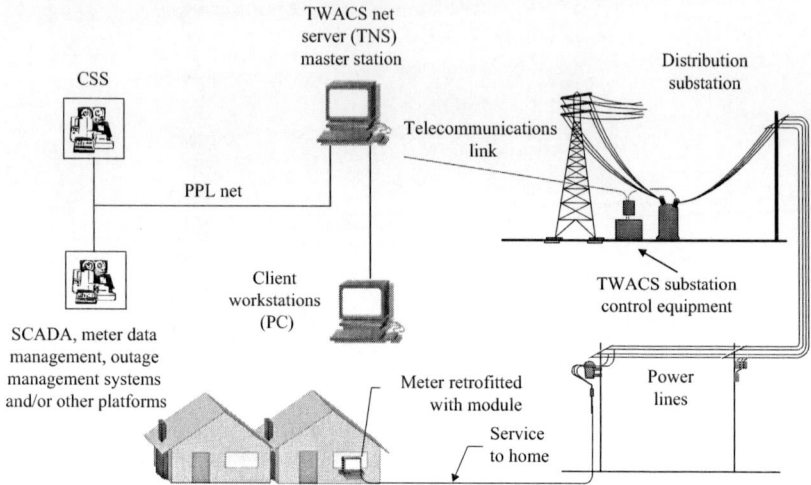

Figure 2.7   Illustration of power line communication (PLC)

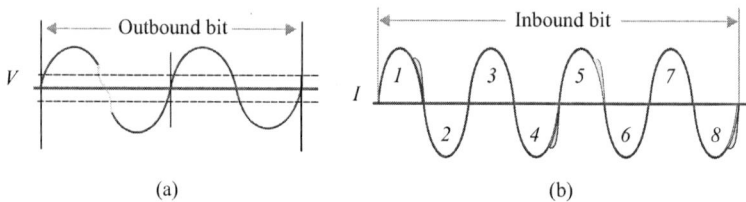

Figure 2.8   (a) Modulation for the outbound signal; (b) modulation for the inbound signal

2.4.2   Dynamic performance improvement

In this section the main functions associated with the analysis of the dynamic performance of the system are considered. These functions are characterized by actions to be taken during faults, unpredicted events, and emergency conditions.

2.4.2.1   Fault location, isolation, and service restoration (FLISR)

The use of automated feeder switching aids in detecting feeder faults, determining the fault location (between two or more switches), isolating the faulted section of the feeder, and providing the possibility of restoring service to "healthy" portions of the feeder.

Service restoration allows to find alternate options to quickly restore power to healthy parts of the system making sure that voltage levels are within accepted ranges and that any overload is avoided. Cold load pickup has to be considered for long outages in order to avoid tripping when re-energizing the feeders. Customer prioritization is carefully considered.

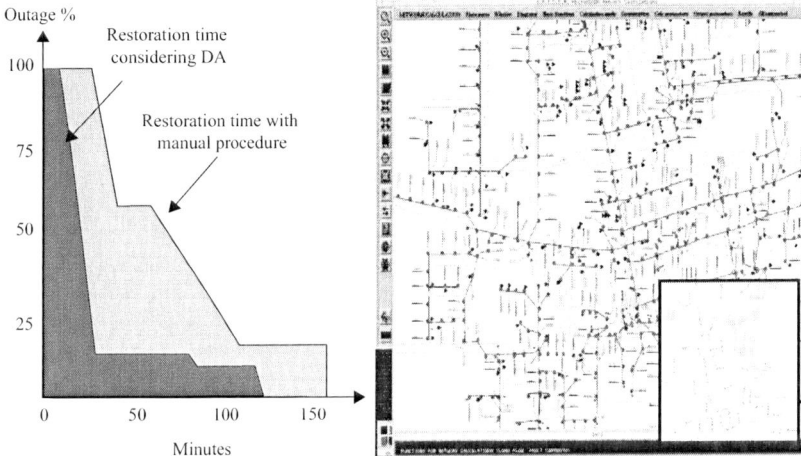

Figure 2.9   Comparison of restoration time with and without DA

Automating this process renders immense benefits as the outage time is greatly reduced, as shown in Figure 2.9.

2.4.2.2   Trouble call system

When a fault occurs, protective relays and fault locators should raise the corresponding alarms that outages have occurred. However, the report of users affected is very important not only to reaffirm the fault occurrence but also to help locating it more quickly.

Utilities now have modern facilities where the calls from users are readily received, classified, and matched with the database of the billing system to identify other users who may be associated with the problem. Thus, they can achieve solution schemes that identify and involve all the customers affected by the event. These facilities are trouble call systems that are manned but assisted with state-of-the-art communication and software capabilities. A typical process of a TCS is illustrated in Figure 2.10, which requires a close relationship with the customer information system (CIS).

2.4.2.3   Alarm triggering

Alarms are triggered by the operation of the protection relays. They send coded information to distribution control centers so that operators are aware of what is happening on the network. Other alarms indicate the state of the power system, for example, voltages at various locations and load flows on the more important circuits. These alarms provide one of the main sources of information flowing in real time into a distribution control center and are normally channeled to one printer in the control room where a hard copy can be produced. The alarm streams are also channeled to the operator's or control engineer's console where they can be displayed on computer screens.

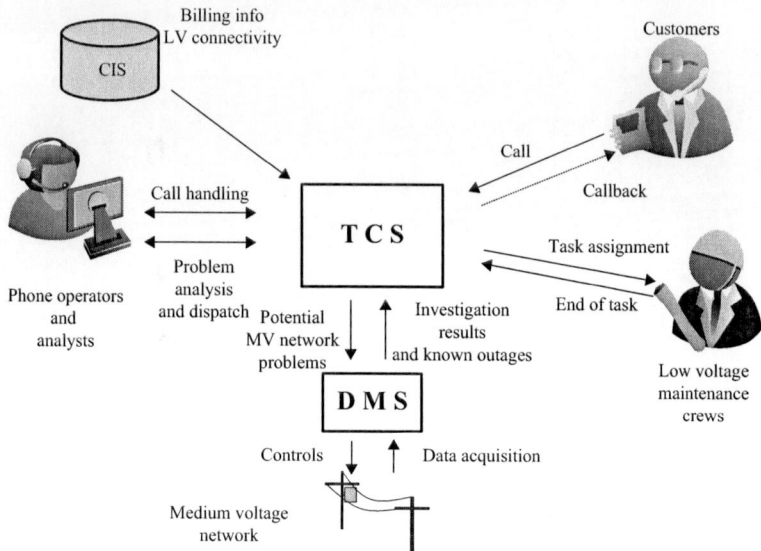

Figure 2.10   Trouble call system

A third avenue for alarm streams is storage in a data logger where the principal function is to retain a history of the alarm streams. This can be used for post-system fault analysis if it is needed later.

It is worth nothing that, with this arrangement, the alarms are not processed. Several events may occur simultaneously, or at a close time proximity with respect to each other, and each incident may trigger many alarms resulting in a large number of alarms flowing into the control center in close succession. The operator would then have to use his judgment to determine what has happened to the system. Subsequent telephone calls from customers may also help to determine the exact location of these incidents such as a blackout in a certain district. The aim of an alarm processor is to help the operator to arrive at a sensible conclusion speedily and to discard redundant information in the alarm streams.

2.4.2.4   Work orders

In order to expedite the coordination of crews responsible for the maintenance or feeder fixing, a proper work order system is essential. After a fault is reported, the system should be capable of automatically locating the most convenient crew and traveling means, and to determine the essential elements required to reestablish service as quickly as possible. This requires a proper coordination with the DMS, the resource scheduler, and the operation manager of the utility, as depicted in Figure 2.11.

2.5   Geographic information system

A geographical information system (GIS) can integrate, store, edit, and display geographically referenced information. A GIS may be linked to the online environment of a DMS. An example is shown in Figure 2.12.

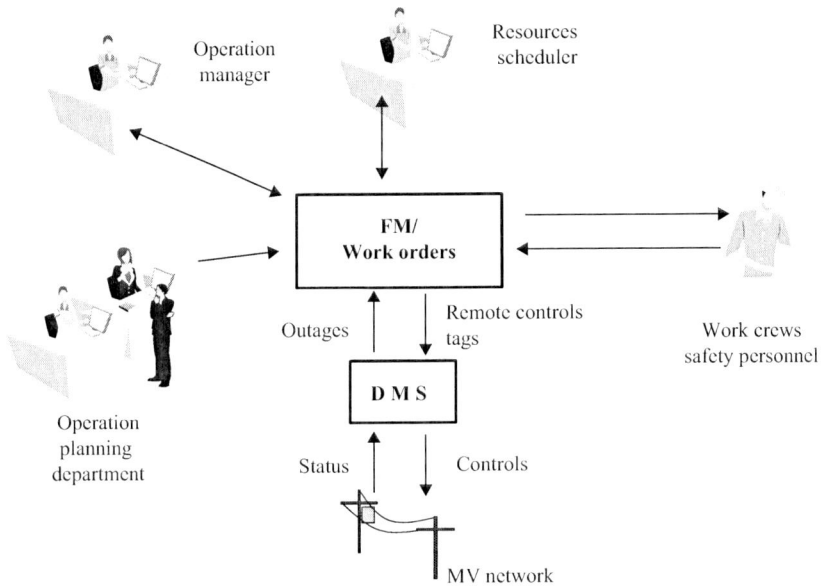

Figure 2.11   Work order illustration

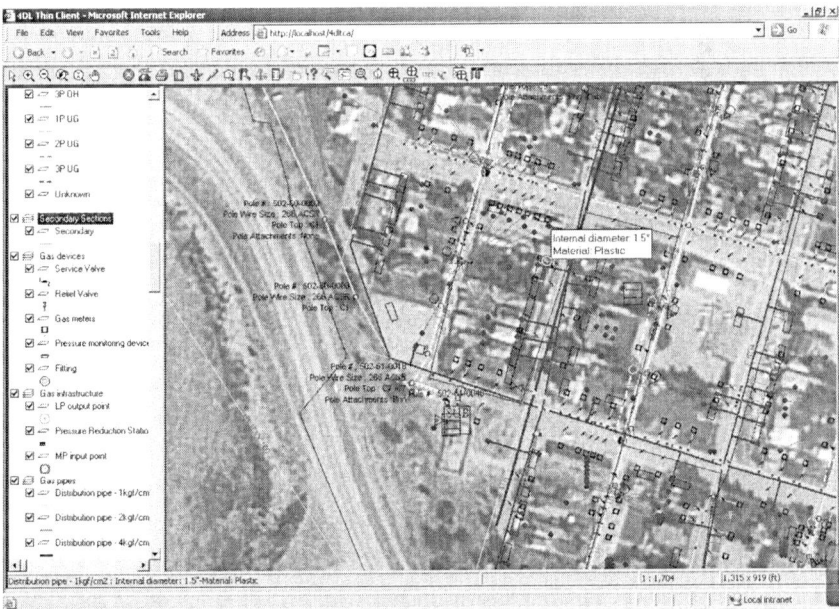

Figure 2.12   Example of a GIS

2.5.1   AM/FM functions

Automated mapping/facility management (AM/FM) is a specific application of the GIS aimed at providing database query and graphical user interface (GUI) for efficient management of distribution system devices. It involves activities related to geographical and electrical maps.

2.5.2   Database management

GIS systems require a computerized mapping system with the capability of storing and displaying the geographic locations of customers, feeder sections, substations, and/or pieces of equipment.

2.6   Communication options

DMS may employ a variety of communication mediums such as cooper wires, fiber optics, VHF and UHF radio, satellites, microwaves, cellular communications, Wi-Fi, and WiMAX.

Rapid advances in wireless, cellular, and satellite systems will soon provide cost-effective solutions for enterprise connectivity. Figure 2.13 illustrates different types of communication.

2.7   Supervisory control and data acquisition

SCADA is designed to achieve control and monitor processes remotely. This system is based on remote data acquisition.

SCADA applications are globally accepted as a real-time power system monitoring and control tool, especially in generation, transmission and distribution

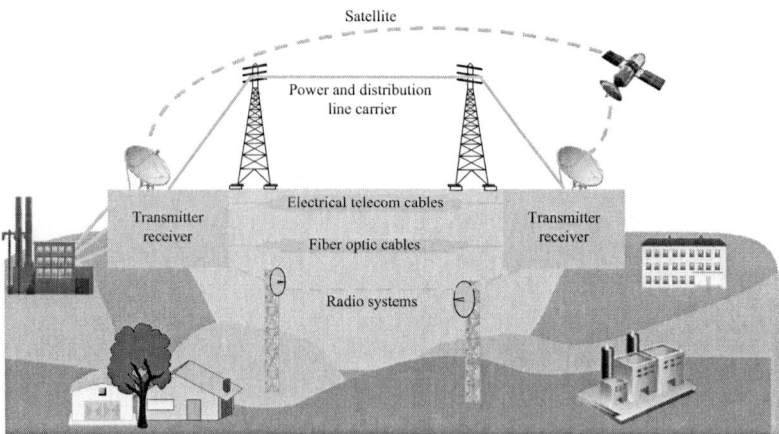

Figure 2.13   Typical communication methods

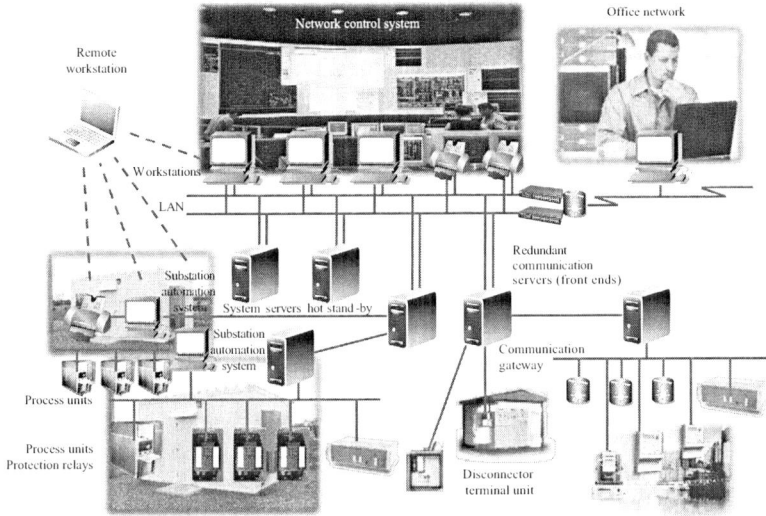

Figure 2.14   SCADA illustration

systems. The RTU collects telemetry analog data and the status of devices in the substations, and communicates control commands. It is installed in a central location such as the control center of the utility with GUI, engineering applications managing history software, and other components.

SCADA describes a technique whereby computers are used to collect real-time data from plant machinery and/or to control plant machinery in real time. Typically this is done through the use of programmable logic controllers (PLCs). Figure 2.14 presents a common SCADA arrangement.

Recent trends in SCADA systems provide greater situational awareness through improved GUI, data presentation and information, intelligent alarm processing, improved integration with other engineering processes and business systems, and enhanced security functions.

2.7.1   SCADA functions

DMS/EMS integration with SCADA is a growing trend. The main functions of SCADA systems are:

* supervisory control,
* data acquisition and processing,
* sequence of events (SOE) registry,
* misoperation revision,
* tagging,
* alarm processing,
* historical information system.

2.7.1.1   Supervisory control

SCADA controls the operation equipment (switches, breakers) in locations remote from the control center. This includes operation, commands, open/close commands, and set point control.

The control should include a selection sequence and verification before operation (select-check-back-before-operate – SCBO). In this case the user selects the equipment to be controlled using SCADA and the operation to be conducted. The operator observes the result of the operation in SCADA. The option to cancel the command should be available. The user is able to manually update the state of all control points that are not telemetry-related.

Commands executed by the user are registered at the time they occur and they are stored in a historic file for future use, and in the historical information system (HIS). Each event is identified by date, time, device name, device ID, and operation time. All failed remote command actions are identified as alarms.

The selected Intelligent Electronic Devices (IEDs) use a process that meets the supervisory functional control requirements. The following factors should be taken into consideration:

- The time delay after the command has been sent or received from a communication point.
- The time delay after a command has been executed before another command can be initiated.
- The backup for multiple control points per command.

Figure 2.15 is a good illustration of a SCADA operating on different pieces of equipment in a distribution system, following orders received from application packages.

Distribution system

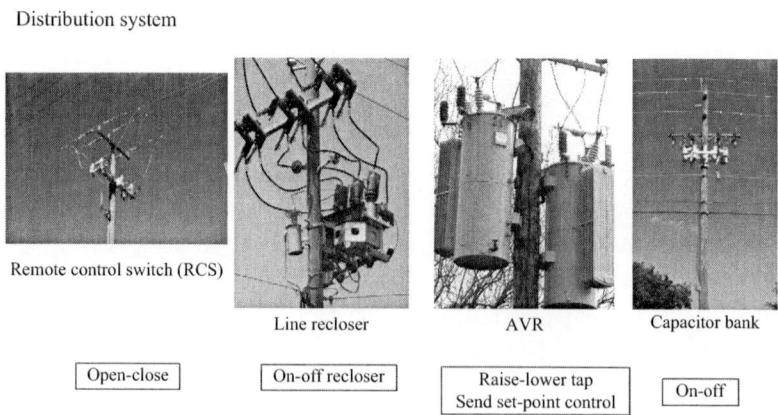

Remote control switch (RCS)

Line recloser   AVR   Capacitor bank

| Open-close | On-off recloser | Raise-lower tap Send set-point control | On-off |

Figure 2.15   SCADA functions: supervisory control

2.7.1.2   Data acquisition and processing

Figure 2.16 depicts a display of a SCADA system showing data acquisition in a distribution utility.

SCADA systems require a reliable connection with databases that have information about the system and the users. Connection with packages having the "intelligence" or the software is essential to performing the appropriate operations. Figure 2.17 shows the interaction of a SCADA with other information packages of the system.

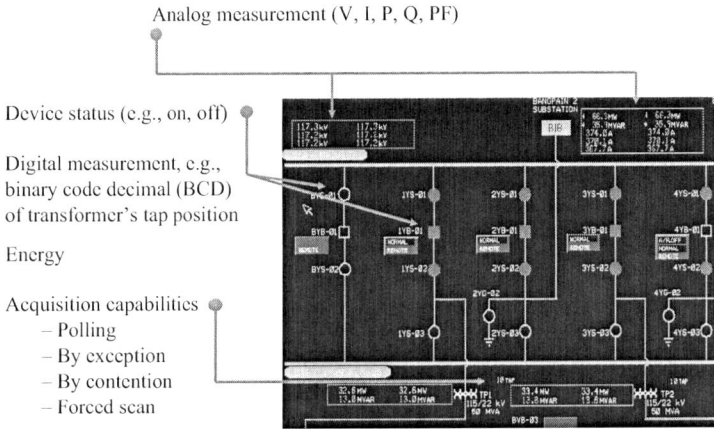

Figure 2.16   SCADA functions: data acquisition

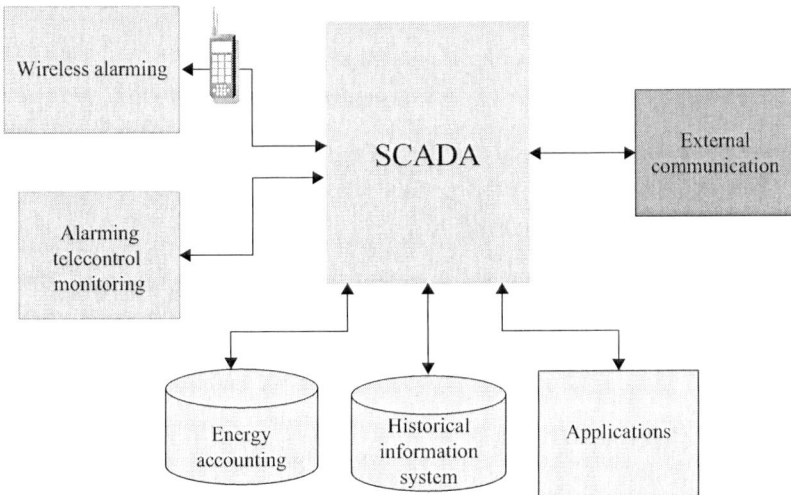

Figure 2.17   Relationship of SCADA with databases

The data acquisition function is employed to supervise the state of the equipment, and measure different equipment parameters and external SCADA systems. These states and measurement data are viewable to the operators and also available for other functions.

The SCADA/DMS/EMS/GMS system master station obtains data from the data concentrators. Data acquisition is conducted through a private WAN dedicated network for IEDs and SCADA/DMS/EMS communications and a wireless data network with IP technology dedicated to IEDs.

These two networks should support TCP/IP or UDP/IP protocols. IEC 60870-5-104 or IEC 61850 protocols are normally used for all substations. The substation data concentrator in addition to obtaining data from auxiliary services should be designed to question the distribution network IEDs. The communication protocol is typically DNP 3.0 or IEC 61850.

The data obtained from each IED includes:

- states (operation state, breaker position, switch position, recloser, alarms, etc.),
- SOE data (high precision event data timestamp),
- metering (currents, voltages, MW, MVAR, frequency, transformer MW and MVAR, transformer tap position, etc.),
- alarms (overvoltage, undervoltage, high frequency, low frequency, etc.),
- input change processing time.

2.7.1.3 Sequence of events (SOE) registry

The distribution system SCADA has the ability to store SOE data for specific events. The SOE information includes relay operation analysis and alarms as part of the equipment operation analysis. The SOE events messaged are similar to the alarm messages, but they will be accurate to the millisecond, for example 17:31:19.509.

The SOE data obtained are stored in chronological order in the Historical Information System (HIS) database and reported individually.

2.7.1.4 Misoperation revision

The misoperation revision consists of the creation and continuous update of the event file. Every 10 sec, a picture of the operator-selected points or the real-time database is stored in the event file which consists of three files: the pre-event, event, and post-event period.

The file size is determined by the time period covered by the recording, the amount of data to be stored, and the available space on the disk. The recommended time frame is 10 min for the pre-event, 15 for the event period, and 15 min for the post-event period.

2.7.1.5 Tagging

The control center SCADA labels each specific event, monitored or not, on a graph. Each label represents the attention of a device using a graphical symbol indicating what the remote control is for such device. Different labels are normally used to classify devices in a priority order that refers to the open and close controls.

2.7.1.6   Alarm processing

The alarms detected by the data acquisition functions are managed in such a way that the predetermined alarm conditions are reported clearly and promptly only for the work station that needs the information.

The alarm management is supported by priority levels. When an alarm is presented, its responsibility area and priority will be displayed.

The alarms and events are saved daily and stored in the HIS for later reference. The alarms are presented in a single line diagram with screen visualization using symbols or color changes.

2.7.1.7   Historical information system

Traditional methods to collect historical data included chart recorders, operator logs, and alarm printers. Trends and historical databases can be constructed with the help of SCADA data. Reports, for example, can include compliance reports, operation reports, asset management, and work order generation.

2.7.2   System architecture

The system architecture has a layout and several components which are presented in the following paragraphs.

2.7.2.1   Master station (control center)

The master station has a distributed software and hardware architecture. Modern master station SCADA systems employ open architecture characteristics allowing interconnection with other systems. Open system standards allow interaction with other products from suppliers.

To ensure openness, the system must meet international standards or industry standards like Microsoft Windows and computer applications related products, IEC 60870-6 (TASE.2) for communications with other control centers, and IEEE 1379 or IEC 61850 for the communications with RTU. IEEE has launched the IEEE Standard 2030 providing alternative approaches and best practices for achieving Smart Grid interoperability.

The main SCADA system elements are shown in Figure 2.18 and should include the following components:

2.7.2.2   Human–machine interface (HMI)

This interface consists of the mimic board and the multi-video display unit. The board map shows a simplified scheme of the power system. This board presents the most detailed representation of the electric system. It is organized to geographically represent the power system, which is useful for observing large disturbances that cover an extended geographical area.

The multi-video display interface is employed to observe the condition of the power system devices in more detail. In modern SCADA systems, multiple screen workstations offer operators easy access to a wide variety of application functions and controls. These workstations combine graphical capabilities with multiple window functions like zoom, detachable menus, and drag-and-drop operations.

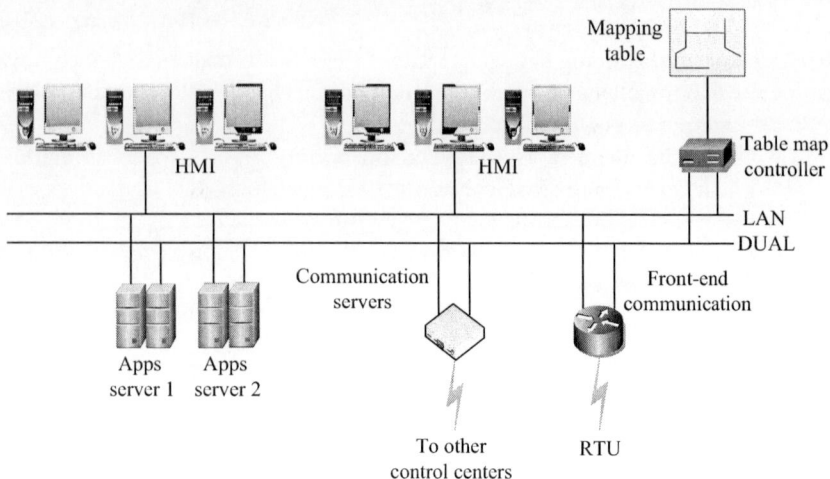

Figure 2.18   Control center general scheme

2.7.2.3   Application servers

- Main SCADA subsystem: This server is mainly used for data processing functions and real-time process control.
- Database subsystem: This server backs up the historic database and other databases.
- Configuration and operation: This server is used for control, operation, and maintenance of the SCADA system.
- Front-end communication: This system is used for data acquisition from the RTU and field equipment. Besides RTU data acquisition functions, it also includes protocol conversion, security check, digital and analog data temporary storage, and analog values and digital state changes.
- External communications server: This server provides interchange with other control centers. A standard protocol such as IEC 60870 (TASE.2) must be used for real-time interchange and data storage.

2.7.2.4   Remote stations

Remote stations are measuring devices or IEDs connected to the plant that are monitored and controlled using an interface.

The remote station could be an RTU or PLC. The RTU is mainly used when communications are difficult. The RTU has digital and analog inputs, and outputs to an LED indicator (selected by channel), with optical insulation for protection against overvoltage or short circuits.

The remote station has RTU that collects the corresponding data for transmission to the control center, and sends commands from the control center to the devices. Additionally, the data concentrators collect data from all IEDs, and, if necessary, provide data interchange with other systems (possibly with other SCADA systems).

2.7.2.5 Architecture selection

The architecture of the overall system control can have different levels of complexity according to the size of the system and the number of users and applications. Not every system needs to have all the elements or the same number of devices. Figures below show different levels starting from simple systems based on RTUs to the more complex ones. Figure 2.19 shows a single master station with multiple RTUs. Figure 2.20 shows a single master station with data concentrator or gateway, and Figure 2.21 shows a multiple master station, LAN/WAN substation connection using routers.

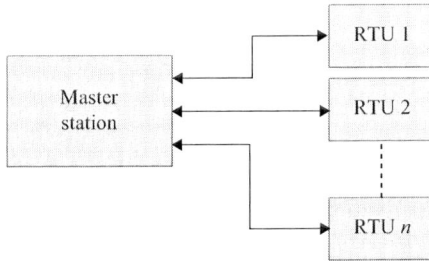

Figure 2.19   Single master station, multiple RTU, radial circuit

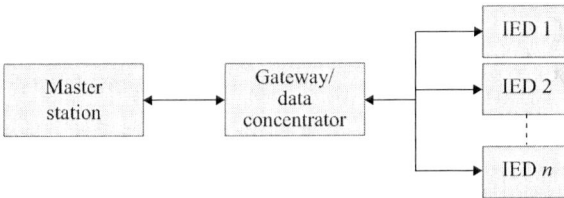

Figure 2.20   Single master station, data concentrator, or gateway

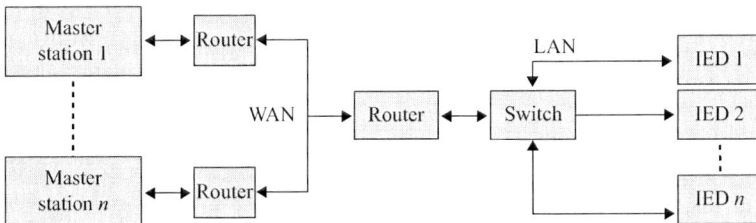

Figure 2.21   Multiple master stations, LAN/WAN substation connection using routers

2.8   Synchrophasors and its application in power systems

Due to the needs of the energy market and in order to better use company assets, key tools are employed. These include communication technology, signal processing, global synchronization diffusion among monitoring, protection, and control applications in the electric power system aiming for efficient operations within the existing electrical, economic, and regulatory margins.

In the past, it was not possible to monitor in real time the relative phase angle in all voltages and intensities in the network due to processing constraints and difficulties in information collection, coordination, and network data synchronization. Changes have been made in new communications solutions, metering, digital data processing, and data synchronization through satellites that accompanied by synchrophasors provide a new scenario to implement a great number of applications that allow to increase network efficiency, tools to detect the power system network conditions, and reliable operation near electrical and physical limits.

In this sense, for example, the idea of real-time system monitoring is recommended to detect stability risk situations to make necessary and accurate decisions limiting the impact on the affected areas avoiding total collapse of voltages in the system, increasing the efficiency of using transmission assets, and reducing investments in new circuits.

New developments in synchronized phasor metering (synchrophasors) are presented as an important evaluation tool for system operators for real-time metering presenting the operator with sufficient information to take appropriate actions in the interconnection system when violations are presented and oscillations could endanger the system stability.

The first phasor measurement unit (PMU) was created in 1988 allowing data synchronization using the clock pulses from a GPS; 7 years later the IEEE 1344 Standard was developed where the GPS synchronized PMU is designated as the synchrophasors. The IEEE 1344 denominated as "the synchrophasor standard" was superseded by the IEEE C37-118 norm in 2005. This norm has been widely accepted as the preferred method for synchrophasor's metering. The norm defines the synchronized phasor samples employed in power system applications and provides methods to quantify metering samples, test to guarantee metering samples, test error limits, and real-time data communication protocols.

2.8.1   Definition

A phasor is the electric network sinusoidal wave representations (voltages or currents) through the projection in the real axis of a vector rotating at angular speed ω at a moment $t = 0$ producing an angle φ with respect to the real axis. The time instant of the rotational vector "picture" defines the maximum values of a reference cosine wave in a way the phasor phase X represents the displacement between the reference wave and the observed cosine wave. In addition to obtaining a mathematical simplification, the system reduces complex electronic system needs and processing capabilities enabling network supervision employing PMUs. Figure 2.22 shows the elaboration of the sinusoidal waveform using its phasorial representation.

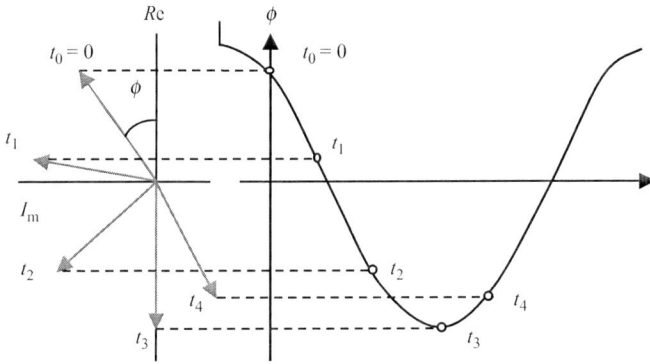

Figure 2.22 Sinusoidal waveform from its phasorial representation

2.8.2 Application of PMUs

In general terms, synchrophasors are phasor samples that are synchronized with UTC (Universal Time Coordinated) time that creates a nominal frequency (50 or 60 Hz) cosine wave as reference. IEEE C37.118 norm from 2005 defines the synchrophasors as a phasor calculated from sampled data using a standard time signal as sampling reference such that phasors sampled in remote locations have a defined common phase relation.

PMUs measure voltage and currents from the power system, using the instrumentation transformers installed in substations, filter and attenuators are employed in the connection of this equipment. Once the phasor is metered, the PMU generates a message with time stamp and phasor data in a format defined in 1995 by the IEEE 1344 norm (replaced in 2005 by IEEE C37.118) in such a way that could be transmitted to a distant location across any available communications link. The format established by the IEEE C37.118 is not required for historic records or monitoring activities, but it is required in real-time control systems.

A simple phasorial network consists of two nodes, one is connected to a PMU linked to a phasorial data concentrator (PDC) in the second node. The communication, as illustrated in Figure 2.23, can be conducted employing any of the technologies currently available as direct wiring, radio network, microwave, telephone, digital radio, or the combination among the ones mentioned. The communication protocol is described in the IEEE C37.118 Standard.

Applications employing synchrophasors have been developed to detect synch loss, multi-area state estimation, oscillation mode identification, voltage stability protection, and dynamic system monitoring. The latest application is real-time wide-area control. PMU offers attractive solutions for protection and control action improvements in modern power systems solving the distributed data time incoherence including a time stamp to the data aligning samples with a common time for processing.

Wide area networks extent across large geographical areas like a country or continent. The main function is focused in network interconnection or terminals

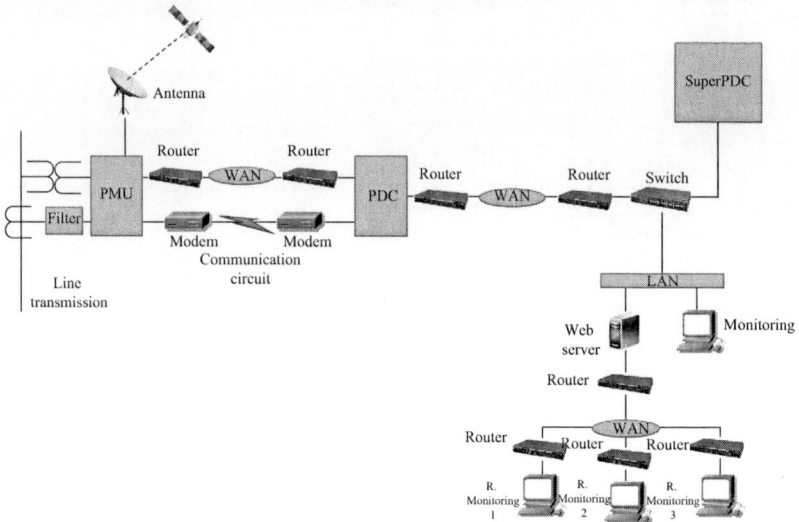

Figure 2.23   PMU's integration with the current communication system

distantly located. The wide area networks operate based on commutation nodes that conduct elements interconnection that allow a continuous large data flow.

The wide area monitoring system's main purpose is to provide system operators with a large information system with analysis tools that increase detection speeds and response time in the event of risky situations allowing the operators to make appropriate decisions efficiently preventing uncontrollable events or cascaded outages. This allows to convert such applications as components of protection and control applications being the first step in Wide Area Monitoring Protection and Control (WAMPC) projects.

The wide area (WAMPC) protection and control schemes have not been widely applied yet and research groups continue to develop new applications generally divided into three groups: power systems monitoring, advanced network protections (wide area), and advanced control schemes.

WAMPC applications comprehend two categories: applications in local substations and applications in system or regional control centers. This classification takes advantage of local intelligence and the substation PMU decision power to conduct corrective actions locally in case power system events occur while maintaining general system control in control centers.

2.8.2.1   Line parameter calculation

Figure 2.24 shows the transmission line model and its parameters. Transmission line impedance parameters can be calculated using PMU on both ends of the line eliminating error sources. PMU measures voltage and current synchrophasors in the terminals: V_S, V_R, I_S and I_R. Line parameters R, X_L, Y_1 and Y_2 can be calculated. Different load and ambient profiles can be obtained from these results calculating

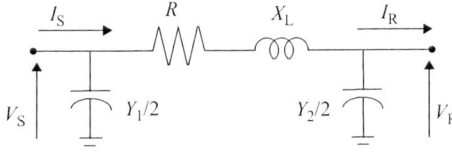

Figure 2.24   Transmission line model

distributed π sections allowing determining line voltage profiles in any transmission conditions.

2.8.2.2   State estimation

This application allows the operator to know relative angle values in different network zones making decisions with the objective of maintaining power system integrity during events that cause unexpected generation or transmission outages since the increase in angle difference represents an increase in static voltage between two points resulting in instability and partial or total blackouts.

2.8.2.3   Transmission line thermal monitoring

Line resistance calculations based on PMU samples offer an economic approximation to thermal monitoring allowing efficient usage of the full transmission capability. By simply obtaining voltage magnitude and angle in the line connection nodes and the current magnitude and angle flowing at the beginning and end points, the line impedance can be obtained. Based on the conductor resistance at 25 °C or 50 °C (manufacturer data), the average conductor temperature can be calculated extrapolating the resistance and temperature relation. With this data and in an economical manner, warnings can be received for overloads, transmission line capacity dynamic control, and/or direct estimation of line depression.

$$Z_1 = \frac{V_S^2 - V_R^2}{I_S \cdot V_R - I_R \cdot V_S} \quad Z_1 = R + jX$$

2.8.2.4   Voltage instability

Voltage instability refers to power systems' capacity to maintain stationary voltages in all busbars after a disturbance; this ability directly depends on a balance between power generated and demanded by loads. The instability is reflected as a continued deterioration of voltage magnitudes in certain busbars or nodes in the system and it could be visualized as the relationship between voltages and demanded power in the busbar; consequently, it can be determined that voltage stability phenomena are directly related to reactive power flow in the network, load behavior in the event of voltage variations, and automatic voltage regulation device actions.

Voltage stability deficiencies represent a high risk for safe and reliable power system operations; consequently, indicators have been established based on certain calculations to determine system state in the event of a possible stability perturbation from the voltage point of view. However, complex mathematical processes needed to evaluate the power system state have been a limiting factor in the implementation of monitoring schemes and voltage instability prevention. Synchrophasors are a solution to the time delay issues with data acquisition since the phasorial metered samples from the PMUs, correctly located, present the system state without involving complex data processing; additionally, the time stamped by the PMU in each sample allows to establish a temporary pattern for each indicator which guarantees that all samples obtained belong to the same system state.

Some indicators applicable to this case are:

- SDI and SDC: Calculations are based on samples metered on transmission element ends in two consecutive time instants and the power increment in one of the terminals.
- ISI: It is based on consecutive voltage and current samples measured at a substation considering the system Thévenin equivalent.
- VSLBI: It applies the maximum power transfer theorem in a calculation based on substation samples and Thévenin equivalent impedance.

2.8.2.5   Power transfer stability

Generator power output increase is only possible up to the maximum electrical power value to be transferred. This power value is referred to as stable state stability limit, and it occurs at 90° maximum angular displacement between the reception busbar and supply busbar. Any attempt to increase the power to be transferred from this maximum transfer point causes a decrease in the electric power to be transferred and consequently a collapse due to the lack of sufficient power to supply the demand, as is shown in Figure 2.25.

The voltage synchrophasor and the transmission line impedance model could be used to illustrate the power transfer curve and the system operational point. When the operational point P_T reaches an unstable condition, preventive actions can be taken.

2.8.2.6   Power oscillations

Significant changes in power transfer cause large voltage variations in the system resulting in angle oscillations which can produce hard oscillations and system instability. To prevent the collapse, synchrophasor-based schemes have been developed to detect imminent unstable operation conditions and inform the system operator; these algorithms monitor oscillation speed to predict instability, hence their implementation in control centers requires data reception at rates higher than 60 Hz with delays close to zero – in any case it can be locally detected by PMUs at substations.

A hard oscillation metering method is the modal analysis which is a signal processing technique that requires samples uniformly spaced; consequently, traditional data acquisition systems are not recommended due to their asynchronous nature. Mode measuring provides valuable information regarding frequency and oscillation damping in the system; it provides dynamic system nature inherent information of the power system.

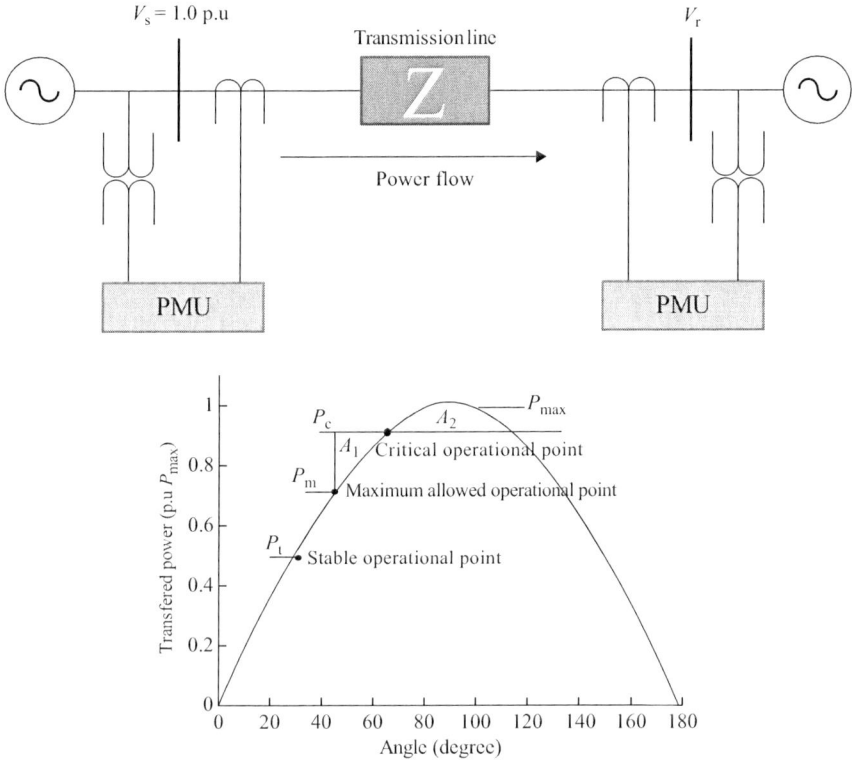

Figure 2.25   Transferred power through a line

2.8.2.7   Mode control governor

When a generator operates in island mode, the governor control system must be in asynchronous mode so that the generation system can control and maintain the frequency in the event of demand changes. Traditional systems that detect synchronism employ frequency samples not aligned in time, so they are susceptible to false synchronism claims when the systems operate at the same frequency but not connected. A more reliable method is to employ angle data in addition to the frequency to detect if an island condition has occurred.

Synchrophasor provides vectorial data with chronologic stamps to determine absolute angles across the power system. In such case, a substation relay with PMU capabilities can communicate angle metering and frequency to the cogeneration control center. With this information, the control system is able to automatically commutate the governor control mode if a separation is detected and it transforms into isochronous mode to regulate frequency.

2.8.2.8   Distributed generation control

Alternate sources generation has converted as an attractive window for new development projects. However, for implementations like photovoltaic to be widely

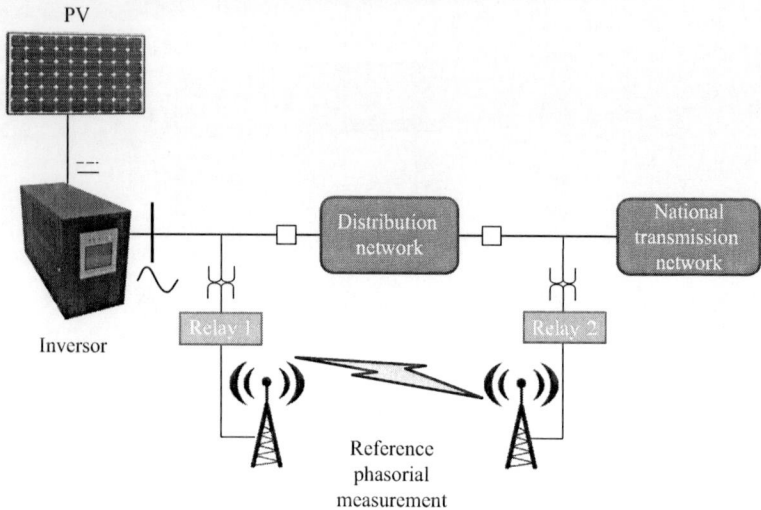

Figure 2.26   Integration of a distributed generation source using PMUs

accepted, their connection to the transmission system must be reliable. When a source is disconnected from the main transmission system, the source must be isolated from the portion of the system where it was separated; a fault in the source trip generates a high risk in personnel safety, in power quality, and in an out-of-phase reclosing.

Traditional methods employing island sensing local metering may not detect an island mode early enough, 2 sec maximum according to IEEE 1547, for all load/ generation conditions. These methods employ local voltage and frequency information verifying if their magnitude is outside predetermined ranges. The inconvenience with current local detection schemes is that they cannot detect a network separation promptly if the difference between power generated and consumed is small.

The synchrophasors provide the means to detect island for almost all load/ generation conditions as they obtain precise metering from all areas. This vision provides a platform for solutions that could maintain the generation on line during transient conditions. Solutions involving two relays, as is shown in Figure 2.26, are available that take phasorial voltage samples and communicate between them (60 times per second). As an example, the second relay can receive remote data from relay one and calculate the angle difference between local and remote, the difference variation with respect to time determines the frequency displacement, and the displacement difference with respect to time indicates the acceleration between terminals.

Chapter 3

Fundamentals of distribution system analysis

Distribution automation requires a deep knowledge of the system to where it is applied. For this, a proper handling of analysis techniques is very important. Distribution system analysis is a part of a broader concept referred to as power system analysis. In distribution system analysis only some fields of the overall picture are studied, and, of course, those referring mainly to feeders and radial systems.

The main topics to be analyzed correspond to the modeling of the elements, the analysis of load flows for ring and radial systems, and short circuit conditions. Load flows are applied for different applications involving not only the analysis of power flows and voltage regulation but also feeder reconfiguration and loss reduction. Short circuit as always is essential in sizing equipment and protective relay setting. Studies of transient and small signal stability are not usually conducted at distribution levels and therefore will not be considered in this book.

3.1 Electrical circuit laws

The most important concepts of electrical circuits are treated in this chapter. The three basic laws of electrical circuits are as follows.

3.1.1 Ohm's law

Ohm's law establishes the relationship between voltage, current, and impedance in electrical circuits as follows:

$$V = IZ \tag{3.1}$$

3.1.2 Kirchhoff's voltage law

For any closed path in a network, Kirchhoff's voltage law (KVL) states that the algebraic sum of the voltages is zero. Some of the voltages will be sources, while other will result from current in passive elements creating a voltage, which is sometimes referred to as a voltage drop. The law applies equally well to circuits driven by constant sources, DC, to time variable sources, v(t) and i(t), and to circuits driven by sources.

3.1.3   Kirchhoff's current law

The connection of two or more circuit elements creates a junction called a node. The junction between two elements is called a simple node and no division of current results. The junction of three or more elements is called a principal node, and here current division does take place. Kirchhoff's current law (KCL) states that the algebraic sum of the currents at a node is zero. It may be stated alternatively that the sum of the currents entering a node is equal to the sum of the currents leaving that node. The basis for the law is the conservation of electric charge.

3.2   Circuit theorems

Circuit theorems are derived from the circuit laws. The three most commonly used for system analysis are as follows.

3.2.1   Thévenin's theorem

This is useful for replacing part of a network which is not of particular interest. Any active network viewed from any two terminals can be replaced by a single driving voltage in series with a single impedance where the Thévenin equivalent voltage is the driving voltage equal to the open circuit voltage between terminals. Likewise, the Thévenin equivalent impedance is equal to the impedance of the network as viewed from the two terminals with all source voltages short-circuited. These concepts are illustrated in the drawings of Figure 3.1.

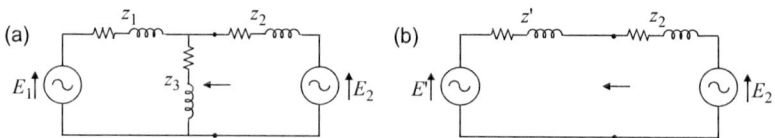

Figure 3.1   (a) Circuit before using the Thévenin's theorem; (b) circuit after using the Thévenin's theorem

where:

$$E' = \frac{z_3}{z_1 + z_3} E_1 \qquad (3.2)$$

and

$$z' = \frac{z_1 z_3}{z_1 + z_3} \qquad (3.3)$$

3.2.2   Star/delta transform

Figure 3.2 shows the circuit equivalent in a star/delta configuration. The expressions to transform from one to the other configuration are the following.

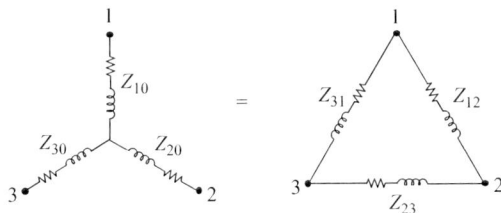

Figure 3.2   Star/delta equivalent

$$Z_{12} = Z_{10} + Z_{20} + \frac{Z_{10}Z_{20}}{Z_{30}} \qquad (3.4)$$

$$Z_{10} = \frac{Z_{12}Z_{31}}{Z_{12} + Z_{23} + Z_{31}} \qquad (3.5)$$

3.2.3   Superposition theorem

In any linear network the current in any branch is equal to the vector sum of the currents caused by each source voltage acting alone with the others short-circuited. An example case is illustrated in Figure 3.3.

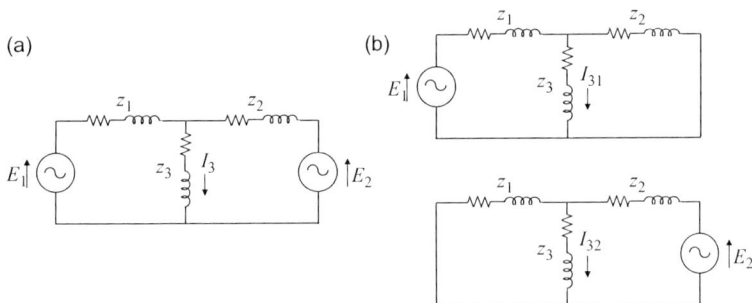

Figure 3.3   (a) Circuit complete; (b) solution using superposition theorem

$$I_3 = I_{31} + I_{32} \qquad (3.6)$$

The superposition theorem is highly used for short circuit calculations in order to include the pre-fault current values that are normally taken from load flow programs.

3.3   Power AC circuits

Let $v(t) = V_m \sin \omega t$ and $i(t) = I_m \sin(\omega t - \theta)$ be the steady-state voltage and current in an AC circuit, here θ is the angle of the impedance of the circuit. The instantaneous power p is then:

$$p(t) = v(t)i(t) = V_m I_m \sin(\omega t)\sin(\omega t - \theta) \qquad (3.7)$$

If T is the period of the voltage, or current, waveform, then the average power P is given by:

$$P = \frac{1}{T}\int_0^T p(t)dt = \frac{1}{2}V_m I_m \cos(\omega t) = VI\cos(\omega t) \tag{3.8}$$

where V and I are the rms voltage and rms current, respectively. The definition now becomes meaningful: the apparent power VI must be multiplied by the factor $\cos\theta$ to obtain the actual average power.

Sometimes, P is referred to as the true power, real power, in-phase power, or active power. Figure 3.4 helps to understand easier this concept (drawn for the case $\theta > 0$), from which the following definitions of (average) power emerge:

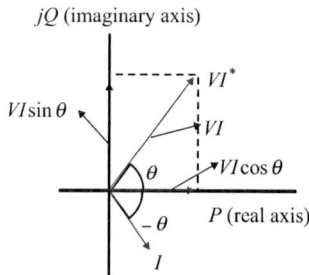

Figure 3.4   Relationship of current, voltage, and power in electrical circuits

Note that while active power is never negative, reactive power can have either sign. The three power units, W, VA, and VAR, are physically equivalent.

All three kinds of power may be derived from a single complex power, defined as VI^* (VA). In fact, since $V\angle 0°$ and $I\angle -\theta°$.

$$VI^* = (V\angle 0°)(I\angle +\theta°) = VI\cos\theta + jVI\sin\theta = P + jQ$$

and so

$$VI = \|VI^*\| P = \text{Re}(VI^*)   Q = \text{Im}(VI^*)$$

Example 3.1 Two voltage sources $V_1 = 110\angle\delta$ and $V_2 = 100\angle 50°$ are connected by a short line of impedance $Z = 2 + j5\Omega$, where δ is the phase angle of source 1. Write a MATLAB program for the system that computes the complex power for each source and the line loss; tabulate the real power and plot P_1, P_2, and P_L versus phase angle δ, if δ ranges from $-38°$ to $32°$ in steps of $2.5°$.

In general terms:

$$I_{12} = \frac{V_1 - V_2}{Z}$$

$$I_{21} = \frac{V_2 - V_1}{Z}$$

$$S_{12} = P_{12} + jQ_{12} = V_1 I_{12}^*$$

$$S_{21} = P_{21} + jQ_{21} = V_2 I_{21}^*$$

$$S_L = P_L + jQ_L = S_{12} + S_{21}$$

The code of MATLAB for the general case:

```
V_1_mag=input('Write the voltage magnitude of V1 [V] = ');
V_1_ang=input('Write the voltage phase angle of V1 (degree) = ');
V_2_mag=input('Write the voltage magnitude of V2 [V] = ');
V_2_ang=input('Write the voltage phase angle of V2 (degree) = ');
var=input('Write the angle variation for V1 (degree) = ');
R= input ('Write the resistance in the line between both sources [ohm] = ');
X= input ('Write the reactance in the line between both sources [ohm] = ');
Z=R+j*X;
angle_1=[V_1_ang-var:2.5:V_1_ang+var]*pi()/180;
n=length(angle_1);
angle_2=ones(1,n)*V_2_ang*180/pi();
V1=V_1_mag.*(cos(angle_1)+j*sin(angle_1));
V2=V_2_mag.*(cos(angle_2)+j*sin(angle_2));

I12=(V1-V2)./Z;
I21=(V2-V1)./Z;

S1=V1.*conj(I12);
P1=real(S1);
Q1=imag(S1);

S2=V2.*conj(I21);
P2=real(S2);
Q2=imag(S2);

SL=S1+S2;
PL=real(SL);
QL=imag(SL);

result=[angle_1'*180/pi(), P1', P2', PL'];
disp(' Delta-1 P1 P2 PL')
disp(result)
plot(angle_1*180/pi(), P1, angle_1*180/pi(), P2, angle_1*180/pi(), PL)
legend('P1','P2','PL')
ylabel('P [W]')
xlabel(' \delta [°]')
```

The results of calculations for Example 3.1 are given in the following list. The graphs for P_1, P_2, and P_L versus phase angle δ are shown in Figure 3.5.

δ	P_1	P_2	P_L
−38,0	−481,751	565,527	83,776
−35,5	−412,364	476,711	64,347
−33,0	−340,602	388,300	47,697
−30,5	−266,605	300,463	33,858
−28,0	−190,511	213,366	22,855
−25,5	−112,466	127,176	14,711
−23,0	−32,618	42,057	9,439
−20,5	48,880	−41,829	7,051
−18,0	131,874	−124,323	7,551
−15,5	216,205	−205,267	10,937
−13,0	301,713	−284,508	17,204
−10,5	388,235	−361,895	26,340
−8,0	475,606	−437,280	38,327
−5,5	563,661	−510,519	53,142
−3,0	652,231	−581,475	70,757
−0,5	741,149	−650,010	91,139
2,0	830,244	−715,995	114,248
4,5	919,346	−779,305	140,042
7,0	1008,288	−839,818	168,470
9,5	1096,898	−897,420	199,478
12,0	1185,009	−952,001	233,009
14,5	1272,453	−1003,457	268,997
17,0	1359,063	−1051,689	307,374
19,5	1444,675	−1096,607	348,067
22,0	1529,125	−1138,125	390,999
24,5	1612,252	−1176,164	436,088
27,0	1693,899	−1210,651	483,249
29,5	1773,911	−1241,520	532,390
32,0	1852,134	−1268,714	583,420

Example 3.2 A balanced three-phase source for the following instantaneous phase voltages:

$$v_{an} = 1250\cos(\omega t)$$

$$v_{bn} = 1250\cos(\omega t - 120°)$$

$$v_{cn} = 1250\cos(\omega t - 240°)$$

This source supplies a balanced load of impedance $Z = 125\angle 30°\,\Omega$ per phase. Using MATLAB, plot the instantaneous powers p_a, p_b, p_c and their sum versus ωt over a range of 0–2π with steps of 0.1. Comment in the nature of the instantaneous power in each phase and the total three-phase real power.

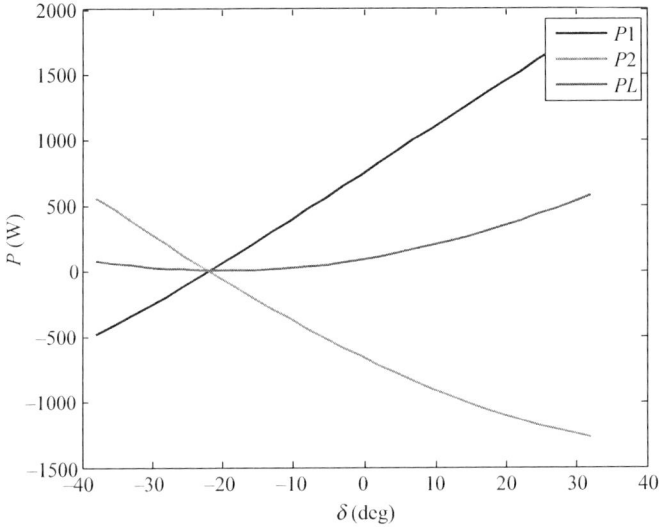

Figure 3.5   Power losses for different voltage angles

In general terms:

$$v_n(t) = V_m \cos(\omega t + \beta)$$

$$i_n(t) = I_m \cos(\omega t + \beta - \theta) = \frac{V_m}{\|Z\|} \cos(\omega t + \beta - \theta)$$

$$p_n(t) = V_m I_m \cos(\omega t + \beta) \cos(\omega t + \beta - \theta)$$

A useful expression is:

$$\cos(A)\cos(B) = \frac{1}{2}\cos(A + B) + \frac{1}{2}\cos(A - B)$$

The new expression of power is:

$$p_n(t) = \frac{V_m I_m}{2}[\cos(2\omega t + 2\beta - \theta) + \cos(\theta)]$$

The expression can be easily calculated and plotted using MATLAB. The following code can apply for this purpose:

```
V_1_m=input('Write the voltage magnitude of V1 [V] = ');
V_1_a=input('Write the voltage phase angle of V1 (degree) = ');
V_2_m=input('Write the voltage magnitude of V2 [V] = ');
V_2_a=input('Write the voltage phase angle of V2 (degree) = ');
V_3_m=input('Write the voltage magnitude of V3 [V] = ');
V_3_a=input('Write the voltage phase angle of V3 (degree) = ');
Za=input('Write the load for phase a [complex number, ohms] = ');
Zb=input('Write the load for phase b [complex number, ohms] = ');
Zc=input('Write the load for phase c [complex number, ohms] = ');
```

```
wt=[0:0.1:2*pi()];
I1_m=V_1_m/abs(Za);
I2_m=V_2_m/abs(Zb);
I3_m=V_3_m/abs(Zc);
P1=(V_1_m*I1_m/2).*(cos((2*wt)+(2*(V_1_a*pi()/180))-
(cos(angle(Za))))+cos(angle(Za)));
P2=(V_2_m*I2_m/2).*(cos((2*wt)+(2*(V_2_a*pi()/180))-
(cos(angle(Zb))))+cos(angle(Zb)));
P3=(V_3_m*I3_m/2).*(cos((2*wt)+(2*(V_3_a*pi()/180))-
(cos(angle(Zc))))+cos(angle(Zc)));


Pt=P1+P2+P3;


plot(wt, P1, wt, P2, wt, P3, wt, Pt)
legend('P1','P2','P3', 'Pt')
ylabel('P [W]')
xlabel(' \omega * t')
```

Figure 3.6 shows the values of P_1, P_2, P_3 and the total Power P_T.

3.4   PU normalization

In power system calculations a normalization of variables called per unit normalization is generally used. It is especially convenient when several transformers are involved.

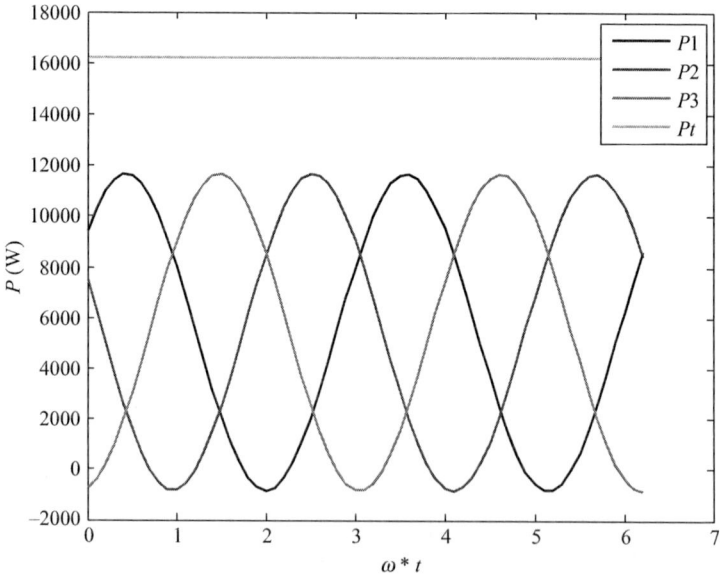

Figure 3.6   Power losses for different voltage angles

The idea is to pick base values for quantities such as voltages, currents, impedances, power, and so on, and to define the quantity in per unit as follows.

$$\text{Quantity in per Unit} = \frac{\text{Actual Quantity}}{\text{Base Value of Quantity}}$$

Base variables are picked to satisfy the same kind of relationship as the variables. Ohm's law establishes that $V = ZI$.

If the equation is expressed by using the base values, it becomes:

$$V_B = Z_B I_B \tag{3.9}$$

Dividing both equations, the following is obtained:

$$\frac{V}{V_B} = \frac{ZI}{Z_B I_B} \tag{3.10}$$

or

$$V_{pu} = Z_{pu} I_{pu} \tag{3.11}$$

Equation (3.10) has the same form as (3.9), which implies that circuit analysis can be done exactly as with (3.9). The pu subscript indicates per unit and is read "per unit."

Like it is done for Ohm's law can also be done in the case of power calculations. For example, corresponding to:

$$S = VI^* \tag{3.12}$$

$$S_B = V_B I_B \tag{3.13}$$

and

$$S_{pu} = V_{pu} \cdot I_{pu}^* \tag{3.14}$$

Note that the two equations (3.9) and (3.13) involve four base quantities V_B, I_B, Z_B, and S_B. The selection of any two base quantities determines the remaining two base quantities. If V_B and S_B are selected, then $I_B = S_B/V_B$ can be replaced in Z_B as follows:

$$Z_B = \frac{V_B}{I_B} = \frac{V_B^2}{S_B} \tag{3.15}$$

The extension to other related variables is routine. For example:

$$S_{pu} = P_{pu} + jQ_{pu} \tag{3.16}$$

where:

$$P_{pu} = \frac{P}{S_B} \quad Q_{pu} = \frac{Q}{S_B}$$

For the impedance:

$$Z_{pu} = R_{pu} + jX_{pu} \tag{3.17}$$

where

$$R_{pu} = \frac{R}{Z_B} \quad X_{pu} = \frac{X}{Z_B}$$

Note that

$$Y_B = \frac{I_B}{V_B} = \frac{1}{Z_B} \tag{3.18}$$

With several items of equipment, with different ratings, it is not usually possible to pick base values so that they are always the same as the nameplate ratings. It is necessary to recalculate the per unit values on the new basis. The key idea is that $Z_{p.u}$ depends on Z_B but, of course, Z_{actual} does not. The relationship between old and new values is:

$$Z_{actual} = Z_{pu}^{old} Z_B^{old} = Z_{pu}^{new} Z_B^{new} \tag{3.19}$$

Then

$$Z_{pu}^{new} = Z_{pu}^{old} \frac{Z_B^{old}}{Z_B^{new}}$$

$$Z_{pu}^{new} = Z_{pu}^{old} \left[\frac{V_B^{old}}{V_B^{new}} \right]^2 \frac{S_B^{new}}{S_B^{old}} \tag{3.20}$$

Note: In applying (3.20), it can be substituted with three-phase and/or line-line values.

Example 3.3 The system shown in Figure 3.7 has the impedances in the bases of the corresponding elements. Find the pu values considering that at the generator, the base values are 11 kV and 50 MVA.

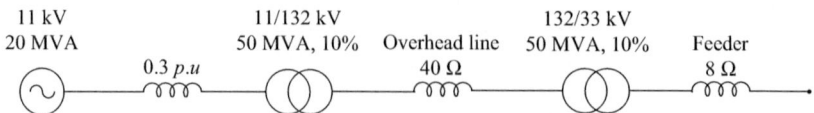

Figure 3.7   Electrical system for example 3.3

At the generator:

Base $kV_B = 11$
Base $MVA_B = 50$

$$Z_B = \frac{(11 \text{ kV})^2}{50 \text{ MVA}} = 2.42 \ \Omega$$

$$I_B = \frac{50 \text{ MVA}}{\sqrt{3} \times 11 \text{ kV}} = 2624.3 \text{ A}$$

Z_{GENpu} on common base:

$$Z_{GENpu} = 0.3 \frac{50 \text{ MVA}}{20 \text{ MVA}} = 0.75 \text{ pu}$$

At the transformers: 0.1 pu

At the line:

Base $kV_B = 132$
Base $MVA_B = 50$

$$Z_B = \frac{(132 \text{ kV})^2}{50 \text{ MVA}} = 348.5 \ \Omega$$

$$I_B = \frac{50 \text{ MVA}}{\sqrt{3} \cdot 132 \text{ kV}} = 218.7 \text{ A}$$

$$Z_{OHLpu} = \frac{40 \ \Omega}{348.5 \ \Omega} = 0.115 \text{ pu}$$

At the feeder:

Base $kVB = 33$
Base $MVAB = 50$

$$Z_B = \frac{(33 \text{ kV})^2}{50 \text{ MVA}} = 21.78 \ \Omega$$

$$I_B = \frac{50 \text{ MVA}}{\sqrt{3} \cdot 33 \text{ kV}} = 874.77 \text{ A}$$

$$Z_{FEEDERpu} = \frac{8 \ \Omega}{21.78 \ \Omega} = 0.367 \text{ pu}$$

3.5   Load flow

Load flows, also called power flows, are mathematical procedures performed on power system networks, which allow the calculation of the voltage magnitudes and angles of all buses in the system. Voltage magnitudes are only known for those buses having generators that allow setting the voltage magnitude if appropriate regulators are available. These buses are referred to as voltage buses. For the strongest bus of the system, the voltage magnitude can be specified and the angle normally defined is equal to zero as this is the reference for the rest of the system buses. This bus is referred to as the slack bus. The other buses were the active and reactive power are defined, are referred to as PQ buses. The bus classification will be mentioned later.

Once the voltage and angles are calculated at all buses, other parameters can easily be determined by means of the expressions already reviewed earlier in this chapter. Those parameters include among others the following: power (total, active, and reactive) transferred, and current including angles, losses, power factors, and loading of the different elements.

Load flows are analyzed by using very efficient algorithms which conduct almost invariably to software packages. This should not though prevent hand calculations, which are appropriate especially for small systems.

Up to the mid-1960s before use of computer, big systems were analyzed with the aid of board simulators. Nowadays there are flexible and affordable software packages that allow comprehensive analysis, and Figure 3.8 illustrates the big changes from the old simulators to the new representations achieved from software packages.

The most known algorithms to implement load flow programs are the Newton–Raphson and Gauss–Seidel methods. These can be used for the solution of H.V. network systems which invariably correspond to transmission systems and big complexes with internal generation. In H.V. systems normally there is a high X/R ratio and the loads are symmetric. Therefore, the analysis assumes three-phase balanced system operation.

These algorithms also can be used in distribution systems, although in these cases convergence could be a problem since the X/R ratio is lower. For these cases then other algorithms are more appropriate. Mathematical algorithms for

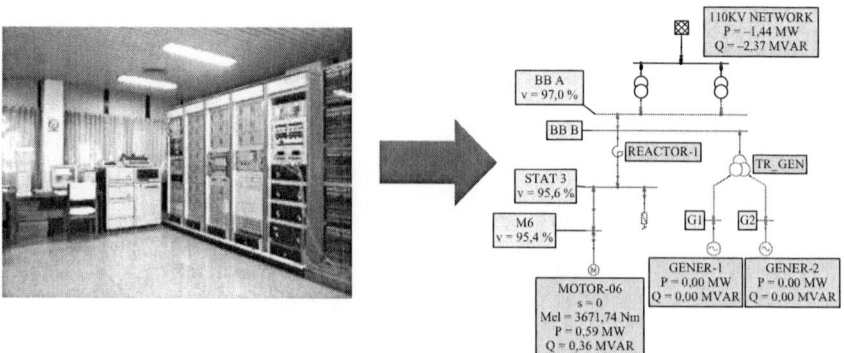

Figure 3.8   Illustration of a power system board simulator

distribution systems are normally simpler than those developed for high-voltage systems because of the radial condition. They will be treated in section 3.6.

In this section the basis and structure of the algorithms for load flows are reviewed, as well as some examples to illustrate their application in a comprehensive case study.

3.5.1   Formulation of the load flow problem

For load flow analysis, all machines at any one bus are treated as a single machine and represented by a single EMF and a series reactance. The basic equations that will be the basis for the nodal analysis are the following:

The current is $I = YV$, which becomes, for the entire system, $I = Y_{\text{bus}}V$
The complex power is $S = VI^*$ and therefore $S^* = V^*I$
In an electrical system Y_{bus} is symmetrical around the principal diagonal.

A simple electrical system with three nodes is shown in the Figure 3.9 to illustrate the concepts. Here each generator is replaced by an equivalent current source and shunt admittance.

At each node it is known that $I_{ik} = Y_{ik}(V_i - V_k)$. Therefore, the following expressions can be obtained from Figure 3.9.

$$I_0 = Y_0 V_0 + Y_{10}(V_0 - V_1)$$
$$I_1 = Y_1 V_1 + Y_{10}(V_1 - V_0) + Y_{12}(V_1 - V_2)$$
$$I_2 = Y_2 V_2 + Y_{12}(V_2 - V_1)$$

Regrouping these terms, the following equations are obtained:

$$I_0 = (Y_0 + Y_{01})V_0 - Y_{01}V_1$$
$$I_1 = -Y_{01}V_0 + (Y_1 + Y_{01} + Y_{12})V_1 - Y_{12}V_2$$
$$I_2 = -Y_{12}V_1 + (Y_2 + Y_{12})V_2$$

In matrix form those equations can be expressed as:

$$\begin{bmatrix} I_0 \\ I_1 \\ I_2 \end{bmatrix} = \begin{bmatrix} Y_0 + Y_{01} & -Y_{01} & 0 \\ -Y_{01} & Y_1 + Y_{01} + Y_{12} & -Y_{12} \\ 0 & -Y_{12} & Y_2 + Y_{12} \end{bmatrix} \begin{bmatrix} V_0 \\ V_1 \\ V_2 \end{bmatrix} \qquad (3.21)$$

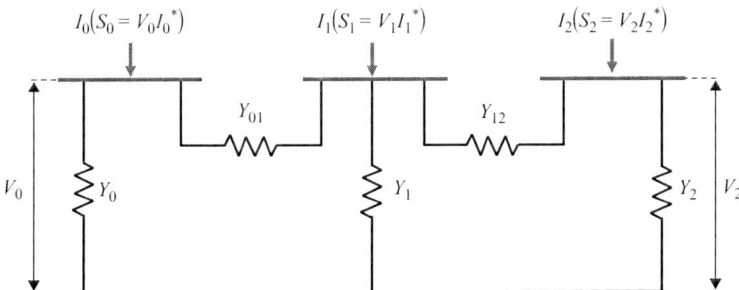

Figure 3.9   Representation of a three-node system with current sources

The equations can be expressed as:

$$\sum Y_{ik} V_k = I_i$$

$$\sum Y_{ik} V_k = \frac{S_i^*}{V_i^*} = I_i \tag{3.22}$$

$$V_i^* \sum Y_{ik} V_k = S_i^*$$

with $i = 1, 2, \ldots, n$; n being the total number of nodes.

3.5.2   Newton–Raphson method

The former equations are nonlinear and require for their solution numerical itera-tive methods. There are a number of them but the most well-known are those of Newton–Raphson and Gauss–Seidel. Here the method of Newton–Raphson will be developed using Taylor's series. The Taylor's series for a function $f(x)$ that has derivatives of all orders at $x = a$ is expressed as:

$$y = f(x) = f(a) + f'(a)(x-a) + \frac{f''(a)}{2!}(x-a)^2 + \ldots + \frac{f^{(n)}(a)}{n!}(x-a)^n \tag{3.23}$$

Neglecting the second-order terms and higher, the expression becomes:

$$y = f(x) = f(a) + f'(a)(x-a)$$

The series converges rapidly for values of x near to a. If x_0 is the initial esti-mate, the expression becomes:

$$y = f(x_0) + f'(x_0)(x_1 - x_0) \tag{3.24}$$

where x_1 is the new value of x which is a closer estimate. This curve crosses the axis at the new value x_1. Thus:

$$0 = f(x_0) + f'(x_0)(x_1 - x_0) \tag{3.25}$$

Normally $x_0 - x_1$ is referred to as Δx. Therefore:

$$\Delta x = -\frac{f(x_0)}{f'(x_0)} \tag{3.26}$$

$$x_1 = x_0 - \Delta x = x_0 - \frac{f(x_0)}{f'(x_0)} \tag{3.27}$$

Figure 3.10 explains this concept.

Furthermore, the Newton–Raphson method is used for solving systems of simultaneous equations. The Taylor series is applied to n nonlinear equations in n unknowns:

$$x_1, x_2, \cdots, x_n$$

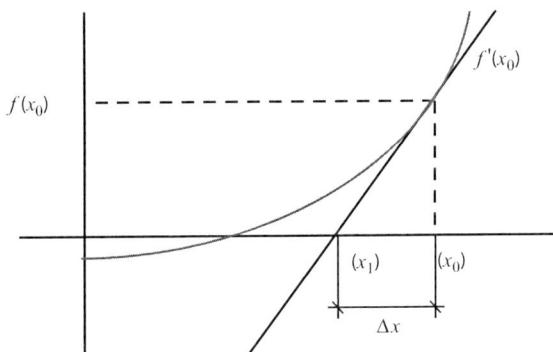

Figure 3.10   Illustration of the Newton–Raphson concept

The differentiable functions are expressed as:

$$f_1(x_1, x_2, \cdots, x_n) = 0$$
$$f_2(x_1, x_2, \cdots, x_n) = 0$$
$$\vdots$$
$$f_n(x_1, x_2, \cdots, x_n) = 0$$

The matrix of partial derivatives is called a Jacobian matrix J. By using the same concept of (3.27), this result can be written as:

$$[X_{k+1}] = [X_k] - [J^{-1}][F(X_k)] \qquad (3.28)$$

The square matrix J is the Jacobian matrix of $F(x)$, whose $(i,k)^{\text{th}}$ element is defined as $\frac{\partial f_i}{\partial x_k}$. This algorithm can be used to solve load flows. In that case, it is required to write the equations as a set $F(x) = 0$. In general it can be written that:

$$x_{k+1} = x_k - \frac{f(x_k)}{f'(x_k)} \qquad (3.29)$$

Example 3.4 Find the solution for the following equation using the expression for quadratic equations and compare the results with Newton–Raphson method:

$$f(x) = 0 = x^2 - 20x + 100$$

This equation is of the form $f(x) = y = ax^2 + bx + c$. The solution applying the mathematical formula is obtained as:

$$x = \frac{-b \pm \sqrt{b^2 - 4ac}}{2a}$$
$$= \frac{20 \pm \sqrt{400 - 400}}{2} = 10$$

Now the Newton–Raphson Method is used and therefore:

$$f'(x) = 2x - 20$$

The first approximation is taken as $x = 12$:

$$f(x) = (12)^2 - 20(12) + 100 = 4$$
$$f'(x) = 2(12) - 20 = 4$$
$$\Delta x_1 = -\frac{f(12)}{f'(12)} = -\frac{4}{4} = -1$$
$$x_1 = x_0 + \Delta x_1 = 12 + (-1) = 11$$

The second iteration:

$$f(x) = (11)^2 - 20(11) + 100 = 1$$
$$f'(x) = 2(11) - 20 = 2$$
$$\Delta x_2 = -\frac{f(11)}{f'(11)} = -\frac{1}{2} = -0.5$$
$$x_2 = x_1 + \Delta x_2 = 11 + (-0.5) = 10.5$$

The third iteration:

$$f(x) = (10.5)^2 - 20(10.5) + 100 = 0.25$$
$$f'(x) = 2(10.5) - 20 = 1$$
$$\Delta x_3 = -\frac{f(10.5)}{f'(10.5)} = -0.25$$
$$x_3 = x_2 + \Delta x_3 = 10.5 + (-0.25) = 10.25$$

The fourth iteration:

$$f(x) = (10.25)^2 - 20(10.25) + 100 = 0.0625$$
$$f'(x) = 2(10.25) - 20 = 0.5$$
$$\Delta x_4 = -\frac{f(11)}{f'(11)} = -\frac{0.0625}{0.5} = -0.125$$
$$x_4 = x_3 + \Delta x_4 = 10.25 + (-0.125) = 10.125$$

The last result is very close to the exact solution. This is shown in Table 3.1.

Table 3.1   Values for example 3.5

Iteration	X_k	X_{k+1}
1	0.0000000	1.0000000
2	1.0000000	1.1029207
3	1.1029207	1.1060571
4	1.1060571	1.1060602

Example 3.5 Use the Newton–Raphson method to resolve the following equation:

$$x = 2 - \sin x$$

with the NRM:

$$f(x) = 0 = x - 2 + \sin x$$

$$f'(x) = 1 + \cos x$$

$$\Delta x = -\frac{f(x)}{f'(x)} = \frac{2 - (x + \sin x)}{1 + \cos x}$$

$$x_{k+1} = x_k + \Delta x$$

$$x_{k+1} = x_k + \frac{2 - (x + \sin x)}{1 + \cos x}$$

3.5.3   Type of buses

The buses are classified in three different categories for load flow analysis as follows:

- Slack buses: In this type of node the values of voltage magnitude and angle are known. The slack generator covers system losses. The angle is used as reference for all voltage angles. Usually this node is set as the supply for the largest generator.
- PV buses: In this type of node, the values of P and V are known. Q is a free variable within limits ($Q_{min} < Q < Q_{max}$). This type of node is used for voltage controlled generators.
- PQ buses: In this type of node, the values of P and Q are known. V is a free variable within limits ($V_{min} < V < V_{max}$).

Figure 3.11 illustrates the different type of buses.

3.5.4   Application of the Newton–Raphson method to solve load flows

The theory seen in the previous sections will be applied to the full solution of load flows with the method of Newton–Raphson.

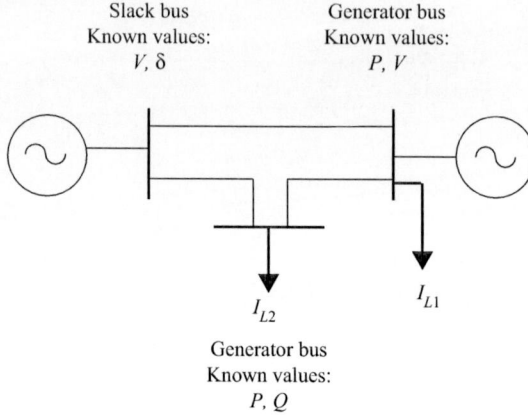

Figure 3.11   *Illustration of types of buses for load flow analysis*

Equation (3.22) can be expressed as follows, indicating the k nodes connected to node i denoted by $k\varepsilon i$.

$$S_i^* = V_i^* \sum_{k\varepsilon i} Y_{ik} V_k \tag{3.30}$$

From (3.30), the following derivation is obtained:

$$S_i^* = V_i \angle{-\theta_i} \sum Y_{ik} V_k \angle{\theta_k}$$

$$S_i^* = V_i \sum Y_{ik} V_k \angle{(\theta_k - \theta_i)}$$

Since $Y_{ik} = G_{ik} + jB_{ik}$ and $S^* = P_i - jQ_i$, then:

$$S_i^* = V_i \sum (G_{ik} + jB_{ik})(V_k \cos\theta_{ik} - jV_k \sin\theta_{ik})$$

$$S_i^* = V_i \sum (G_{ik} \cos\theta_{ik} + B_{ik} \sin\theta_{ik})V_k$$

$$+ jV_i \sum (B_{ik} \cos\theta_{ik} - G_{ik} \sin\theta_{ik})V_k \tag{3.31}$$

Now the values of P and Q calculated are the following:

$$P_i(\text{calculated}) = V_i \sum (G_{ik} \cos\theta_{ik} + B_{ik} \sin\theta_{ik})V_k$$

$$P_i = V_i^2 G_{ii} + V_i \sum_{k\varepsilon i,\, k\neq i} (G_{ik} \cos\theta_{ik} + B_{ik} \sin\theta_{ik})V_k$$

$$Q_i(\text{calculated}) = V_i \sum (G_{ik} \sin\theta_{ik} - B_{ik} \cos\theta_{ik})V_k$$

$$Q_i = -V_i^2 B_{ii} + V_i \sum_{k\varepsilon i,\, k\neq i} (G_{ik} \sin\theta_{ik} - B_{ik} \cos\theta_{ik})V_k$$

$$\Delta P_i^j = P_i(\text{specified}) - P_i^j(\text{calculated})$$

$$\Delta Q_i^j = Q_i(\text{specified}) - Q_i^j(\text{calculated})$$

with $i = 1, 2, \ldots, n-1$, where P_1, jQ_1 specified are net power at bus i in pu. ΔP and ΔQ are calculated from the equations of P and Q as follows:

$$X^{r+1} = X^r - \Delta X$$
$$\Delta X = J^{-1} F(X^r)$$

$$\begin{bmatrix} \Delta\theta \\ \Delta V \end{bmatrix} = [J^{-1}] \begin{bmatrix} \Delta P \\ \Delta Q \end{bmatrix}$$

$$[J]_{\text{Jacobian}} = \begin{bmatrix} H & N \\ M & L \end{bmatrix}$$

$$\begin{bmatrix} \Delta P \\ \Delta Q \end{bmatrix} = [J] \begin{bmatrix} \Delta\theta \\ \dfrac{\Delta V}{V} \end{bmatrix} = \begin{bmatrix} H & N \\ M & L \end{bmatrix} \begin{bmatrix} \Delta\theta \\ \dfrac{\Delta V}{V} \end{bmatrix}$$

$$H_{ii} = \frac{\partial \Delta P_i}{\partial \theta_i} \qquad H_{ik} = \frac{\partial \Delta P_i}{\partial \theta_k}$$

$$N_{ii} = V_i \frac{\partial \Delta P_i}{\partial V_i} \qquad N_{ik} = V_k \frac{\partial \Delta P_i}{\partial \theta_k}$$

$$M_{ii} = \frac{\partial \Delta Q_i}{\partial \theta_i} \qquad M_{ik} = \frac{\partial \Delta Q_i}{\partial \theta_k}$$

$$L_{ii} = V_i \frac{\partial \Delta Q_i}{\partial V_i} \qquad L_{ik} = V_k \frac{\partial \Delta Q_i}{\partial V_k}$$

The following trigonometric functions have to be considered:

$$\frac{\partial \cos \theta_{ik}}{\partial \theta_i} = -\sin \theta_{ik} \qquad \frac{\partial \cos \theta_{ik}}{\partial \theta_k} = \sin \theta_{ik}$$

$$\frac{\partial \sin \theta_{ik}}{\partial \theta_i} = \cos \theta_{ik} \qquad \frac{\partial \sin \theta_{ik}}{\partial \theta_k} = -\cos \theta_{ik}$$

By applying the trigonometric functions, the values of the Jacobian become:

$$H_{ii} = \frac{\partial \Delta P_i}{\partial \theta_i} = -V_i \sum_{k \in i, k \neq i} (G_{ik} \sin \theta_{ik} - B_{ik} \cos \theta_{ik}) V_k$$

$$= -Q_i^{\text{calculated}} - |V_i|^2 B_{ii}$$

$$H_{ik} = \frac{\partial \Delta P_i}{\partial \theta_k} = V_i(G_{ik} \sin \theta_{ik} - B_{ik} \cos \theta_{ik})V_k$$

$$M_{ii} = \frac{\partial \Delta Q_i}{\partial \theta_i} = V_i \sum_{k\varepsilon i, k \neq i} (G_{ik} \cos \theta_{ik} + B_{ik} \sin \theta_{ik})V_k = P_i^{\text{calculated}} - |V_i|^2 G_{ii}$$

$$M_{ik} = \frac{\partial \Delta Q_i}{\partial \theta_k} = -V_i (G_{ik} \cos \theta_{ik} + B_{ik} \sin \theta_{ik})V_k$$

$$N_{ii} = V_i \frac{\partial \Delta P_i}{|\partial V_i|} = P_i^{\text{calculated}} + |V_i|^2 G_{ii} = M_{ii} + 2|V_i|^2 G_{ii}$$

$$N_{ik} = V_k \frac{\partial \Delta P_i}{|\partial V_k|} = -M_{ik}$$

$$L_{ii} = V_i \frac{\Delta Q_i}{|\partial V_i|} = Q_i^{\text{calculated}} - |V_i|^2 B_{ii} = -H_{ii} - 2|V_i|^2 B_{ii}$$

$$L_{ik} = V_k \frac{\Delta \theta}{|\partial V|_k} = H_{ik}$$

The algorithm of Newton–Raphson method can be implemented following the chart shown in Figure 3.12. It is schematic but illustrates the procedure. The equations used are those shown in the previous paragraph.

3.5.5   Decoupling method

Brian Stott proposed a very clever procedure to solve load flows, known as the Decoupled method taking advantage of an inherent characteristic of power systems which is strong interdependence between active power and bus voltage angles, and between reactive power and voltage magnitudes. Decoupling makes reference to the separate solution of the P-θ and Q-V problems. These facts can be appreciated in Figure 3.13 showing a two-node system.

This method doesn't need to invert the Jacobian J, which is its most important advantage. However, the method is slow to converge.

Assuming $\bar{V}_1$ as reference (1 pu and $\theta_1 = 0$) and $\bar{Z} = jX$, since $R \ll z$, then:

$$\bar{V}_2 = \bar{V}_1 - \bar{I}\bar{Z} = \bar{V}_1 - \bar{I}(jX)$$

It has to be taken into account that:

$$\bar{I} = \frac{(P - jQ)}{\bar{V}_1^*}$$

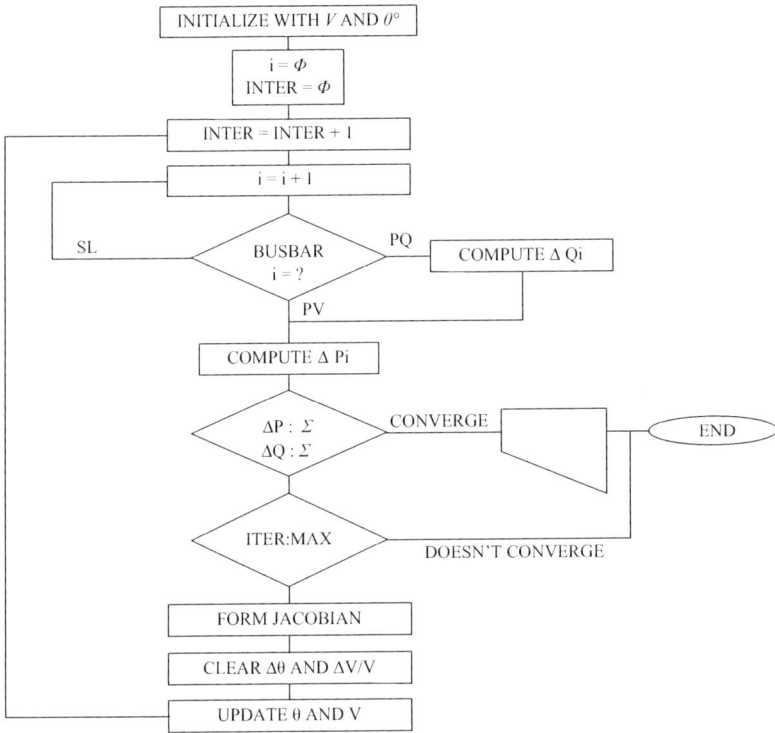

Figure 3.12 Flow chart for load flow analysis using Newton–Raphson algorithm

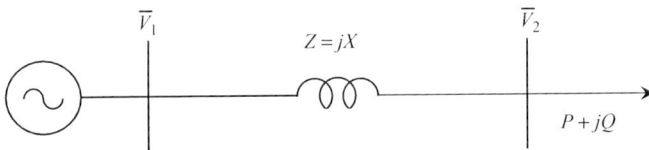

Figure 3.13 Two bus system with load at the end bus

Since $\bar{V}_1 = 1\angle 0° = \bar{V}_1^*$, therefore:

$$\bar{V}_2 = \bar{V}_1 - \frac{(P - jQ)}{\bar{V}_1^*}(jX) = \frac{|\bar{V}_1|^2 - (P - jQ)(jX)}{\bar{V}_1^*} = 1 - XQ - jXP$$

It can be seen from the phasor diagram that a change in P (ΔP) changes the angle by $\Delta\theta$ but has a small effect in the voltage magnitude, as shown in Figure 3.14. On the other hand, a change in Q (ΔQ) changes the voltage by ΔV but has a small effect in the angle, as shown in Figure 3.15.

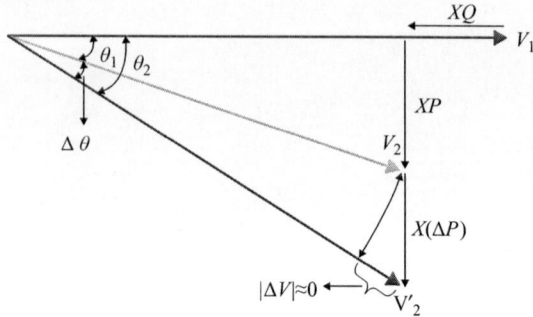

Figure 3.14   Illustration of the relationship between P and θ

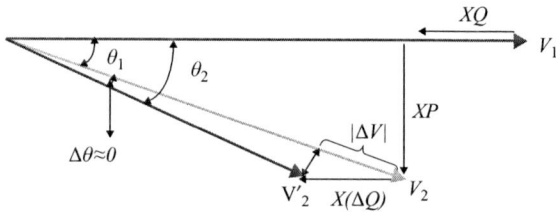

Figure 3.15   Illustration of the relationship between Q and V

Algebraically the above approximate relation may be stated as:

$$\Delta P = H \Delta \theta$$

$$\Delta Q = L' \frac{\Delta V}{V}$$

Therefore, applying the decoupling phenomena to an electrical system with several nodes, the sub-matrices N and J in equation can be neglected resulting in:

$$
\begin{bmatrix} \Delta P \\ \Delta Q \end{bmatrix} = \begin{bmatrix} H & \\ & L \end{bmatrix} * \begin{bmatrix} \Delta \theta \\ \Delta V/V \end{bmatrix}
$$

$$\Delta P_i = \left[\sum_{k=1, k \neq i} V_i V_k (G_{ik} \sin \theta_{ik} - B_{ik} \cos \theta_{ik}) \Delta \theta_k \right] - (Q_i^{\text{calculated}} + V_i^2 B_{ii}) \Delta \theta_i$$

$$\Delta Q_i = \left[\sum_{k=1, k \neq i} V_i V_k (G_{ik} \sin \theta_{ik} - B_{ik} \cos \theta_{ik}) \frac{|\Delta V_k|}{V_k} \right] - (Q_i^{\text{calculated}} + V_i^2 B_{ii}) \frac{|\Delta V_k|}{V_i}$$

θ_{ik} is small and so $\cos\theta_{ik} = 1.0$ and $\sin\theta_{ik} = 0.0$, $G_{ik}\sin\theta_{ik} \ll B_{ik}\cos\theta_{ik}$. As a product of engineering studies, it is also found that $Q_i \ll V_i^2 B_{ii}$. Therefore, Q_i can be neglected:

$$\Delta P_i = \left[\sum_{k=1,k\neq i} -V_i V_k B_{ik}\Delta\theta_k \right] - (V_i^2 B_{ii})\Delta\theta_i$$

$$\frac{\Delta P_i}{V_i} = \left[\sum_{k=1,k\neq i} -V_k B_{ik}\Delta\theta_k \right] - (V_i B_{ii})\Delta\theta_i$$

$$\Delta Q_i = \left[\sum_{k=1,k\neq i} -V_i V_k B_{ik}\frac{|\Delta V_k|}{V_k} \right] - (V_i^2 B_{ii})\frac{|\Delta V_k|}{V_i}$$

$$\frac{\Delta Q_i}{V_i} = \left[\sum_{k=1,k\neq i} -V_i V_k B_{ik}\frac{|\Delta V_k|}{V_i V_k} \right] - (V_i^2 B_{ii})\frac{|\Delta V_k|}{V_i^2} = \left[\sum_{k=1,k\neq i} -B_{ik}|\Delta V_k| \right] - (B_{ii})|\Delta V_k|$$

Further approximation can be made by setting voltage on the right-hand side of equation to unit results in:

$$\frac{\Delta P_i}{V_i} = -\left[\sum_{k=1,k\neq i} B_{ik}\Delta\theta_k + (\Delta\theta_i B_{ii}) \right] = -B'\Delta\theta$$

$$\frac{\Delta Q_i}{V_i} = -\left[\sum_{k=1,k\neq i} B_{ik}|\Delta V_k| + (B_{ii}|\Delta V_k|) \right] = -B''|\Delta V|$$

The application of this simplification, which solves separately the P-θ and Q-V, renders a 4:1 saving on storage of the Jacobian matrix. This results in more iterations, which is compensated by saving in inversion time and shorter individual iteration time.

Example 3.6 For the system shown in Figure 3.16, calculate the voltages and powers at Meadow and Harrisburg nodes for the first two iterations using the Newton–Raphson method.

Bear in mind the following information:
Base power: 100 MVA
Convergence value: 0.1 MVA
Reactive power limits for bus 1: −5 MVAR and 10 MVAR

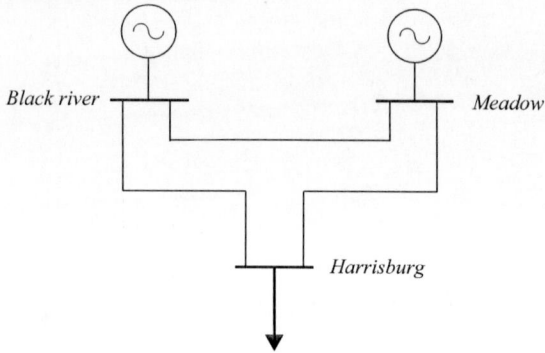

Figure 3.16   Power system for example 3.6

The voltages and active/reactive power at each node are indicated in Table 3.2. The line impedances are indicated in Table 3.3.

Table 3.2   Values of voltages for example 3.6

Node	Name	Voltage	Power
0	Black River	1.01	–
1	Meadow	1.00	$P_G = 55$ MW
2	Harrisburg	–	$P_C = 105$ MW, $Q_C = 65$ MVAR

Table 3.3   Values of impedances for example 3.6

Lines	Impedances in pu
0–1	$0.015 + 0.035j$
0–2	$0.015 + 0.055j$
1–2	$0.0075 + 0.0175j$

$$y_{01} = \frac{1}{0.015 + j0.035} = 10.34 - j24.14$$

$$y_{02} = \frac{1}{0.015 + j0.055} = 4.62 - j16.92$$

$$y_{12} = \frac{1}{0.0075 + j0.0175} = 20.69 - j48.28$$

The bus admittance matrix is:

$$Y = \begin{bmatrix} y_{01} + y_{02} & -y_{01} & -y_{02} \\ -y_{01} & y_{01} + y_{12} & -y_{12} \\ -y_{02} & -y_{12} & y_{02} + y_{12} \end{bmatrix}$$

$$Y = \begin{bmatrix} 14.96 - j41.06 & -10.34 + j24.14 & -4.62 + j16.92 \\ -10.34 + j24.14 & 31.03 - j72.42 & -20.69 + j48.28 \\ -4.62 + j16.92 & -20.69 + j48.28 & 25.31 - j65.20 \end{bmatrix}$$

With the data given, Table 3.4 can be formed.

Table 3.4 Initial input data for example 3.6

Node	Type	Data	Unknown
0	Slack	$V_0 = 1.01$ $\theta_0 = 0.0$	P_0 Q_0
1	PV	$V_1 = 1.00$ $P_1 = 0.55$	θ_1 Q_1
2	PQ	$P_2 = -1.05$ $Q_2 = -0.65$	θ_2 V_2

The power equations injected at each node are:

$$P_i^{cal} = V_i \sum_j V_j \left(G_{ij} \cos \theta_{ij} + B_{ij} \sin \theta_{ij} \right)$$

$$Q_i^{cal} = V_i \sum_j V_j \left(G_{ij} \sin \theta_{ij} - B_{ij} \cos \theta_{ij} \right)$$

$$\Delta P = P_i^{sp} - P_i^{cal} = 0$$

$$\Delta Q = Q_i^{sp} - Q_i^{cal} = 0$$

$$P_i^{sp} = P_{Gi} - P_{Ci} \text{ and } Q_i^{sp} = Q_{Gi} - Q_{Ci}$$

For the first iteration:

$$V^0 = \begin{bmatrix} 1.01 \\ 1.00 \\ 1.00 \end{bmatrix} \quad \theta^0 = \begin{bmatrix} 0 \\ 0 \\ 0 \end{bmatrix}$$

$$P_1^{(0)} = \left(V_1^2 G_{11} \right) + \left(V_0 (G_{10} \cos \theta_{10} + B_{10} \sin \theta_{10}) + V_2 (G_{12} \cos \theta_{12} + B_{12} \sin \theta_{12}) \right) V_1$$

$$P_1^{(0)} = \left((1.0)^2 (31.03) \right) + \left(\begin{array}{l} (1.01)(-10.34 \cos(0) + 24.14 \sin(0)) + \\ (1.0)(-20.69 \cos(0) + 48.28 \sin(0)) \end{array} \right) (1.0)$$

$$P_1^{(0)} = -0.103$$

$$P_2^{(0)} = \left(V_2^2 G_{22}\right) + \left(V_0(G_{20}\cos\theta_{20} + B_{20}\sin\theta_{20}) + V_1(G_{21}\cos\theta_{21} + B_{21}\sin\theta_{21})\right)V_2$$

$$P_2^{(0)} = \left((1.0)^2(25.31)\right) + \left(\begin{array}{l}(1.01)(-4.62\cos(0) + 16.92\sin(0)) + \\ (1.0)(-20.69\cos(0) + 48.28\sin(0))\end{array}\right)(1.0)$$

$$P_2^{(0)} = -0.046$$

$$Q_2^{(0)} = \left(V_2^2 B_{22}\right) + \left(V_0(G_{20}\sin\theta_{20} + B_{20}\cos\theta_{20}) + V_1(G_{21}\sin\theta_{21} + B_{21}\cos\theta_{21})\right)V_2$$

$$Q_2^{(0)} = -\left((1.0)^2(-65.20)\right) + \left(\begin{array}{l}(1.01)(4.62\sin(0) - 16.92\cos(0)) + \\ (1.00)(20.69\sin(0) - 48.28\cos(0))\end{array}\right)(1.0)$$

$$Q_2^{(0)} = -0.169$$

Therefore:

$$0.55 - P_1^{\text{cal}} = 0$$

$$-1.05 - P_2^{\text{cal}} = 0$$

$$-0.65 - Q_2^{\text{cal}} = 0$$

The Jacobian for the first iteration is:

$$J = \begin{bmatrix} \dfrac{\partial P_1}{\partial \theta_1} & \dfrac{\partial P_1}{\partial \theta_2} & \dfrac{\partial P_1}{\partial |V_2|} \\[2ex] \dfrac{\partial P_2}{\partial \theta_1} & \dfrac{\partial P_2}{\partial \theta_2} & \dfrac{\partial P_2}{\partial |V_2|} \\[2ex] \dfrac{\partial Q_2}{\partial \theta_1} & \dfrac{\partial Q_2}{\partial \theta_2} & \dfrac{\partial Q_2}{\partial |V_2|} \end{bmatrix}$$

Then, the following results for ΔP and ΔQ are obtained:

$$\Delta P = \begin{bmatrix} - \\ 0.653 \\ -1.004 \end{bmatrix} \quad \Delta Q = \begin{bmatrix} - \\ - \\ -0.481 \end{bmatrix}$$

The Jacobian components are:

$$\frac{\partial P_1}{\partial \theta_1} = -(1.00)\left(\begin{array}{l}(-10.34\sin(0) - 24.14\cos(0))(1.01) + \\ (-20.69\sin(0) - 48.28\cos(0))(1.00)\end{array}\right) = 72.66$$

$$\frac{\partial P_1}{\partial \theta_2} = \frac{\partial P_2}{\partial \theta_1} = (1.00)(-20.69\sin(0) - 48.28\cos(0))(1.00) = -48.28$$

$$\frac{\partial P_2}{\partial \theta_2} = -(1.00)\left(\begin{array}{c} (-4.62\sin(0) - 16.92\cos(0))(1.01)+ \\ (-20.69\sin(0) - 48.28\cos(0))(1.01) \end{array} \right) = 65.37$$

$$\frac{\partial Q_2}{\partial \theta_1} = -(1.00)(-20.69\cos(0) + 48.28\sin(0))(1.00) = 20.69$$

$$\frac{\partial Q_2}{\partial \theta_2} = (1.00)\left(\begin{array}{c} (-4.62\cos(0) + 16.92\sin(0))(1.01)+ \\ (-20.69\cos(0) + 48.28\sin(0))(1.00) \end{array} \right) = -25.36$$

$$\frac{\partial P_2}{\partial |V_2|} = (1.00)(-20.69\cos(0) + 48.28\sin(0))(1.00) = -20.69$$

$$\frac{\partial P_2}{\partial |V_2|} = \frac{\partial Q_2}{\partial \theta_2} + 2|V_2|^2 G_{22} = -25.36 + 2|1|^2 25.31 = 25.26$$

$$\frac{\partial Q_2}{\partial |V_2|} = -\frac{\partial P_2}{\partial \theta_2} - 2|V_2|^2 B_{22} = -65.37 + 2|1|^2(65.20) = 65.03$$

$$J = \begin{bmatrix} 72.66 & -48.28 & -20.69 \\ -48.28 & 65.37 & 25.26 \\ 20.69 & -25.36 & 65.03 \end{bmatrix}$$

$$b = \begin{bmatrix} \Delta P_1 \\ \Delta P_2 \\ \Delta Q_2 \end{bmatrix} = \begin{bmatrix} 0.654 \\ -1.004 \\ -0.481 \end{bmatrix}$$

Solving for $J\Delta x = b$, with $x = [\begin{array}{ccc} \Delta\theta_1 & \Delta\theta_2 & \Delta V_2/V_2 \end{array}]T$:

$$\Delta x = \begin{bmatrix} -0.0030 \\ -0.0131 \\ -0.0116 \end{bmatrix}$$

$$\theta_1^{(1)} = \theta_1^{(0)} + \Delta\theta_1^{(0)} = 0 - 0.0030 = -0.0030 = -0.172°$$

$$\theta_2^{(1)} = \theta_2^{(0)} + \Delta\theta_2^{(0)} = 0 - 0.0131 = -0.0116 = -0.751°$$

$$V_2^{(1)} = V_2^{(0)}\left(1 + \frac{\Delta V_2^{(0)}}{V_2^{(0)}}\right) = 1.00(1 - 0.0116) = 0.988$$

$$V^{(1)} = \begin{bmatrix} 1.01 \\ 1.00 \\ 0.988 \end{bmatrix} V \qquad \theta^{(1)} = \begin{bmatrix} 0 \\ -0.0030 \\ -0.0131 \end{bmatrix} \text{rads}$$

Then:

$$P_0^{(1)} = \left((1.01)^2 (14.97) \right)$$

$$+ \left(\begin{array}{l} (1.00)(-10.35\cos(0.0030) + 24.14\sin(0.0030)) + \\ (0.988)(-4.62\cos(0.0131) + 16.93\sin(0.0131)) \end{array} \right) (1.01)$$

$$P_0^{(1)} = 15.271 - 10.38 - 4.389 = 0.502$$

$$P_1^{(1)} = \left((1.00)^2 (31.04) \right)$$

$$+ \left(\begin{array}{l} (1.01)(-10.35\cos(-0.0030) + 24.14\sin(-0.0030)) + \\ (0.988)(-20.69\cos(0.0101) + 48.28\sin(0.0101)) \end{array} \right) (1.00)$$

$$P_1^{(1)} = 31.04 - 10.527 - 19.959 = 0.554$$

$$P_2^{(1)} = \left((0.988)^2 (25.31) \right)$$

$$+ \left(\begin{array}{l} (1.01)(-4.62\cos(-0.0131) + 16.93\sin(-0.0131)) + \\ (1.00)(-20.69\cos(-0.0101) + 48.28\sin(-0.0101)) \end{array} \right) (0.988)$$

$$P_2^{(1)} = 24.706 - 4.831 - 20.922 = -1.047$$

$$Q_0^{(1)} = -\left((1.01)^2 (-41.07) \right)$$

$$+ \left(\begin{array}{l} (1.00)(-10.35\sin(0.003) - 24.14\cos(0.003)) + \\ (0988)(-4.62\sin(0 + 0.0131) - 16.93\cos(0.0131)) \end{array} \right) (1.01)$$

$$Q_0^{(1)} = 41.896 - 24.413 - 16.953 = 0.53$$

$$Q_1^{(1)} = -\left((1.00)^2 (-72.42) \right)$$

$$+ \left(\begin{array}{l} (1.01)(-10.35\sin(-0.003) - 24.14\cos(-0.003)) + \\ (0.988)(-20.69\sin(0.0101) - 48.28\cos(0.0101)) \end{array} \right) (1.00)$$

$$Q_1^{(1)} = 72.42 - 24.35 - 47.905 = 0.165$$

$$Q_2^{(1)} = -\left((0.988)^2 (-65.21) \right)$$

$$+ \left(\begin{array}{l} (1.01)(-4.62\sin(-0.0131) - 16.93\cos(-0.0131)) + \\ (1.00)(-20.69\sin(-0.0101) - 48.28\cos(-0.0101)) \end{array} \right) (0.988)$$

$$Q_2^{(1)} = 63.654 - 16.832 - 47.492 = -0.67$$

$$P_i^{cal} = \begin{bmatrix} 0.502 \\ 0.554 \\ -1.047 \end{bmatrix} \quad Q_i^{cal} = \begin{bmatrix} 0.530 \\ 0.165 \\ -0.670 \end{bmatrix}$$

$$\Delta P = \begin{bmatrix} - \\ -0.004 \\ -0.003 \end{bmatrix} \quad \Delta Q = \begin{bmatrix} - \\ - \\ 0.02 \end{bmatrix}$$

These are the results for the first iteration. The residues exceed the limit of 10^{-1}. Furthermore, the maximum reactive power that the generator 1 could produce is 10 MVAR and the Q_i^{cal} obtained is 16.5 MVAR. It makes that the generator 1 is out of the limits. For that reason, the node 1 changes to a PQ node on the next iteration, maintaining the reactive power in $Q_i^{cal} = 0.165$ pu. For this change, the Jacobian that needs to be found on the second iteration would be the corresponding to two PQ nodes.

The same process is followed for the next iterations until the algorithm converges. Final results are obtained using a software package and are presented in Figure 3.17. They correspond to the same pu values that are given in the problem formulation. However, they are shown in kV, MW, and MVAR since the package used required inputs in real magnitudes.

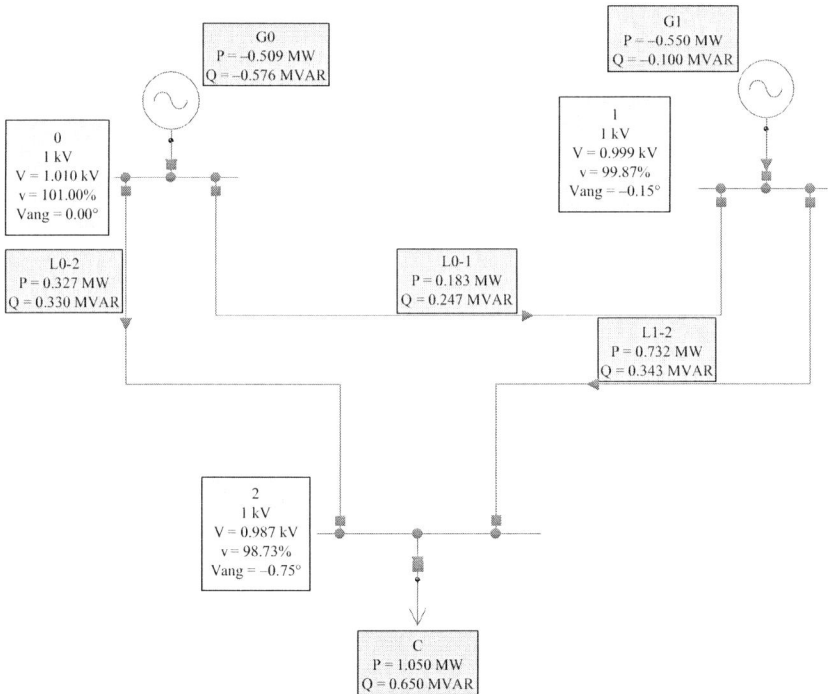

Figure 3.17   Load flow results in pu for the system of example 3.6

In order to present the results with realistic values, the buses are assumed with voltage levels of 115 kV, and the base magnitudes as 100 MVA and 115 kV. The line impedances are also expressed in the same values. By making these adjustments, the new load flow obtained is presented in Figure 3.18 which of course has equivalent values of those of the previous figure.

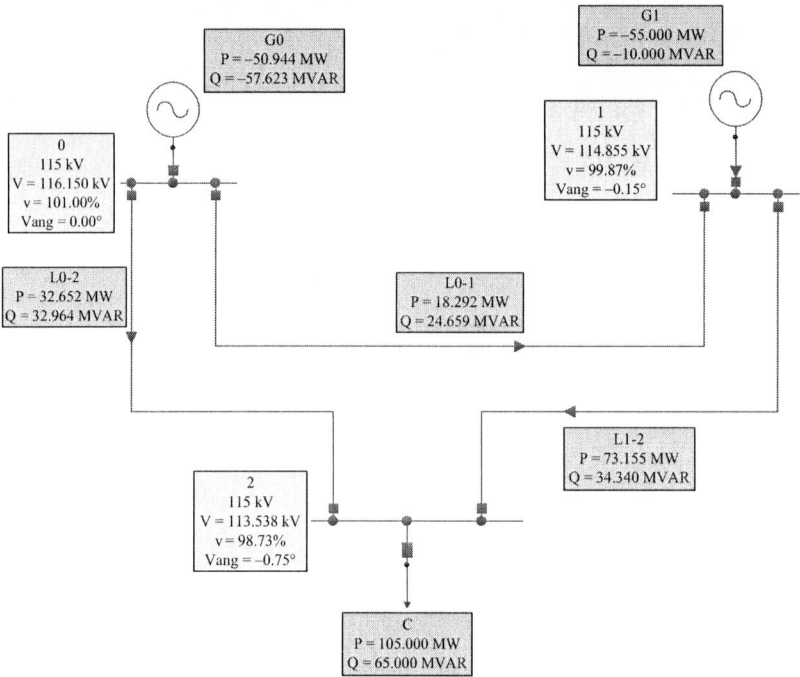

Figure 3.18   Load flow results in real magnitudes for the system of example 3.6

Example 3.7 An example taken from a power system at 230/115/34.5/13.2 is used here to illustrate the application of load flow. Figure 3.19 shows the single line of the system. Rated bus voltages are given in Table 3.5. Tables 3.6 and 3.7 give the HV and LV transformer data. Table 3.8 gives the generator data. Bus and line data are given in Tables 3.9 and 3.10.

Load flow analysis is required to assess the voltage levels and determine remedial actions where required.

Load flows will be run to analyze the behavior of the system under normal conditions. If violations occur, measures to overcome them have to be proposed.

The load flow for the base condition was run using a software package. The results are given in graphical form in Figure 3.20 for a portion of the system from which it is clear that the voltages at buses Willow 34.5 kV and Willow 13.2 kV are

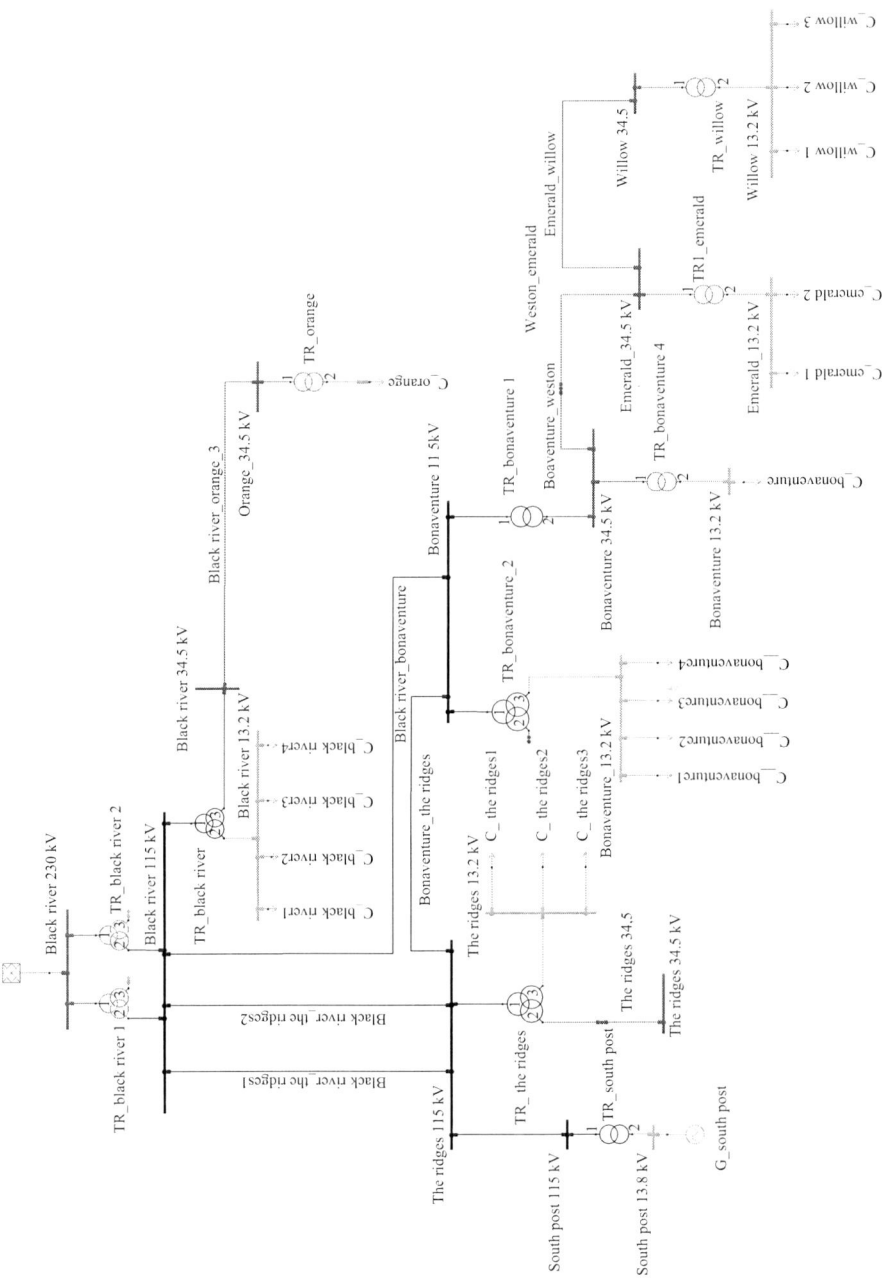

Figure 3.19 Single-line diagram for example 3.7

Table 3.5   Rated bus voltages for example 3.7

Bus	V_n (kV)	Bus	V_n (kV)
THE RIDGES 115 kV	115	BONAVENTURE_13.2 kV	13.2
THE RIDGES 13.2 kV	13.2	BONAVENTURE2_1	34.5
THE RIDGES 34.5 KV	34.5	EMERALD	34.5
BLACK RIVER	13.2	EMERALD 1	13.1
BLACK RIVER 115 kV	115	ORANGE_34.5 kV	34.5
BLACK RIVER 13.2 kV	13.2	ORANGE13.2 kV	13.2
BLACK RIVER 230 kV	230	SOUTH POST 115 kV	115
BLACK RIVER 34.5 kV	34.5	SOUTH POST 13.8 kV	13.8
BLACK RIVER 13.8 kV	13.8	THE RIDGES1 34.5 kV	34.5
BONAVENTURE 115 kV	115	WESTON	34.5
BONAVENTURE 13.2 kV	13.2	WILLOW 13.2 kV	13.2
BONAVENTURE 34.5 kV	34.5	WILLOW 34.5 kV	34.5

0.91 and 0.902, respectively, which are below the allowed range of $\pm 5\%$. The following actions will be considered to overcome the problem:

- installation of capacitors at Willow 13.2 kV bus,
- operation of the tap changer of the transformer connecting buses Willow 34.5 kV and Willow 13.2 kV, and
- adding another line between nodes Emerald 34.5 kV–Willow 34.5 kV.

One of the remedial actions is the installation of capacitors. A capacitor of 3.36 MVAR has to be installed at Willow 13.2 kV bus in order to push the voltage to 1.0 pu. The results are shown in Figure 3.21, and the results indicate that the new voltages at buses Willow 34.5 kV and Willow 13.2 kV are 0.963 and 1.0, respectively, which demonstrates the validity of the approach.

Other action that can be considered is the operation of the tap changer of the transformer connecting buses Willow 34.5 kV and Willow 13.2 kV. As shown in Figure 3.22 a tap setting at 7 is required to take the voltage at the low voltage side to 1 pu. In this case although the voltage at the bus Willow 34.5 kV is not 1.00, the improvement is clear.

Finally the option of adding another line between nodes Emerald 34.5 kV– Willow 34.5 kV was considered. The results are shown in Figure 3.23. Here the voltage of Willow 34.5 kV improves but still does not fit within the accepted range. The voltage of Willow 13.2 kV is also below the accepted range. Actually this option is mainly considered when the initial line is overloaded.

Other actions can be analyzed in addition to those mentioned before, like increasing the wire size of the transmission line connecting buses Emerald 34.5 kV and Willow 34.5 kV if this is practical, or adding series shunt capacitors. Normally, to find the optimal case, a number of cases have to be run and the solution eventually could be a combination of several actions where economic reasons should be considered too.

Table 3.6 HV transformer data for power system of example 3.7

Transformer	S_{r12} (MVA)	S_{r23} (MVA)	S_{r31} (MVA)	V_{r1} (kV)	V_{r2} (kV)	V_{r3} (kV)	Z_{cc12} (%)	Z_{cc23} (%)	Z_{cc31} (%)	Vector group	Tap (oper)	Tap (min)	Tap (nom)	Tap (max).	ΔV (%)
TR_THE RIDGES	30	30	15	115.0	34.5	13.2	8.8	8.8	8.8	YNd11yn0	4	−7	0	7	−1.25
TR_BLACK RIVER 1	90	90	90	230.0	115.0	13.8	10.0	12.0	25.0	YNyn0d5	1	−10	0	14	−1.20
TR_BLACK RIVER 2	90	90	90	230.0	115.0	13.8	10.0	12.0	25.0	YNyn0d5	1	−10	0	14	−1.20
TR_BLACK RIVER	12	12	12	115.0	34.5	13.8	7.6	3.3	11.3	YNd1yn2	3	−4	0	4	−1.25
TR_BONAVENTURE_2	22	22	22	115.0	34.5	13.8	5.0	7.0	9.6	Dyn5yn5	0	−7	0	7	−1.25

Table 3.7   MV transformer data for power system of example 3.7

Transformer	S_r (MVA)	V_{r1} (kV)	V_{r2} (kV)	Z_{cc} (1) (%)	Vector group	Tap (oper)	Tap (mín)	Tap (nom)	Tap (max)	ΔV (%)
TR_ORANGE	3	34.5	13.2	4.9	Dyn5	0	−3	0	3	−1.25
TR_BONA-VENTURE 1	30	115.0	34.5	9.0	YNyn0	10	1	7	15	−1.25
TR1_EMERALD	10	34.5	13.8	6.1	Dyn5	2	−5	0	5	−1.25
TR_WILLOW	10	34.5	13.8	10.3	Dyn11	0	−10	0	10	−1.25
TR_BONA-VENTURE 4	10	34.5	13.8	12.5	Dyn5	0	−3	0	3	−1.25
TR_SOUTH POST	50	115.0	13.8	8.0	YNd1	2	−1	0	3	2.5

Table 3.8   Generator data for power system of example 3.7

Generator	S_r (MVA)	V_r (kV)	$\cos \phi$ (MW)	P_G (MVAR)	Q_G
G_SOUTH POST	68.8	13.8	0.8	15	13

Table 3.9   Bus data for power system of example 3.7

Load	P (MW)	Q (MVAR)	S (MVA)	I (kA)	$\cos \phi$
C_ORANGE	1.0	0.6	1.2	0.1	0.9
C_EMERALD 1	3.7	2.7	4.6	0.2	0.8
C_EMERALD 2	3.1	2.3	3.8	0.2	0.8
C_BONAVENTURE	2.0	1.0	2.2	0.1	0.9
C_BLACK RIVER 2	1.3	1.0	1.6	0.1	0.8
C_BLACK RIVER 1	3.5	2.6	4.3	0.2	0.8
C_BLACK RIVER 3	0.1	0.1	0.1	0.0	0.9
C_BLACK RIVER 4	0.0	0.0	0.0	0.0	0.0
C_WILLOW 2	1.1	0.8	1.3	0.1	0.8
C_WILLOW 1	1.1	0.8	1.3	0.1	0.8
C_WILLOW 3	1.1	0.8	1.3	0.1	0.8
C_BONAVENTURE 4	4.7	3.5	5.9	0.3	0.8
C_BONAVENTURE 3	4.3	3.2	5.4	0.2	0.8
C_BONAVENTURE 2	1.5	1.1	1.9	0.1	0.8
C__BONAVENTURE 1	4.0	2.9	5.0	0.2	0.8
C_THE RIDGES 1	1.1	0.8	1.4	0.1	0.8
C_THE RIDGES 3	5.2	3.9	6.5	0.3	0.8
C_THE RIDGES 2	4.9	3.6	6.1	0.3	0.8

Table 3.10 Line data for power system of example 3.7

Line	Long (km)	R(1) (Ohm/ km)	X(1) (Ohm/ km)	C(1) (µF/ km)	B(1) (µS/ km)	R(0) (Ohm/ km)	X(0) (Ohm/ km)	B(0) (µS/ km)	I,max (down- stream) A	Bus to bus	
L_BLACK RIVER_ORANGE_3	9.6	0.4	0.5	0.0	4.6	0.5	1.9	2.1	340	BLACK RIVER 34.5 kV	ORANGE_34.5 kV
BOAVENTURE_WESTON	2.5	0.2	0.4	0.0	4.4	0.5	1.7	2.4	434	WESTON	BONAVENTURE 34.5 kV
WESTON_EMERALD	1.6	0.2	0.4	0.0	4.4	0.5	1.7	2.4	434	WESTON	EMERALD
EMERALD_WILLOW	20.2	0.2	0.4	0.0	4.4	0.5	1.7	2.4	434	EMERALD	WILLOW 34.5
BLACK RIVER_THE RIDGES1	12.0	0.1	0.5	0.0	3.4	0.2	0.9	2.4	0	THE RIDGES 115 kV	BLACK RIVER 115 kV
L_THE RIDGES_SOUTH POST115 kV	0.1	0.1	0.5	0.0	3.4	0.2	0.9	2.4	0	SOUTH POST 115 kV	THE RIDGES 115 kV
EMERALD_WILLOW2	20.2	0.2	0.4	0.0	4.4	0.5	1.7	2.4	434	EMERALD	WILLOW 34.5
L_THE RIDGES 34.5	0.3	0.2	0.4	0.0	4.4	0.5	1.7	2.4	434	THE RIDGES1 34.5 kV	THE RIDGES 34.5 kV
BLACK RIVER_BONAVENTURE	10.0	0.1	0.5	0.0	3.4	0.2	0.9	2.4	0	BLACK RIVER 115 kV	BONAVENTURE 115 kV
BLACK RIVER_THE RIDGES2	12.0	0.1	0.5	0.0	3.4	0.2	0.9	2.4	0	BLACK RIVER 115 kV	THE RIDGES 115 kV
BONAVENTURE_THE RIDGES	9.1	0.1	0.5	0.0	1.7	0.4	1.6	0.0	0	BONAVENTURE 115 kV	THE RIDGES 115 kV

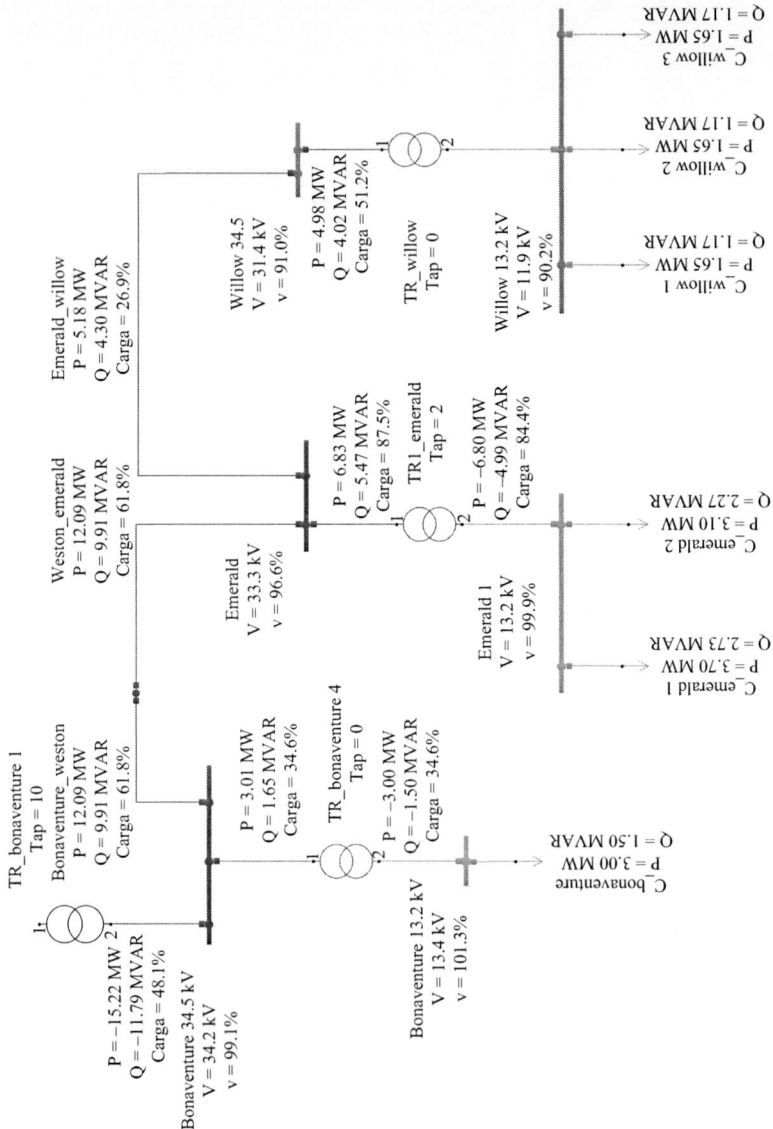

Figure 3.20   Portion of the power system of example 3.7

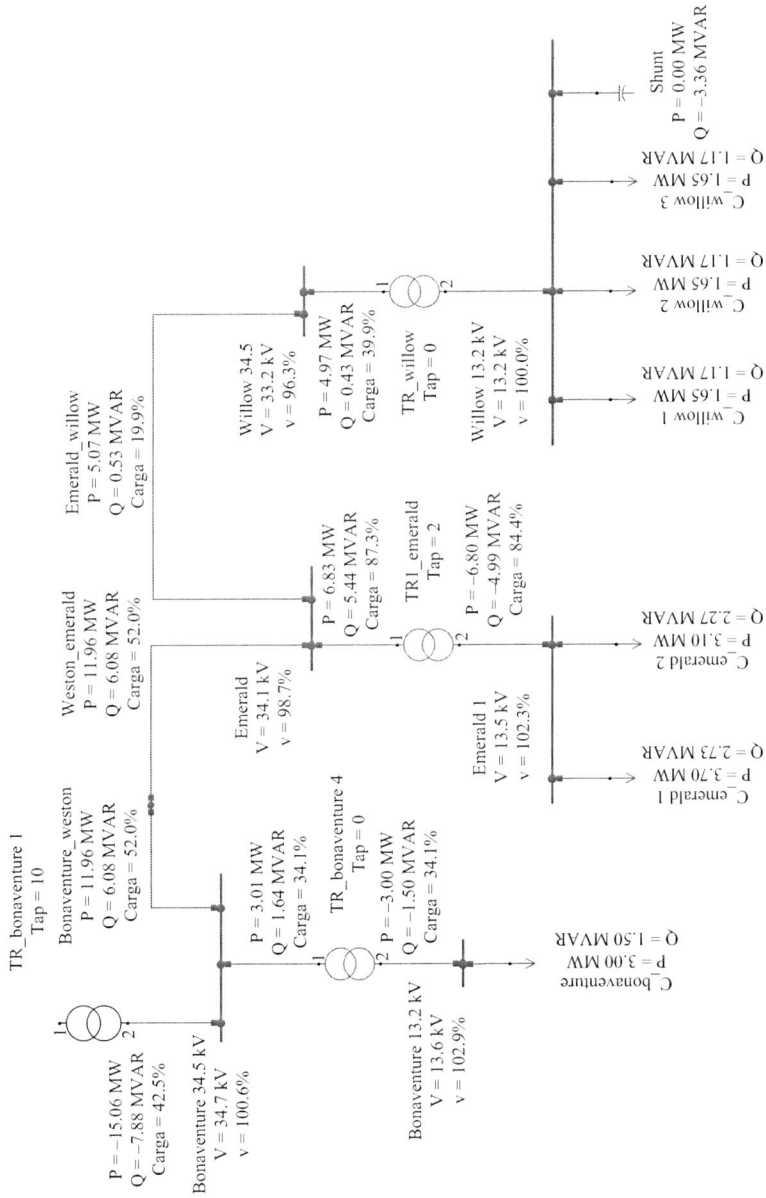

TR_bonaventure 1
Tap = 10

Bonaventure_weston
P = 11.96 MW
Q = 6.08 MVAR
Carga = 52.0%

Weston_emerald
P = 11.96 MW
Q = 6.08 MVAR
Carga = 52.0%

Emerald_willow
P = 5.07 MW
Q = 0.53 MVAR
Carga = 19.9%

Shunt
P = 0.00 MW
Q = –3.36 MVAR

P = –15.06 MW
Q = –7.88 MVAR
Carga = 42.5%

Bonaventure 34.7 kV
V = 34.7 kV
v = 100.6%

P = 3.01 MW
Q = 1.64 MVAR
Carga = 34.1%

TR_bonaventure 4
Tap = 0

P = –3.00 MW
Q = –1.50 MVAR
Carga = 34.1%

Bonaventure 13.2 kV
V = 13.6 kV
v = 102.9%

C_bonaventure
P = 3.00 MW
Q = 1.50 MVAR

Emerald
V = 34.1 kV
v = 98.7%

P = 6.83 MW
Q = 5.44 MVAR
Carga = 87.3%

TR1_emerald
Tap = 2

P = –6.80 MW
Q = –4.99 MVAR
Carga = 84.4%

Emerald 1
V = 13.5 kV
v = 102.3%

C_emerald 1
P = 3.70 MW
Q = 2.73 MVAR

C_emerald 2
P = 3.10 MW
Q = 2.27 MVAR

Willow 34.5
V = 33.2 kV
v = 96.3%

P = 4.97 MW
Q = 0.43 MVAR
Carga = 39.9%

TR_willow
Tap = 0

Willow 13.2 kV
V = 13.2 kV
v = 100.0%

C_willow 1
P = 1.65 MW
Q = 1.17 MVAR

C_willow 2
P = 1.65 MW
Q = 1.17 MVAR

C_willow 3
P = 1.65 MW
Q = 1.17 MVAR

Figure 3.21   Bus voltage results considering a capacitor bank at 13.2 kV Willow bus

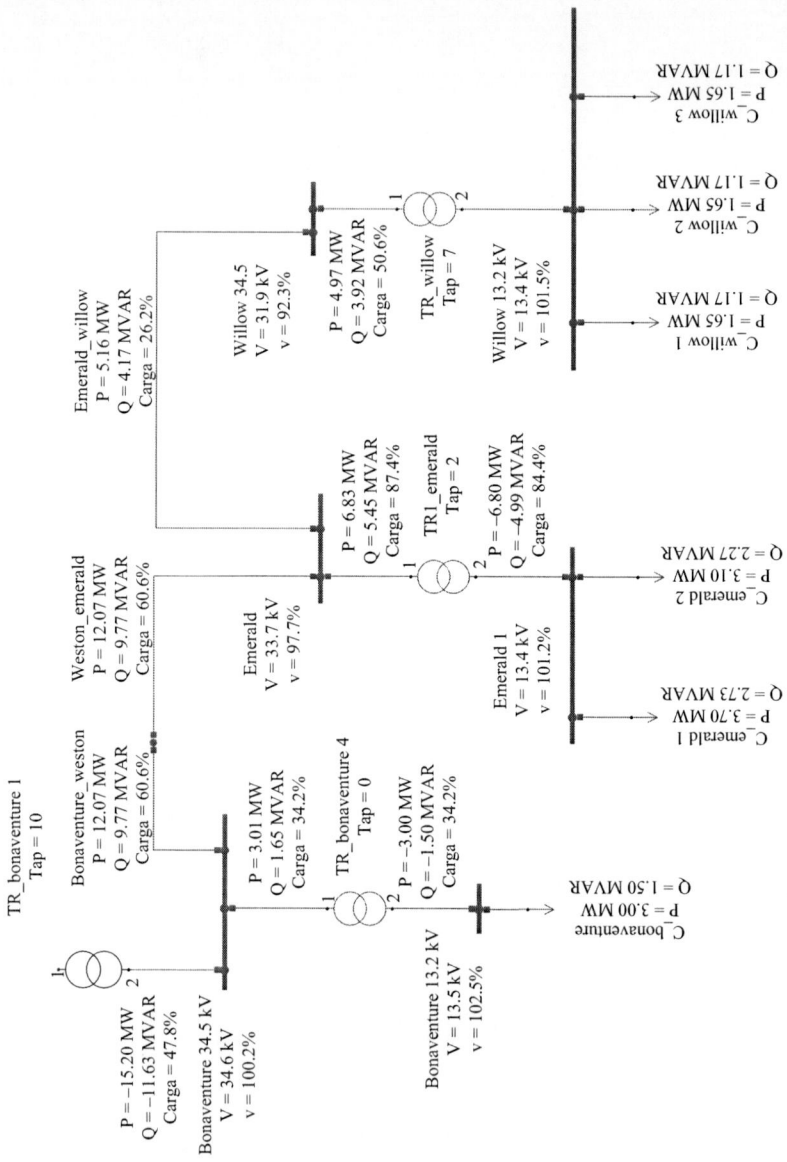

Figure 3.22   Bus voltage results of example 3.7 considering tap changer operation

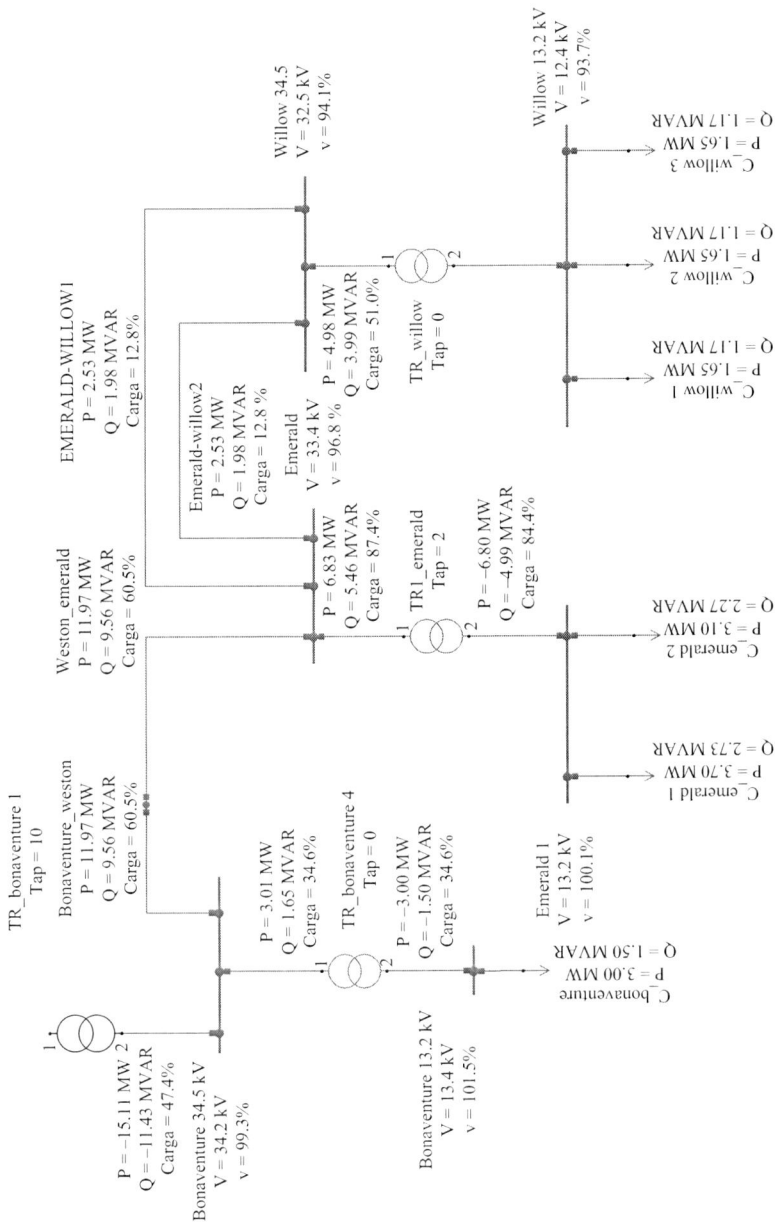

Figure 3.23   Bus voltage results of example 3.7 considering a new line

3.6   Radial load flow concepts

Power systems are analyzed by efficient algorithms which conduct almost invariably to software packages. This should not though prevent hand calculations, which are appropriate especially for small systems. Up to the mid-1960s before use of computer, big systems were analyzed with the aid of board simulations.

Nowadays there are flexible and affordable software packages that allow comprehensive analysis. In this section the basis and structure of the two main tools are covered focusing on distribution systems. These tools are load flow and short circuit calculations.

Standard algorithms like of Newton–Raphson or Gauss–Seidel can be used for the solution of distribution systems. However, it is better for these cases to use programs designed for radial systems for two main reasons:

- Programs designed for power systems normally assume a high X/R ratio.
- A program developed specifically for distribution systems will be more efficient and simpler than those developed for high-voltage systems.

This method has low memory requirement and good accuracy. On the other hand, it is simple to be implemented and has good convergence speed.

3.6.1   Theoretical background

The simplest way to start the analysis of radial feeders is by examining Figure 3.24.

$$\bar{V}_s - \bar{V}_r = \bar{I}(R + jX) \tag{3.32}$$

where:

$\quad\quad$ s $\quad$ Source node
$\quad\quad$ r $\quad$ Receiving node
$\quad\quad$ V_s $\quad$ Voltage magnitude in source node
$\quad\quad$ V_r $\quad$ Voltage magnitude in receiving node
$\quad$ P, Q $\quad$ Active and reactive load
$\quad$ R, X $\quad$ Branch resistance and reactance

It is known that the vector value of current is defined by the following expression:

$$\bar{I} = \frac{P - jQ}{\bar{V}_r^{\,*}} \tag{3.33}$$

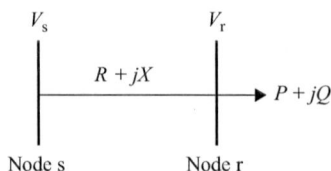

Figure 3.24   Two-node systems

Then:

$$(V_s\angle\theta_s) - (V_r\angle\theta_r) = \frac{(P - jQ)(R + jX)}{(V_r\angle - \theta_r)} \tag{3.34}$$

$$(V_s\angle\theta_s)(V_r\angle - \theta_r) - V_r^2 = (PR + QX) + j(PX - QR) \tag{3.35}$$

By separating the real and imaginary parts:

$$V_s V_r \cos(\theta_s - \theta_r) - V_r^2 = PR + QX \tag{3.36}$$

$$V_s V_r \sin(\theta_s - \theta_r) = PX - QR \tag{3.37}$$

By squaring (3.36) and (3.37):

$$V_s^2 V_r^2 \cos^2(\theta_s - \theta_r) = \left(V_r^2 + (PR + QX)\right)^2 \tag{3.38}$$

$$V_s^2 V_r^2 \sin^2(\theta_s - \theta_r) = (PX - QR)^2 \tag{3.39}$$

$$\cos^2(\theta_s - \theta_r) + \sin^2(\theta_s - \theta_r) = 1 \tag{3.40}$$

And adding up (3.38) and (3.39) the following expression is obtained:

$$V_s^2 V_r^2 = V_r^4 + 2V_r^2(PR + QX) + P^2R^2 + 2PXQR + Q^2X^2 \\ + P^2X^2 - 2PXQR + Q^2R^2 \tag{3.41}$$

By grouping and reducing terms:

$$V_s^2 \cdot V_r^2 = V_r^4 + 2V_r^2(PR + QX) + P^2(R^2 + X^2) + Q^2(R^2 + X^2) \tag{3.42}$$

$$V_s^2 V_r^2 = V_r^4 + 2V_r^2(PR + QX) + (P^2 + Q^2)(R^2 + X^2) \tag{3.43}$$

and finally:

$$V_r^4 + \left(2(PR + QX) - V_s^2\right)V_r^2 + (P^2 + Q^2)(R^2 + X^2) = 0 \tag{3.44}$$

The solution for V_r at the bus load of Figure 3.19 is obtained by solving directly (3.44), which is independent of the angle. The solution is obtained easily since the equation is of quadratic form. P and Q are exact equivalents for the network connected to node r.

Active and reactive power losses can be calculated using the magnitudes for I^2R and I^2X that are obtained from the current magnitude given by the following expression:

$$I = \frac{\sqrt{P^2 + Q^2}}{V_r} \tag{3.45}$$

$$L_p = \frac{R(P^2 + Q^2)}{V_r^2} \tag{3.46}$$

$$L_q = \frac{X(P^2 + Q^2)}{V_r^2} \qquad (3.47)$$

where:

L_p Active losses in the branch
L_q Reactive losses in the branch

3.6.2   Distribution network models

3.6.2.1   Balanced three-phase models

Three-phase models are represented by their resistance and reactance in pu. In most cases the capacitance is neglected unless the line is very long.

All loads including capacitors for reactive compensation are represented by their active (P_0) and reactive (Q_0) components at 1.0 pu voltage. The effect of voltage variations is represented as:

$$P = P_0 V^k \qquad (3.48)$$

$$Q = Q_0 V^k \qquad (3.49)$$

where:

V Voltage magnitude
$k = 0$ Loads of constant power
$k = 1$ Loads of constant current
$k = 2$ Loads of constant impedance

The value of k can vary according to the load characteristics.

3.6.2.2   Unbalanced three-phase network

The model is based on the following considerations: Any distribution line is made up of one, two, or three phases and the neutral conductor, which is considered to be connected to ground and therefore at ground potential. This is a valid approximation in most cases. The three-phase loads are considered to be made up of three single-phase loads connected in star.

3.6.3   Nodes and branches identification

To implement the software for radial load flow and feeder reconfiguration, it is necessary to have an adequate coding for every element to permit a proper handling of the data. The methodology proposed considers the radiality of the feeders where any branch always has power flow in one direction. In this method, the name of each branch is always associated with the names of the respective end nodes. There is not a restriction to the names applied to the nodes and therefore they can be numbers, letters or a combination of both. However any name should only be used for one node in the entire distribution network.

The methodology proposed considers the radiality of the feeders where any branch always has power flow in one direction. The name of each branch is always

associated with the names of the respective end nodes. After any feeder is configured, it is possible to identify the path to get to the source from any node. Two
columns of the configured feeder are used. This process starts by linking the node
under consideration, which is in the right hand side column, to its opposite which is
in the left-hand side column.

With the branches of a feeder from a database arranged in a two-column list,
the process to determine the configuration of any feeder starts by picking up the
node opposite the branch leaving the source node. Once this node is identified, the
node or nodes at the end of the branch or branches leaving the referred node are in
turn identified. Likewise, the rest of the nodes of the feeder can be identified until
the entire feeder is configured again in a two-column arrangement. This method
does not require that the elements of the initial database be ordered, which makes
the process very flexible and quick.

After any feeder is configured, it is possible to identify the path to get to the
source from any node. To achieve this, the two columns of the configured feeder
are used. This process starts by linking the node under consideration, which is in
the right hand side column, to its opposite which is in the left-hand side column.
Then a new link is established between this last node and the next one upwards
which has the same name located at the right hand side column. The process continues until the source node is reached.

If any reconfiguration takes place, the method can easily identify the reconfigured feeders provided that clear indications are given in respect to the branches or
links that have been opened and closed. Care must be taken to ensure that any
feeder maintains its radiality and its connection to the source.

3.6.4   Illustration of nodes and branches identification

In order to illustrate the advantage of the method, Figure 3.25 shows two circuits
that need to be configured, and Figure 3.26 shows the corresponding branch identification in a two column arrangement. The branches that have operating switches
are indicated with an asterisk. The initial database shows that the branches are set
out in a random manner. The second database shows the configurations of the two
circuits. The last database shows the configurations after the switches with stars
have operated.

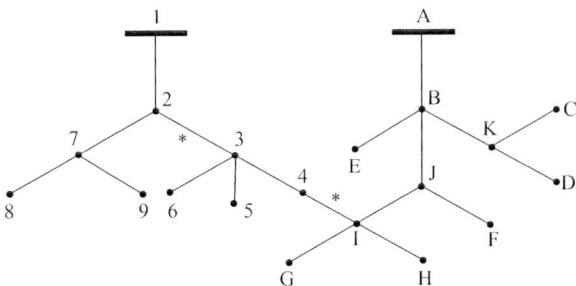

Figure 3.25   Two feeder system with numbers and letters nomenclature

1. Database

C	K
7	8
B	K
2	7
H	I
6	3
E	B
5	3
F	J
1	2
9	7
B	J
3	4
NC * 3	2
K	D
A	B
I	G
I	J
NO * 4	I

2. Initial configuration

i) Circuit with numbers

SN	RN
1	2
2	3
2	7
3	6
3	4
3	5
7	9
7	8

ii) Circuit with letters

A	B
B	E
B	K
B	J
K	C
K	D
J	I
J	F
I	G
I	H

3. Reconfigured circuits

Closing 4 – 1
Opening 2 – 3

i) Circuit with numbers

1	2
2	7
7	9
7	8

ii) Circuit with letters

A—B
B E
B K
B J
K C
K D
J I
J F
I 4
I H
I G
4 3
3 5
3 6

The line shows the path between the load connected to node 5 and the source node

Note : SN: Sending node
RN: Receiving node
NC: Normally closed
NO: Normally opened

Figure 3.26   Two-feeder identification for the example system

To illustrate the advantages of the node numbering system it should be noted that the nodes of the first feeder were coded with numbers and those of the second feeder with letters. The main advantages of this method are:

- A simple and unique identification for any branch regardless of the feeder to which it belongs.
- The ability to allow the path from any node to the source node to be established easily.
- The input of branches can be carried out in any order.
- The input of branches does not need to identify which end is a source or which is a load.

3.6.5   Algorithm to develop radial load flow

The process is started from the farthest node from the source knowing the equivalent P and Q values. For the nodes upstream, the equivalent power loads are obtained from the P and Q values assigned to the nodes and are increased by the losses across the line that are obtained with the corresponding equations. Once the equivalent power at the source is obtained, the downstream process is carried out to calculate the voltages of the nodes. The overall process is finished when a convergence value for voltage or power is attained.

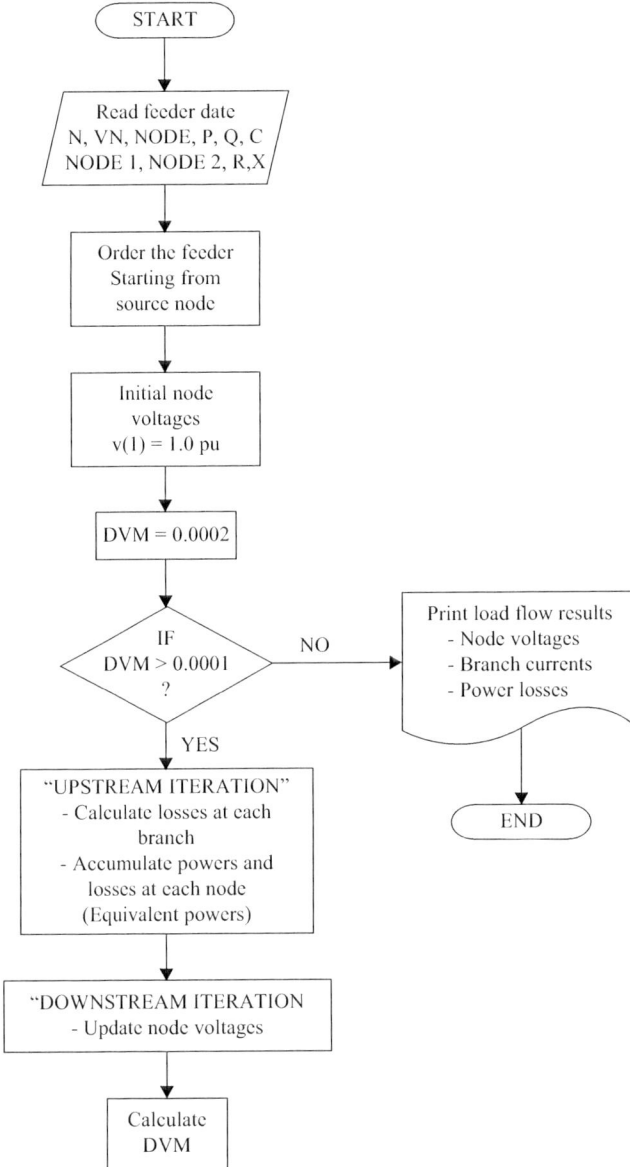

Figure 3.27   Load flow for radial systems

An algorithm is shown in Figure 3.27. This algorithm is built up as indicated in the following steps:

- Read network data including line impedances, source voltages, and power values at the nodes.
- Assign voltages to the nodes for the initial start and calculate equivalent powers depending on voltage magnitudes.

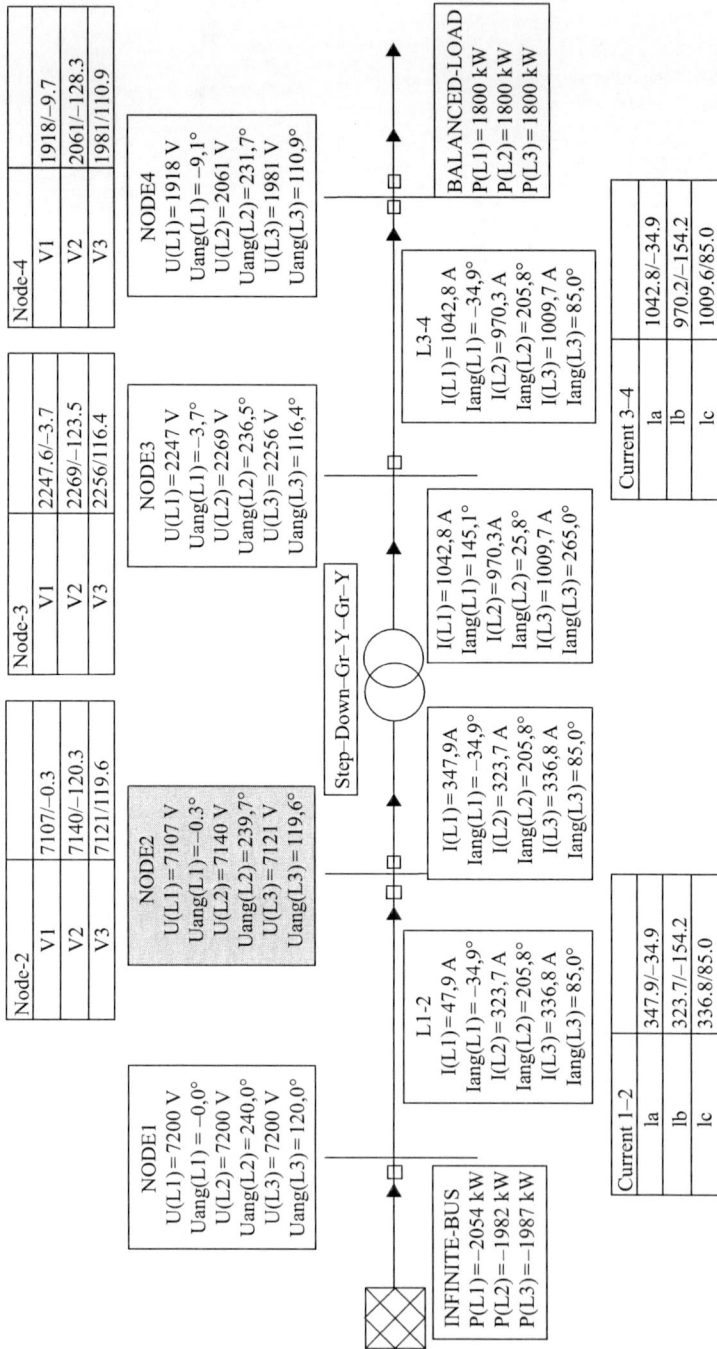

Node-4		
V1	1918/-9.7	
V2	2061/-128.3	
V3	1981/110.9	

NODE4
U(L1)=1918 V
Uang(L1)=-9,1°
U(L2)=2061 V
Uang(L2)=231,7°
U(L3)=1981 V
Uang(L3)=110,9°

BALANCED-LOAD
P(L1)=1800 kW
P(L2)=1800 kW
P(L3)=1800 kW

Node-3		
V1	2247.6/-3.7	
V2	2269/-123.5	
V3	2256/116.4	

NODE3
U(L1)=2247 V
Uang(L1)=-3,7°
U(L2)=2269 V
Uang(L2)=236,5°
U(L3)=2256 V
Uang(L3)=116,4°

L3-4
I(L1)=1042,8 A
Iang(L1)=-34,9°
I(L2)=970,3 A
Iang(L2)=205,8°
I(L3)=1009,7 A
Iang(L3)=85,0°

Current 3-4	
Ia	1042.8/-34.9
Ib	970.2/-154.2
Ic	1009.6/85.0

Node-2		
V1	7107/-0.3	
V2	7140/-120.3	
V3	7121/119.6	

NODE2
U(L1)=7107 V
Uang(L1)=-0,3°
U(L2)=7140 V
Uang(L2)=239,7°
U(L3)=7121 V
Uang(L3)=119,6°

Step-Down-Gr-Y-Gr-Y

I(L1)=1042,8 A
Iang(L1)=145,1°
I(L2)=970,3A
Iang(L2)=25,8°
I(L3)=1009,7 A
Iang(L3)=265,0°

I(L1)=347,9A
Iang(L1)=-34,9°
I(L2)=323,7 A
Iang(L2)=205,8°
I(L3)=336,8 A
Iang(L3)=85,0°

L1-2
I(L1)=47,9 A
Iang(L1)=-34,9°
I(L2)=323,7 A
Iang(L2)=205,8°
I(L3)=336,8 A
Iang(L3)=85,0°

Current 1-2	
Ia	347.9/-34.9
Ib	323.7/-154.2
Ic	336.8/85.0

NODE1
U(L1)=7200 V
Uang(L1)=-0,0°
U(L2)=7200 V
Uang(L2)=240,0°
U(L3)=7200 V
Uang(L3)=120,0°

INFINITE-BUS
P(L1)=-2054 kW
P(L2)=-1982 kW
P(L3)=-1987 kW

Figure 3.28 Grounded Wye–Grounded Wye step-down transformer with balanced load

- Calculate equivalent powers at each node by taking the P and Q values associated with each one and adding the power losses through the respective branches. This is the upstream routine.
- Update voltages at each node starting from the source voltage with the new power equivalents. This is the upstream routine.
- Update losses with the new voltages in the upstream process and afterwards voltage levels until the convergence criteria are satisfied.
- The convergence criteria could be based on the loss values in the upstream routine or on the voltage updating in the downstream routine.

An example case is shown in Figure 3.28.

Proposed exercises

1. Two voltage sources $V_1 = 450\angle-2°$ and $V_2 = 440\angle48°$ are connected by a line of impedance $Z = 1.44 + j3$ Ω. Calculate the complex power for each source and the line loss.
2. Use any calculating software package (i.e., MATLAB) to elaborate the Newton–Raphson code to solve the problem in example 3.6.

Chapter 4
Short circuit calculation

Calculation of short circuit values is essential in power system analysis. The results are used in a number of applications like sizing of breakers and other elements of power systems, designing grounding grids, setting protective equipment, evaluating Total Harmonic Distortion (THD) of harmonic currents, and performing arc flash analysis.

4.1 Nature of short circuit currents

When calculating short circuit currents, it is necessary to take into account two factors which could result in the currents varying with time:

- the presence of the DC component
- the behavior of the generator under short circuit conditions

The circuit R-L in Figure 4.1 helps to illustrate this concept.

$$e(t) = L\frac{di(t)}{dt} + Ri(t) \tag{4.1}$$

The previous expression for the short circuit current is a differential equation, whose solution has two parts:

$$i(t) = i_h(t) + i_p(t) \tag{4.2}$$

where:

$i_h(t)$ Solution to the homogeneous equation corresponding to the transient period

$i_p(t)$ Solution to the particular equation corresponding to the steady-state period

By the use of differential equation theory, the complete solution can be expressed in the following form:

$$i(t) = \frac{V_{max}}{Z}\left[\sin(\omega t + \alpha - \theta) - \sin(\alpha - \theta)e^{-\left(\frac{R}{L}\right)t}\right] \tag{4.3}$$

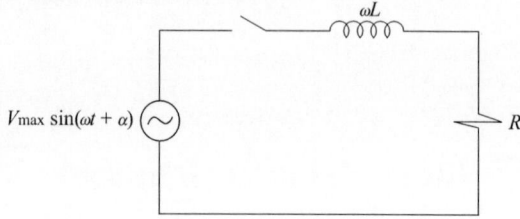

Figure 4.1   L-R circuit

$$Z = \sqrt{(R^2 + \omega^2 L^2)} \tag{4.4}$$

The first term varies sinusoidally and is called the AC component. The second term decreases exponentially and is called the DC component.

α is the closing angle which defines the point on which the fault occurs. θ is defined as $\tan^{-1}\left(\frac{\omega L}{R}\right)$.

The effective value of the total asymmetric short circuit current can be obtained from the following expression:

$$I_{\text{rms asym}} = \sqrt{I_{\text{rms}}^2 + I_{\text{DC}}^2} \tag{4.5}$$

The solution for the particular equation has the following form:

$$i_p(t) = A\sin(\omega t + \alpha) + B\cos(\omega t + \alpha)$$

$$\frac{di_p(t)}{dt} = A\omega\cos(\omega t + \alpha) - B\omega\sin(\omega t + \alpha)$$

The general expression for the solution:

$$e_p(t) = L\frac{di_p(t)}{dt} + Ri_p(t)$$

$$V_m\sin(\omega t + \alpha) = AR\sin(\omega t + \alpha) + BR\cos(\omega t + \alpha)$$
$$+ AL\omega\cos(\omega t + \alpha) - BL\omega\sin(\omega t + \alpha)$$

By direct comparison:

$$V_m = AR - B\omega L$$
$$0 = BR + A\omega L$$
$$B = -A\frac{\omega L}{R}$$

Replacing this expression:

$$V_m = AR + A\frac{\omega^2 L^2}{R}$$

$$V_\text{m}R = AR^2 + A\omega^2 L^2$$

$$A = \frac{R}{R^2 + \omega^2 L^2} V_\text{m}$$

$$B = -\frac{\omega L}{R^2 + \omega^2 L^2} V_\text{m}$$

$$i_\text{p}(t) = V_\text{m} \frac{R}{R^2 + \omega^2 L^2} \sin(\omega t + \alpha) - V_\text{m} \frac{\omega L}{R^2 + \omega^2 L^2} \cos(\omega t + \alpha)$$

The following consideration can be used (see Figure 4.2 for a better understanding):

$$\cos\theta = \frac{R}{\sqrt{R^2 + \omega^2 L^2}}$$

$$\sin\theta = \frac{\omega L}{\sqrt{R^2 + \omega^2 L^2}}$$

The expression becomes:

$$i_\text{p}(t) = \frac{V_\text{m}}{\sqrt{R^2 + \omega^2 L^2}} \left[\frac{R}{\sqrt{R^2 + \omega^2 L^2}} \sin(\omega t + \alpha) - \frac{\omega L}{\sqrt{R^2 + \omega^2 L^2}} \cos(\omega t + \alpha) \right]$$

$$i_\text{p}(t) = \frac{V_\text{m}}{\sqrt{R^2 + \omega^2 L^2}} [\sin(\omega t + \alpha)\cos\theta - \cos(\omega t + \alpha)\sin\theta]$$

$$i_\text{p}(t) = \frac{V_\text{m}}{\sqrt{R^2 + \omega^2 L^2}} [\sin(\omega t + \alpha - \theta)]$$

The solution for the homogeneous equation has the following form:

$$Ri_\text{h}(t) + L\frac{di_\text{h}(t)}{dt} = 0$$

$$i_\text{h}(t) = Ke^{-\frac{R}{L}t}$$

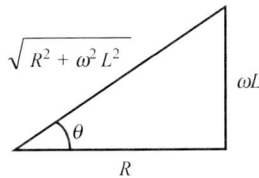

Figure 4.2   Impedance components

(a)

$$\alpha - \theta = 0$$

(b)

$$\alpha - \theta = -\frac{\pi}{2}$$

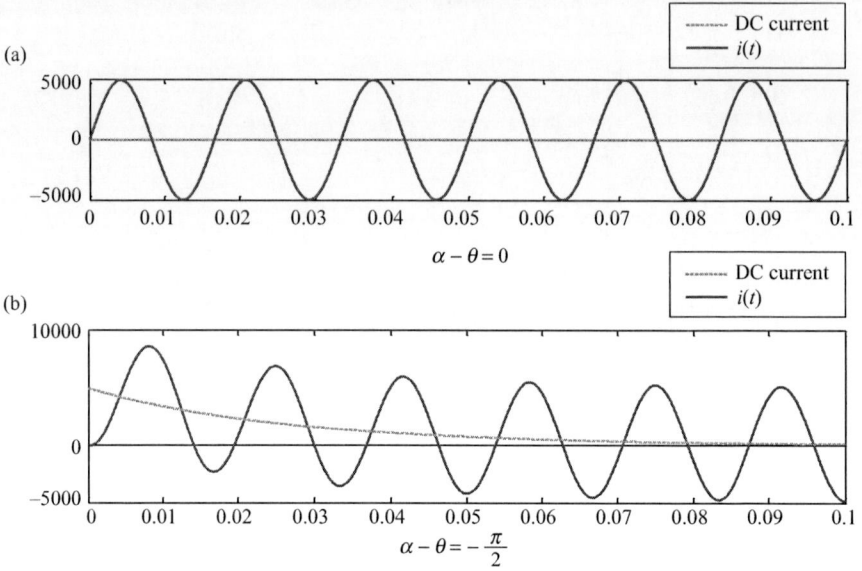

Figure 4.3   Variation of fault current due to the DC component

Since $i(t) = i_h(t) + i_p(t)$, then:

$$i(t) = Ke^{-\frac{R}{L}t} + \frac{V_m}{\sqrt{R^2 + \omega^2 L^2}}[\sin(\omega t + \alpha - \theta)]$$

For the initial condition, when $t = 0^+ \Rightarrow i(t) = 0$:

$$0 = K + \frac{V_m}{\sqrt{R^2 + \omega^2 L^2}}[\sin(\alpha - \theta)]$$

$$K = -\frac{V_m}{\sqrt{R^2 + \omega^2 L^2}}[\sin(\alpha - \theta)]$$

Finally:

$$i(t) = -\frac{V_m}{\sqrt{R^2 + \omega^2 L^2}}[\sin(\alpha - \theta)]e^{-\frac{R}{L}t} + \frac{V_m}{\sqrt{R^2 + \omega^2 L^2}}[\sin(\omega t + \alpha - \theta)]$$

$$i(t) = \frac{V_m}{\sqrt{R^2 + \omega^2 L^2}}\left[\sin(\omega t + \alpha - \theta) - \sin(\alpha - \theta)e^{-\frac{R}{L}t}\right]$$

Figure 4.3 shows how the current can vary with the difference between α and θ.

Example 4.1 Write a code in MATLAB to calculate the fault current during an event for the first 0.05 sec, and plot the AC and DC components of the fault current, and the total current $i(t)$, for the circuit shown in Figure 4.1.

The code for solving the problem can be as follows:

```
V=input('Write the Voltage pre-fault (in V): ');
R=input('Write the equivalent resistance of the system (in ohms): ');
L=input('Write the equivalent inductance of the system (in hernies): ');
f=input('Write the nominal frequency (in hertz): ');
alpha=input('Write the angle when the fault occurred (in degree): ');

Vm=V/sqrt(3);
angle=alpha*(pi()/180);
w=2*pi()*f;
theta=atan(w*L/R);
Z=sqrt((R^2)+((w*L)^2));

t=[0:0.0001:0.05]';

i_ac=(Vm/Z).*(sin(w.*t+angle-theta));
i_dc=(-Vm/Z)*(sin(angle-theta).*exp(-(R/L).*t));
i=i_ac+i_dc;

result=[t, i_ac, i_dc, i];
disp(' ')
disp(' t[s] iAC(t)[A] iDC(t)[A] i(t)')
disp(result)

plot(t,i_ac,'-g')
hold on
plot(t,i_dc,'-r')
hold on
plot(t,i,'-b')
xlabel('Time (sec)')
ylabel('Current (Amp)')
title('Asymmetrical Fault Current')
hleg1 = legend('AC Current','DC Current','Total
Current','Location','NorthEastOutside');
```

Figure 4.4 shows the plot of the AC, DC, and total currents, corresponding to Example 4.1 if the parameters of the circuit are $R = 1.5\,\Omega$, $L = 0.032$ H, and $\alpha = 30°$.

When a fault occurs close to the terminals of rotating machinery, a decaying AC current is produced similar in pattern to that flowing when an AC voltage is applied to an R-L circuit as discussed previously in this section. Here, the decaying pattern is due to the fact that the magnetic flux in the windings of rotating machinery cannot change instantaneously because of the nature of the magnetic

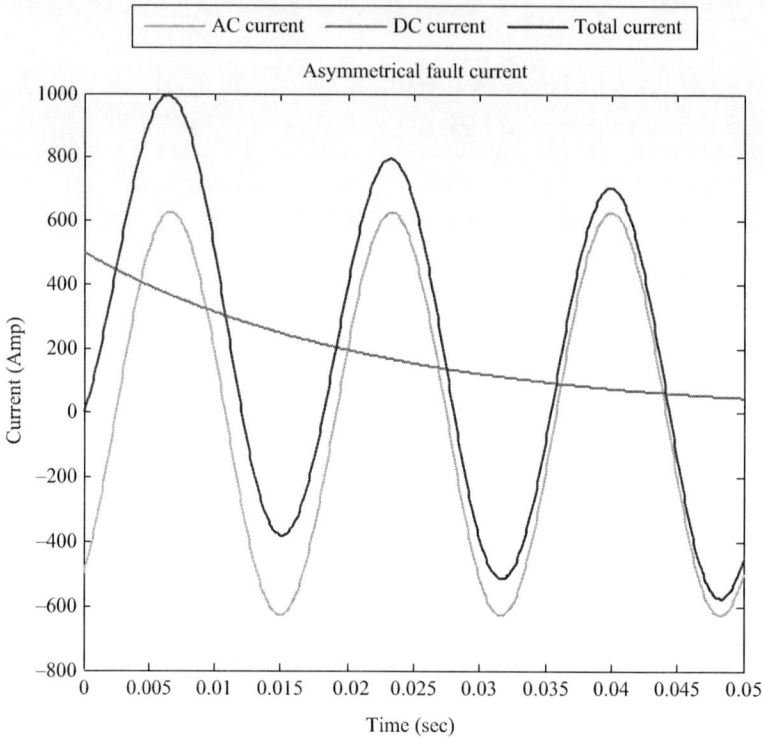

Figure 4.4   Results for example 4.1

circuits involved. The reduction in current from its value at the onset, due to the gradual decrease in the magnetic flux caused by the reduction of the m.m.f. of the induction current, can be seen in Figure 4.5. This effect is known as armature reaction.

The physical situation that is presented to a generator, and which makes the calculations quite difficult, can be interpreted as a reactance which varies with time. Notwithstanding this, in the majority of practical applications it is possible to take account of the variation of reactance in only three stages without producing significant errors. In Figure 4.6 it will be noted that the variation of current with time, $I(t)$, comes close to the three discrete levels of current, I'', I', and I, the subtransient, transient, and steady-state currents, respectively. The corresponding values of direct axis reactance are denoted by X_d'', X_d' and X_d, and the typical variation with time for each of these is illustrated in Figure 4.7.

Short circuit levels vary considerably during a fault, taking into account the rapid drop of the current due to the armature reaction of the synchronous machines and the fact that extinction of an electrical arc is never achieved instantaneously. Therefore, short circuit currents have to be calculated carefully in order to obtain the correct value for the respective applications.

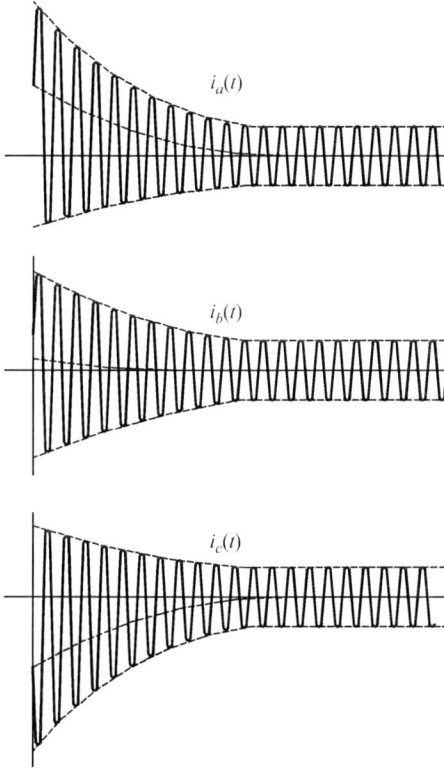

Figure 4.5 Transient SC current at generator terminals

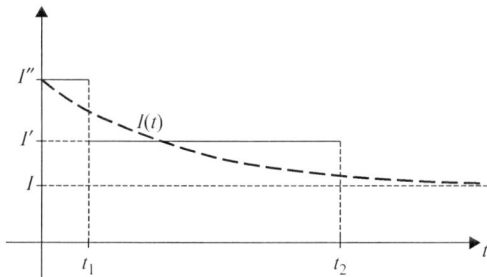

Figure 4.6 Variation of current with time during a fault at generator terminals

The following paragraphs refer to the short circuit currents which are specifically used for the selection of interrupting equipment and protection relay settings – the so-called normal duty rating. ANSI/IEEE Standards C37 and IEC 6090 refer to four duty types defined as first cycle or momentary, peak, interrupting or breaking, and time-delayed or steady-state currents.

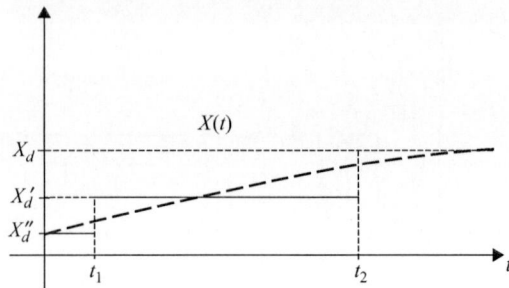

Figure 4.7   Variation of generator reactance during a fault

First cycle currents, also called momentary currents, are the currents that present one half of a cycle after fault initiation. In European Standards these values are indicated by I''_k. These are the currents that are sensed by circuit breaker protection equipment when a fault occurs and are therefore also called close and latch currents. They are calculated with DC offset but no AC decrement in the sources, and use the machine subtransient reactances. Peak currents correspond to the maximum currents during the first cycle after the fault occurs and differ from the first cycle currents that are totally asymmetrical rms currents.

Interrupting currents, also known as contact parting currents, are the values that have to be cleared by interrupting equipment. In European standards, these values are called breaking currents and typically are calculated in the range of 3–5 cycles. These currents contain DC offset and some decrement of the AC current. Time-delayed or steady-state short circuit currents correspond to the values obtained between 6 and 30 cycles. These currents should not contain DC offset, and synchronous and induction contributions should be neglected and transient reactances or higher values should be used in calculating the currents.

Reactance values to be used for the different duties are reproduced in Figure 4.8, based on IEEE Standard 399-1990. For each case, asymmetrical or symmetrical rms values can be defined depending on whether the DC component is included or not. The peak values are obtained by multiplying the rms values by $\sqrt{2}$.

The asymmetrical values are calculated as the square root of the sum of the squares of the DC component and the rms value of the AC current, that is:

$$I_{rms} = \sqrt{I_{DC}^2 + I_{AC}^2} \tag{4.6}$$

4.2   Calculation of fault duty values

The momentary current is used when specifying the closing current of switchgear. Typically, the AC and DC components decay to 90% of their initial values after the first half cycle. From this, the value of the rms current would then be:

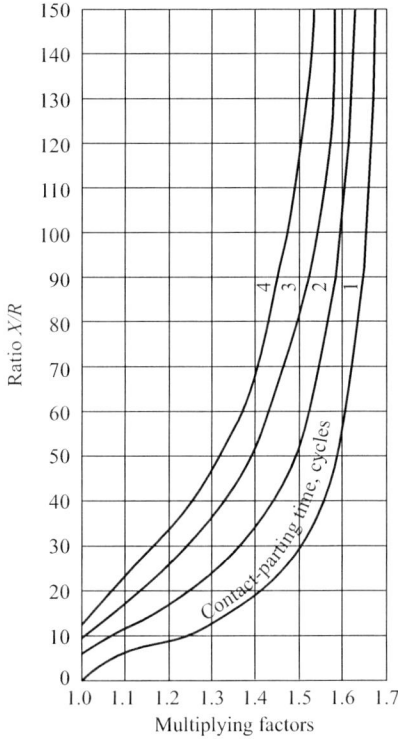

Figure 4.8 Multiplying factors for three-phase and L-L ground faults (from IEEE 551-2006)

$$I_{\text{rms.asym.closing}} = \sqrt{I^2_{\text{DC}} + I^2_{\text{AC.rms.sym}}}$$

$$= \sqrt{(0.9\sqrt{2}V/X''_d)^2 + (0.9V/X''_d)^2}$$

$$= 1.56V/X''_d = 1.56 I_{\text{rms.sym}} \qquad (4.7)$$

Usually a factor of 1.6 is used by manufacturers and in international standards so that, in general, this value should be used when carrying out similar calculations.

The peak value is obtained by arithmetically adding together the AC and DC components. It should be noted that, in this case, the AC component is multiplied by a factor of 2. Thus:

$$I_{\text{peak}} = I_{\text{DC}} + I_{\text{AC}}$$

$$= (0.9\sqrt{2}V/X''_d) + (0.9\sqrt{2}V/X''_d) \qquad (4.8)$$

$$= 2.55 I_{\text{rms.sym}}$$

When considering the specification for the switchgear opening current, the so-called rms value of interrupting current is used in which, again, the AC and DC components are taken into account, and therefore:

$$I_{\text{rms.asym.int}} = \sqrt{I_{\text{DC}}^2 + I_{\text{AC.rms.sym.int}}^2}$$

Replacing the DC component by its exponential expression gives:

$$I_{\text{rms.asym.int}} = \sqrt{\left(\sqrt{2}I_{\text{rms.sym.int}}e^{-(R/L)t}\right)^2 + I_{\text{rms.sym.int}}^2}$$

$$= I_{\text{rms.sym.int}}\sqrt{2e^{-2(R/L)t} + 1} \tag{4.9}$$

The expression $(I_{\text{rms.asym.int}}/I_{\text{rms.sym.int}})$ has been drawn for different values of X/R, and for different switchgear contact separation times, in ANSI Standard C37.5-1979. The multiplying factor graphs are reproduced in Figure 4.8.

As an illustration of the validity of the curves for any situation, consider a circuit breaker with a total contact separation time of two cycles – one cycle due to the relay and one related to the operation of the circuit breaker mechanism.

If the frequency, f, is 60 Hz and the ratio X/R is given as 50, with $t = 2$ cycles $= 0.033$ sec, then $(X/R) = (\omega L/R) = 50$.

Thus $(L/R) = (50/\omega) = (50/2\pi f) = 0.132$. Therefore:

$$\frac{I_{\text{asym}}}{I_{\text{sym}}} = \sqrt{2e^{2(-R/L)t} + 1} = \sqrt{2e^{2\left(-0.033/0.132\right)} + 1} = 1.49$$

Example 4.2 A three-phase, 500 MVA, 13.8 kV synchronous generator has the following parameters:

$X_d = 1.20$ pu $X_d' = 0.15$ pu $X_d' - 3pt' = 0.09$ pu
$T_d' = 0.60$ sec $T_d' - 3pt' = 0.035$ sec $T_a = 0.13$ sec

A three-phase short circuit occurs at the terminals of the machine while it's operating at rated voltage and no-load. Assume that the air-gap voltage is 100% of the rated voltage; that is, $E_{\text{air-gap}} = 1.10\,V_{\text{rated}}$. Determine the instantaneous value of current for a completely displaced wave.

The total AC component of the short circuit current may be expressed as follows:

$$i_{\text{AC}} = \left[(i_d'' - i_d')e^{-t/T_d''} + (i_d' - i_d)e^{-t/T_d'} + i_d\right]\cos\omega t$$

Or

$$i_{AC} = \left[\sqrt{2} \left(\frac{V_{rated}}{X_d''} - \frac{V_{rated}}{X_d'} \right) e^{-t/\tau_d''} + \sqrt{2} \left(\frac{V_{rated}}{X_d'} - \frac{E_{air-gap}}{X_d} \right) e^{-t/\tau_d'} \right.$$

$$\left. + \sqrt{2} \left(\frac{E_{air-gap}}{X_d} \right) \right] \cos\omega t$$

The total DC component of the short circuit current may be expressed as follows:

$$i_{DC} = i_{DC,MAX} e^{-t/\tau_a}$$

$$i_{DC,MAX} = -i_d'' = -\sqrt{2} \left(\frac{V_{rated}}{X_d''} \right)$$

$$i_a(t) = [6.28e^{-28.57t} + 8.13e^{-1.667t} + 1.3]\cos 120\pi t - 15.71e^{-7.69t}$$

The expression obtained is plotted in Figure 4.9.

4.3   Fault calculation for symmetrical faults

The calculation of short circuit values for symmetrical faults is the simplest. The system has to be modeled by using only single phase values.

It is important to emphasize that symmetrical faults produce the largest short circuit values for power and sometimes for currents when systems are not solidly grounded.

Example 4.3 For the system of example 3.3, calculate short circuit current value if a fault occurs at the end of the feeder (Figure 4.10).

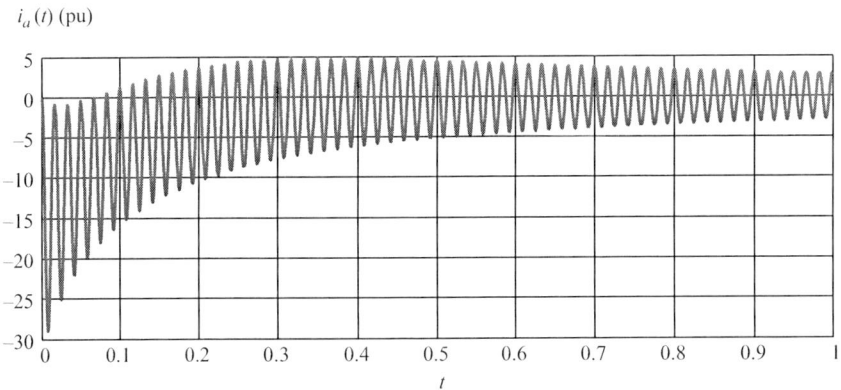

Figure 4.9   Short circuit total current

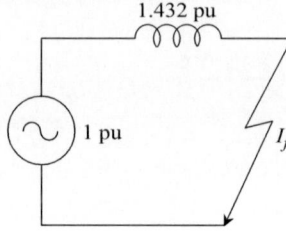

Figure 4.10   Equivalent impedance for example 3.3

$$Z_f = Z_{GEN} + Z_{t1} + Z_{OHL} + Z_{t2} + Z_{FEEDER} = 1.432 \text{ pu}$$

$$I_f = \frac{1}{1.432} = 0.698 \text{ pu}$$

$$I_{11 \text{ kV}} = 0.698 \text{ pu} \times 2624 \text{ A} = 1831.55 \text{ A}$$

$$I_{132 \text{ kV}} = 0.698 \text{ pu} \times 218.7 \text{ A} = 152.65 \text{ A}$$

$$I_{33 \text{ kV}} = 0.698 \text{ pu} \times 874.77 \text{ A} = 610.6 \text{ A}$$

4.4   Symmetrical components

Symmetrical faults, that is three-phase faults and three-phase-to-earth faults, with symmetrical impedances to the fault, leave the electrical system balanced and therefore can be treated by using a single-phase representation. This symmetry is lost during asymmetric faults – line-to-earth, line-to-line, and line-to-line-to-earth – and in these cases a method of analyzing the fault that provides a convenient means of dealing with the asymmetry is required.

On June 28, 1918, Charles L. Fortescue presented at the 34th Annual Convention of the AIEE, in Atlantic City, NJ a very famous paper entitled "Method of Symmetrical Co-ordinates Applied to the Solution of Polyphase Networks." The excerpt of this paper is shown in Figure 4.11. This method allowed the representation of any unbalanced system with *n* phases, into *n* systems of balanced networks. The method was later called symmetrical components method. The main use of this method was of course in the analysis of unbalanced three-phase systems.

While the method can be applied to any unbalanced polyphase system, the theory is summarized here for the case of an unbalanced three-phase system.

When considering a three-phase system, each vector quantity, voltage or current, is replaced by three components so that a total of nine vectors uniquely represent the values of the three phases. As is shown in Figure 4.12, the three-system balanced phasors are designated as:

1. positive-sequence components, which consist of three phasors of equal magnitude, spaced 120° apart, and rotating in the same direction as the phasors in the power system under consideration, that is, the positive direction;

*Presented at the 34th Annual Convention of
the American Institute of Electrical Engineers,
Atlantic City, N. J., June 28, 1918.*

METHOD OF SYMMETRICAL CO-ORDINATES APPLIED TO THE SOLUTION OF POLYPHASE NETWORKS

BY C. L. FORTESCUE

ABSTRACT OF PAPER

In the introduction a general discussion of unsymmetrical systems of co-planar vectors leads to the conclusion that they may be represented by symmetrical systems of the same number of vectors, the number of symmetrical systems required to define the given system being equal to its degrees of freedom. A few trigonometrical theorems which are to be used in the paper are called to mind. The paper is subdivided into three parts, an abstract of which follows. It is recommended that only that part of Part I up to formula (33) and the portion dealing with star-delta transformations be read before proceeding with Part II.

Part I deals with the resolution of unsymmetrical groups of numbers into symmetrical groups. These numbers may represent rotating vectors of systems of operators. A new operator termed the sequence operator is introduced which simplifies the manipulation. Formulas are derived for three-phase circuits. Star-delta transformations for symmetrical co-ordinates are given and expressions for power deduced. A short discussion of harmonics in three-phase systems is given.

Part II deals with the practical application of this method to symmetrical rotating machines operating on unsymmetrical circuits. General formulas are derived and such special cases, as the single-phase induction motor, synchronous motor-generator, phase converters of various types, are discussed.

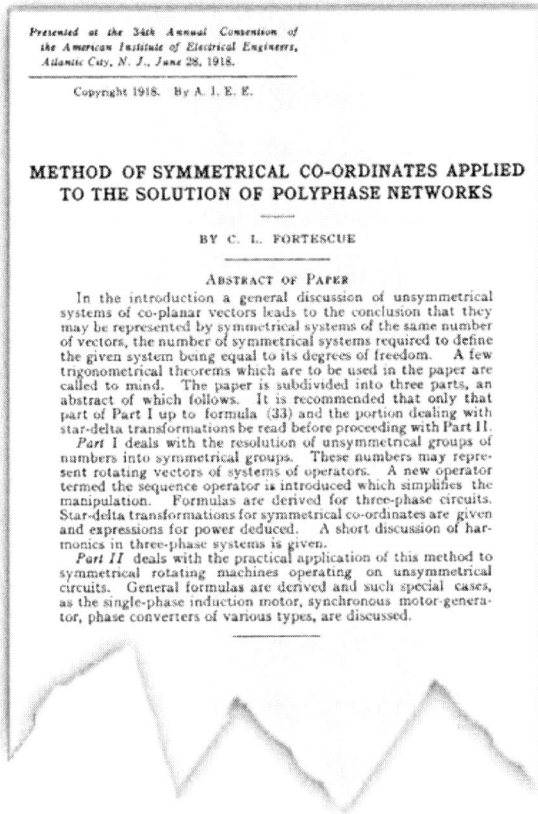

Figure 4.11   Excerpt of the famous paper by Charles L. Fortescue

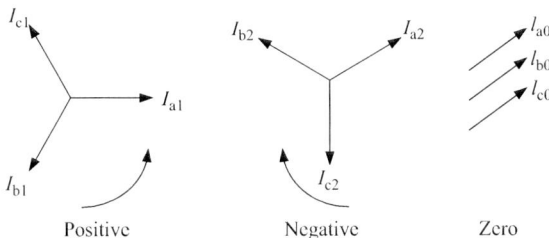

Figure 4.12   Illustration of symmetrical components

2. negative-sequence components, which consist of three phasors of equal magnitude, spaced 120° apart, rotating in the same direction as the positive-sequence phasors but in the reverse sequence;

3. zero-sequence components, which consist of three phasors equal in magnitude and in phase with each other, rotating in the same direction as the positive-sequence phasors.

It is possible to define all the phases in terms of phase a, since the method allows to obtain three balanced systems, as shown in Figure 4.13.

Currents values of any three-phase system, I_a, I_b, and I_c, can be represented as:

$$I_a = I_{a0} + I_{a1} + I_{a2}$$
$$I_b = I_{b0} + I_{b1} + I_{b2}$$
$$I_c = I_{c0} + I_{c1} + I_{c2}$$

Replacing the sequence component values, the following equations are obtained:

$$I_a = I_{a0} + I_{a1} + I_{a2}$$
$$I_b = I_{a0} + a^2 I_{a1} + a I_{a2}$$
$$I_c = I_{a0} + a I_{a1} + a^2 I_{a2}$$

Therefore, the following matrix relationship can be established:

$$\begin{bmatrix} I_a \\ I_b \\ I_c \end{bmatrix} = \begin{bmatrix} 1 & 1 & 1 \\ 1 & a^2 & a \\ 1 & a & a^2 \end{bmatrix} \begin{bmatrix} I_{a0} \\ I_{a1} \\ I_{a2} \end{bmatrix}$$

Inverting the matrix of coefficients:

$$\begin{bmatrix} I_{a0} \\ I_{a1} \\ I_{a2} \end{bmatrix} = \frac{1}{3} \begin{bmatrix} 1 & 1 & 1 \\ 1 & a & a^2 \\ 1 & a^2 & a \end{bmatrix} \begin{bmatrix} I_a \\ I_b \\ I_c \end{bmatrix}$$

From the above matrix it can be deduced that:

$$I_{a0} = \frac{1}{3}(I_a + I_b + I_c)$$

$$I_{a1} = \frac{1}{3}(I_a + a I_b + a^2 I_c)$$

$$I_{a2} = \frac{1}{3}(I_a + a^2 I_b + a I_c)$$

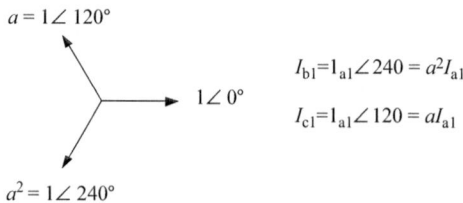

$$a = 1\angle 120°$$
$$1\angle 0°$$
$$I_{b1} = 1_{a1}\angle 240 = a^2 I_{a1}$$
$$I_{c1} = 1_{a1}\angle 120 = a I_{a1}$$
$$a^2 = 1\angle 240°$$

Figure 4.13   *Magnitudes of the positive-sequence network*

In three-phase systems the neutral current is equal to $I_n = (I_{a0} + I_{b0} + I_{c0})$ and, therefore, $I_n = 3I_{a0}$.

Voltage values of any three-phase system, V_a, V_b, and V_c, can be represented as:

$$V_a = V_{a0} + V_{a1} + V_{a2}$$
$$V_b = V_{b0} + V_{b1} + V_{b2}$$
$$V_c = V_{c0} + V_{c1} + V_{c2}$$

Replacing the sequence component values, the following equations are obtained:

$$V_a = V_{a0} + V_{a1} + V_{a2}$$
$$V_b = V_{a0} + a^2 V_{a1} + a V_{a2}$$
$$V_c = V_{a0} + a V_{a1} + a^2 V_{a2}$$

Therefore, the following matrix relationship can be established:

$$\begin{bmatrix} V_a \\ V_b \\ V_c \end{bmatrix} = \begin{bmatrix} 1 & 1 & 1 \\ 1 & a^2 & a \\ 1 & a & a^2 \end{bmatrix} \begin{bmatrix} V_{a0} \\ V_{a1} \\ V_{a2} \end{bmatrix}$$

Inverting the matrix of coefficients:

$$\begin{bmatrix} V_{a0} \\ V_{a1} \\ V_{a2} \end{bmatrix} = \frac{1}{3} \begin{bmatrix} 1 & 1 & 1 \\ 1 & a & a^2 \\ 1 & a^2 & a \end{bmatrix} \begin{bmatrix} V_a \\ V_b \\ V_c \end{bmatrix}$$

From the above matrix it can be deduced that:

$$V_{a0} = \frac{1}{3} (V_a + V_b + V_c)$$

$$V_{a1} = \frac{1}{3} (V_a + a V_b + a^2 V_c)$$

$$V_{a2} = \frac{1}{3} (V_a + a^2 V_b + a V_c)$$

In three-phase systems the neutral current is equal to $V_n = (V_{a0} + V_{b0} + V_{c0})$ and, therefore, $V_n = 3V_{a0}$.

An example of this is shown in Figure 4.14 for an arbitrary unbalanced three-phase system.

4.4.1 Importance and construction of sequence networks

The impedance of a circuit in which only positive-sequence currents are circulating is called the positive-sequence impedance and, similarly, those in which only negative- and zero-sequence currents flow are called the negative- and zero-sequence impedances. These sequence impedances are designated Z_1, Z_2, and Z_0, respectively and are used in calculations involving symmetrical components. Since generators are designed to supply balanced voltages, the generated voltages are of

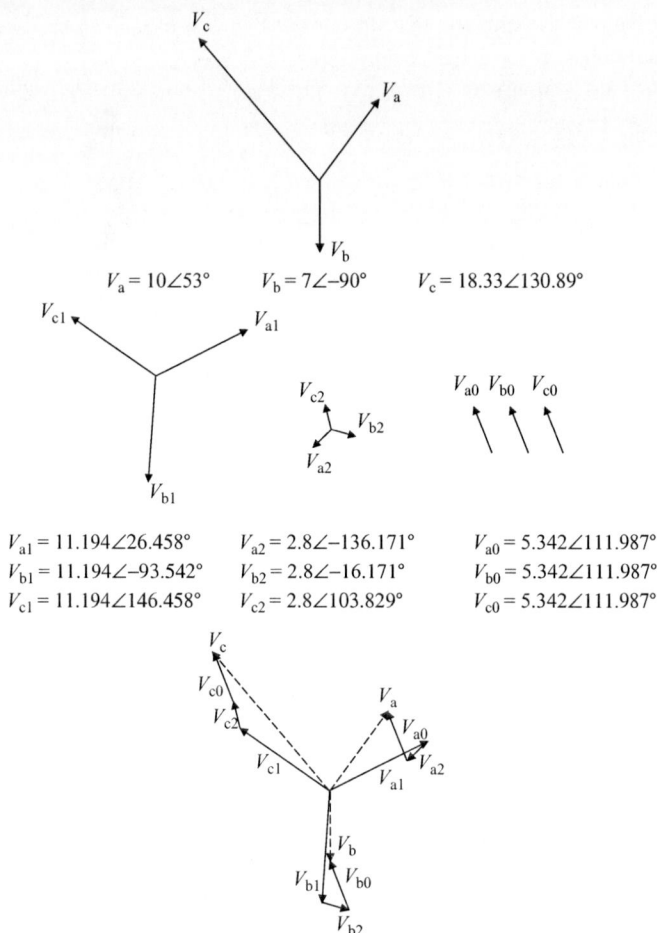

$V_a = 10\angle 53°$ $V_b = 7\angle -90°$ $V_c = 18.33\angle 130.89°$

$V_{a1} = 11.194\angle 26.458°$ $V_{a2} = 2.8\angle -136.171°$ $V_{a0} = 5.342\angle 111.987°$
$V_{b1} = 11.194\angle -93.542°$ $V_{b2} = 2.8\angle -16.171°$ $V_{b0} = 5.342\angle 111.987°$
$V_{c1} = 11.194\angle 146.458°$ $V_{c2} = 2.8\angle 103.829°$ $V_{c0} = 5.342\angle 111.987°$

Figure 4.14   Symmetrical components of an unbalanced three-phase system

positive sequence only. Therefore, the positive-sequence network is composed of an emf source in series with the positive-sequence impedance. The negative- and zero-sequence networks do not contain emfs but only include impedances to the flow of negative- and zero-sequence currents, respectively.

The positive- and negative-sequence impedances of overhead line circuits are identical, as are those of cables, being independent of the phase if the applied voltages are balanced. The zero-sequence impedances of lines differ from the positive- and negative-sequence impedances since the magnetic field creating the positive- and negative-sequence currents is different from that for the zero-sequence currents. The following ratios may be used in the absence of detailed information. For a single circuit line, $Z_0/Z_1 = 2$ when no earth wire is present and 3.5 with an earth wire. For a double circuit line $Z_0/Z_1 = 5.5$. For underground cables Z_0/Z_1 can be taken as 1 to 1.25 for single core, and 3 to 5 for three-core cables.

For transformers, the positive- and negative-sequence impedances are equal because in static circuits these impedances are independent of the phase order, provided that the applied voltages are balanced. The zero-sequence impedance is either the same as the other two impedances, or infinite, depending on the transformer connections. The resistance of the windings is much smaller and can generally be neglected in short circuit calculations.

When modeling small generators and motors it may be necessary to take resistance into account. However, for most studies only the reactances of synchronous machines are used. Three values of positive reactance are normally quoted – subtransient, transient, and synchronous reactances, denoted by X_d'', X_d', and X_d. In fault studies the subtransient and transient reactances of generators and motors must be included as appropriate, depending on the machine characteristics and fault clearance time. The subtransient reactance is the reactance applicable at the onset of the fault occurrence. Within 0.1 sec the fault level falls to a value determined by the transient reactance and then decays exponentially to a steady-state value determined by the synchronous reactance.

In connecting sequence networks together, the reference busbar for the positive- and negative-sequence networks is the generator neutral which, in these networks, is at earth potential, so only zero-sequence currents flow through the impedances between neutral and earth. The reference busbar for zero-sequence networks is the earth point of the generator. The current which flows in the impedance Z_n between the neutral and earth is three times the zero-sequence current. Figure 4.15 illustrates the sequence networks for a generator. The zero-sequence network carries only zero-sequence current in one phase which has an impedance of $Z_0 = 3Z_n + Z_{e0}$.

The voltage and current components for each phase are obtained from the equations given for the sequence networks. The equations for the components of voltage, corresponding to the a phase of the system, are obtained from Figure 4.15 as follows:

$$V_{a1} = E_a - I_{a1}Z_1$$
$$V_{a2} = -I_{a2}Z_2$$
$$V_{a0} = -I_{a0}Z_0$$

where:

E_a No-load voltage to earth of the positive-sequence network
Z_1 Positive-sequence impedance of the generator
Z_2 Negative-sequence impedance of the generator
Z_0 Zero-sequence impedance of the generator (Z_{g0}) plus three times the impedance to earth

The above equations can be applied to any generator which carries unbalanced currents and are the starting point for calculations for any type of fault. The same approach can be used with equivalent power systems or to loaded generators, E_a then being the voltage behind the reactance before the fault occurs.

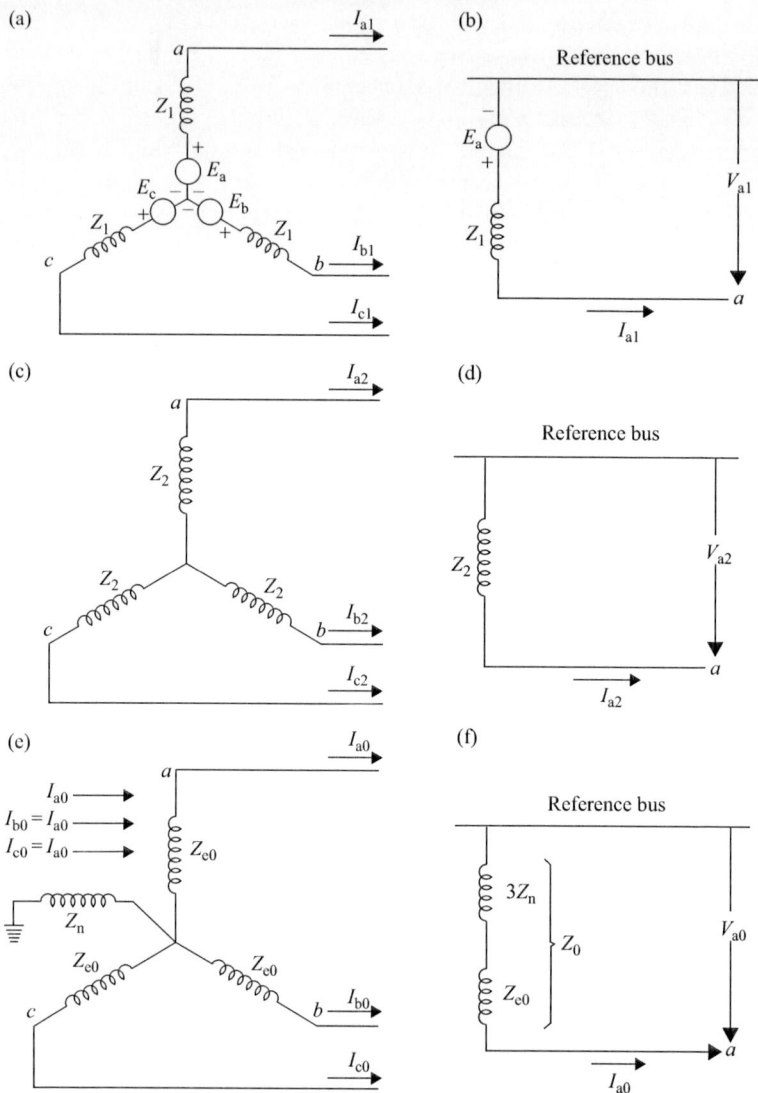

Figure 4.15   Representation of positive-, negative-, and zero-sequence impedances

4.4.2   Calculation of asymmetrical faults using symmetrical components

The positive-, negative-, and zero-sequence networks, carrying currents I_1, I_2, and I_0, respectively, are connected together in a particular arrangement to represent a given unbalanced fault condition. Consequently, in order to calculate fault levels using the method of symmetrical components, it is essential to determine the individual sequence impedances and combine these to make up the correct sequence networks. Then, for each type of fault, the appropriate combination of sequence networks is formed in order to obtain the relationships between fault currents and voltages.

4.4.2.1   Line-to-earth fault

The conditions for a solid fault from line a to earth are represented by the equations $I_b = 0$, $I_c = 0$, and $V_a = 0$. As in the previous equations, it can easily be deduced that $I_{a1} = I_{a2} = I_{a0} = E_a/(Z_1 + Z_2 + Z_0)$. Therefore, the sequence networks will be connected in series, as indicated in Figure 4.16(a). The current and voltage conditions are the same when considering an open-circuit fault in phases b and c, and thus the treatment and connection of the sequence networks will be similar.

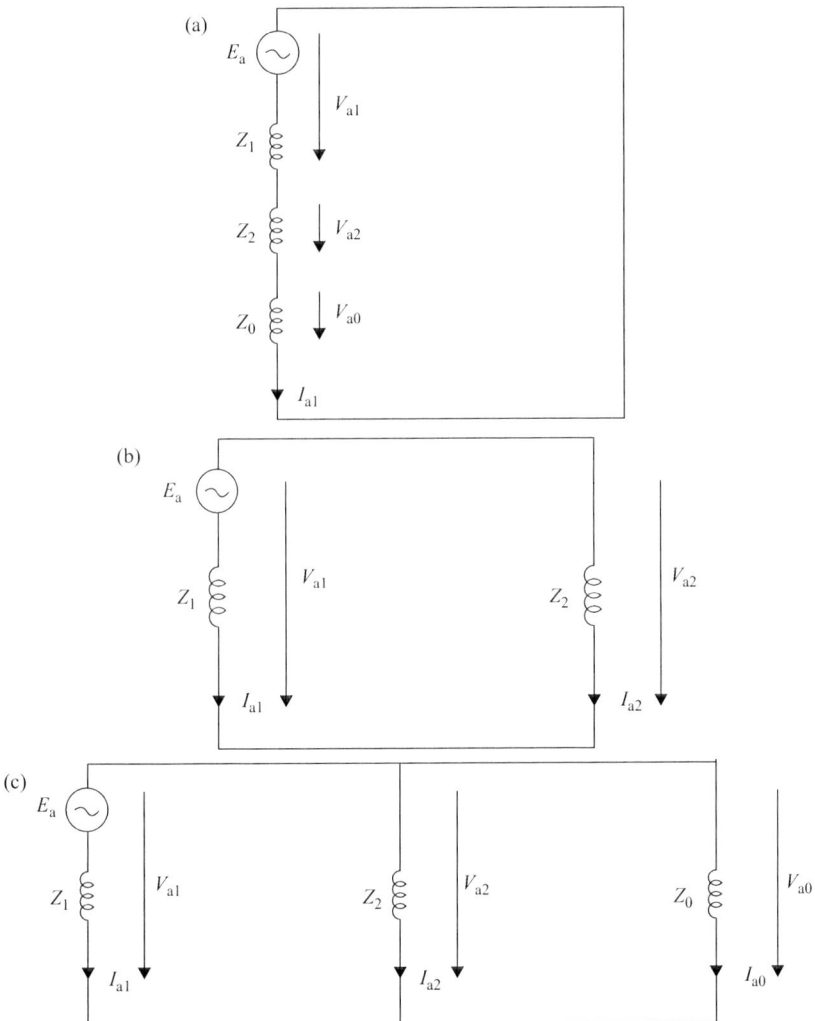

Figure 4.16   Representation of sequence networks

4.4.2.2   Line-to-line fault

The conditions for a solid fault between lines b and c are represented by the equations $I_a = 0$, $I_b = -I_c$ and $V_b = V_c$. Equally it can be shown that $I_{a0} = 0$ and $I_{a1} = E_a/(Z_1 + Z_2) = -I_{a2}$. For this case, with no zero-sequence current, the zero-sequence network is not involved and the overall sequence network is composed of the positive- and negative-sequence networks in parallel as indicated in Figure 4.16(b).

4.4.2.3   Line-to-line-to-earth fault

The conditions for a fault between lines b and c and earth are represented by the equations $I_a = 0$ and $V_b = V_c = 0$. From these equations it can be proved that:

$$I_{a1} = \frac{E_a}{Z_1 + \dfrac{Z_0 Z_2}{Z_0 + Z_2}}$$

The three sequence networks are connected in parallel as shown in Figure 4.16(c).

4.4.3   Equivalent impedances for a power system

When it is necessary to study the effect of any change on the power system, the system must first of all be represented by its corresponding sequence impedances. The equivalent positive- and negative-sequence impedances can be calculated directly from:

$$Z = \frac{V^2}{P}$$

where:

  Z Equivalent positive- and negative-sequence impedances
  V Nominal phase-to-phase voltage
  P Three-phase short circuit power

The equivalent zero-sequence of a system can be derived from the expressions of sequence components referred to for a single-phase fault, that is, $I_{a1} = I_{a2} = I_{a0} = V_{LN}/(Z_1 + Z_2 + Z_0)$, where V_{LN} is the line-to-neutral voltage.

For lines and cables the positive- and negative-sequence impedances are equal. Thus, on the basis that the generator impedances are not significant in most distribution network fault studies, it may be assumed that overall $Z_2 = Z_1$, which simplifies the calculations. Thus, the above formula reduces to $I_a = 3I_{a0} = 3V_{LN}/(2Z_1 + Z_0)$, where V_{LN} is the line-to-neutral voltage and $Z_0 = (3V_{LN}/I_a) - 2Z_1$.

4.4.4   Supplying the current and voltage signals to protection systems

In the presence of a fault the current transformers (CTs) circulate current proportional to the fault current to the protection equipment without distinguishing between the vectorial magnitudes of the sequence components. Therefore, in the majority of cases, the relays operate on the basis of the corresponding values of fault current and/or voltages, regardless of the values of the sequence components. It is very important to emphasize that, given this, the advantage of using symmetrical components is that they facilitate the calculation of fault levels even though the relays in the majority of cases do not distinguish between the various values of the symmetrical components.

In Figure 4.17 the positive- and negative-sequence values of current and voltage for different faults are shown together with the summated values of current and voltage. Relays usually only operate using the summated values in the right-hand columns. However, relays are available which can operate with specific values of some of the sequence components. In these cases there must be methods for obtaining these components, and this is achieved by using filters which produce the mathematical operations of the resultant equations to resolve the matrix for voltages and for currents. Although these filters can be constructed for electromagnetic elements, the growth of electronics has led to their being used increasingly in logic circuits. Among the relays which require this type of filter in order to operate are those used in negative-sequence and earth-fault protection.

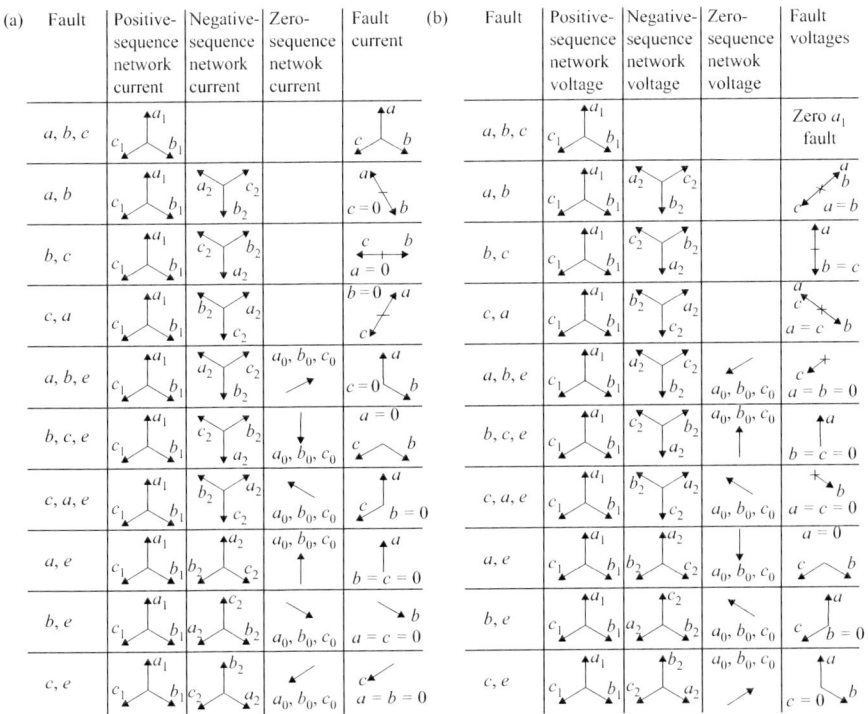

(a)

Fault	Positive-sequence network current	Negative-sequence network current	Zero-sequence netwok current	Fault current
a, b, c				
a, b				$c = 0$
b, c				$a = 0$
c, a				$b = 0$
a, b, e			a_0, b_0, c_0	$c = 0$
b, c, e			a_0, b_0, c_0	$a = 0$
c, a, e			a_0, b_0, c_0	$b = 0$
a, e			a_0, b_0, c_0	$b = c = 0$
b, e			a_0, b_0, c_0	$a = c = 0$
c, e			a_0, b_0, c_0	$a = b = 0$

(b)

Fault	Positive-sequence network voltage	Negative-sequence network voltage	Zero-sequence netwok voltage	Fault voltages
a, b, c				Zero a_1 fault
a, b				$a = b$
b, c				$b = c$
c, a				$a = c$
a, b, e			a_0, b_0, c_0	$a = b = 0$
b, c, e			a_0, b_0, c_0	$b = c = 0$
c, a, e			a_0, b_0, c_0	$a = c = 0$
a, e			a_0, b_0, c_0	$a = 0$
b, e			a_0, b_0, c_0	$b = 0$
c, e			a_0, b_0, c_0	$c = 0$

Figure 4.17   Sequence currents and voltages for different types of faults

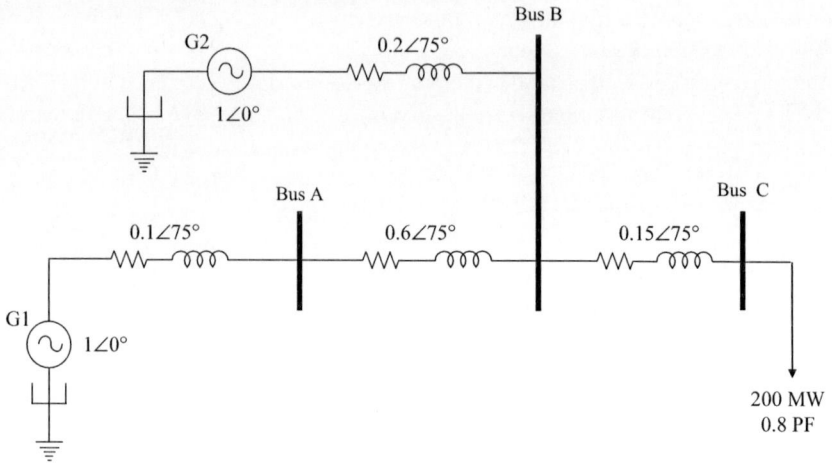

Figure 4.18   Single line diagram for example 4.4

Example 4.4

Notes

- All the impedances shown in Figure 4.18 are per unit positive-sequence impedances at the base of 115 kV and 100 MVA.
- For lines (branches between bus A and bus B and also between bus B and bus C), the zero-sequence impedance (Z_0) is 2.5 times the positive-sequence impedance (Z_1).
- For generators (source impedances behind busses A and B), both the zero-sequence impedance (Z_0) and positive-sequence impedance (Z_1) are equal.
- G1 and G2 are perfect voltage sources generating 1.0 pu voltage with zero-source impedances. These voltage sources and their series impedances are Thévenin equivalent sources behind bus A and bus B, respectively.

1. For the power system shown in Figure 4.18, calculate the fault currents for (i) symmetrical three-phase fault and (ii) single line-to-ground fault (SLG) at bus C.
2. For the three-phase fault, calculate the positive-sequence contributions from G1 and G2, and for the SLG fault, calculate zero-sequence contributions from G1 and G2.

1. Calculations for symmetrical L-L-L fault at bus C:
   The fault impedance is neglected here.
   Positive-sequence network for L-L-L fault calculations at bus C is as follows (Figure 4.19).
   Positive-sequence fault current for symmetrical L-L-L fault would be:

$$I_{f1} = \frac{E}{Z_{eq}} = \frac{1\angle 0}{0.3055\angle 75} = 3.2733\angle -75 \text{ pu}$$

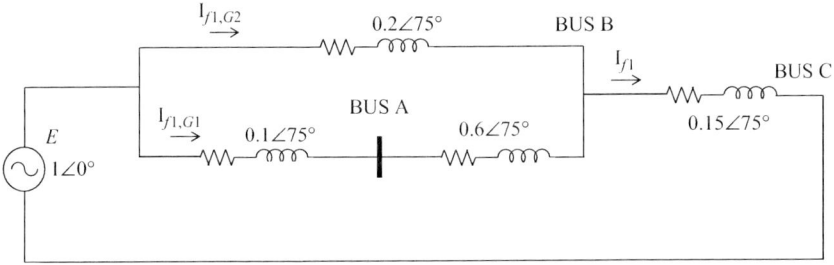

Figure 4.19 Positive-sequence diagram for example 4.4

Positive-sequence fault current contributions for G1 and G2 using current divider rule:

$$I_{f1,G1} = \frac{0.2\angle75}{0.2\angle75 + 0.7\angle75} \times 3.2733\angle{-75} \text{ pu} = 0.7274\angle{-75} \text{ pu}$$

$$I_{f1,G2} = I_{f1} - I_{f1,G1} = 2.5459\angle{-75} \text{ pu}$$

Fault currents in three phases:

$$I_{base} = \frac{100 \times 10^6}{\sqrt{3} \times 115 \times 10^3} = 502.0437 \text{ A}$$

$$I_{f,A} = (3.272\angle{-75} \text{ pu})(502.0437) = 1642\angle{-75} \text{ A}$$

Similarly, for other two phases:

$$I_{f,B} = 1642\angle{-195} \text{ A}$$

$$I_{f,C} = 1642\angle45 \text{ A}$$

2. Calculations for S-L-G fault at bus C (Figure 4.20):

$$I_{f1} = I_{f2} = I_{f0} = \frac{E}{Z_{eq1} + Z_{eq2} + Z_{eq0}}$$

Note: For lines, zero-sequence impedance (Z_0) is 2.5 times the positive-sequence impedance (Z_1), whereas, for equivalent generators both the zero-sequence impedance (Z_0) and positive-sequence impedance (Z_1) are equal.

$$Z_{eq1} = Z_{eq2} = [(0.1\angle75 + 0.6\angle75)\|0.2\angle75] + 0.15\angle75 = 0.3055\angle75 \text{ pu}$$

$$Z_{eq0} = [(0.1\angle75 + 1.5\angle75)\|0.2\angle75] + 0.375\angle75 = 0.55272\angle75 \text{ pu}$$

Sequence fault currents for S-L-G fault would be:

$$I_{f1} = I_{f2} = I_{f0} = \frac{E}{Z_{eq1} + Z_{eq2} + Z_{eq0}} = \frac{1\angle0}{2(0.3055\angle75) + 0.552\angle75}$$

$$= 0.8592\angle{-75} \text{ pu}$$

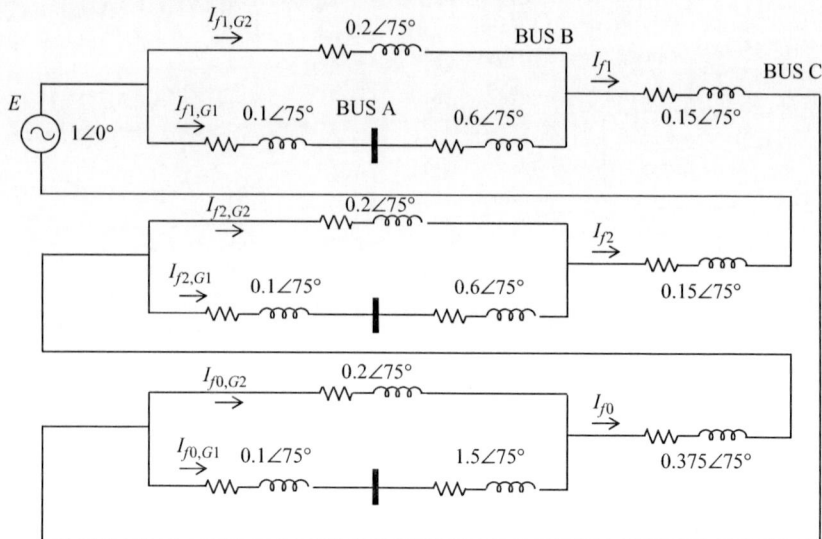

Figure 4.20   Sequence network connection for a single phase fault of example 4.4

Zero-sequence fault current contributions for G1 and G2 using current divider rule:

$$I_{f0,G1} = \frac{0.2\angle75}{(0.2\angle75) + 1.6\angle75}0.8592\angle-75 \text{ pu} = 0.095466\angle-75 \text{ pu}$$

$$I_{f0,G2} = I_{f0} - I_{f0,G1} = 0.8592\angle-75 - 0.095466\angle-75 = 0.7637\angle-75 \text{ pu}$$

Fault currents in three phases (Figure 4.21):

$$I_{base} = \frac{100 \times 10^6}{\sqrt{3} \times 115 \times 10^3} = 502.0437 \text{ A}$$

$$I_{f,A} = I_{f1} + I_{f2} + I_{f0} = 3(0.8592\angle-75 \text{ pu}) = 2.5775\angle-75 \text{ pu}$$

$$I_{f,A} = (2.5775\angle-75 \text{ pu})(502.0437) = 1294.0176\angle-75 \text{ A}$$

$$I_{f,B} = I_{f,C} = 0 \text{ A}$$

Example 4.5 For example 3.7 calculate the short circuit currents for a three-phase faults at busbar 115 kV The Ridges. Notice that the plant connected at South Post is considered out for maintenance.

Figure 4.22 shows the reduced diagram when the three-phase fault occurred. Only the equivalent system before bus Black River 230 kV, the three-winding transformers between Black River 230 kV bus and Black River 115 kV bus, and the lines between Black River 115 kV bus and The Ridges 115 kV bus have an effect on this fault.

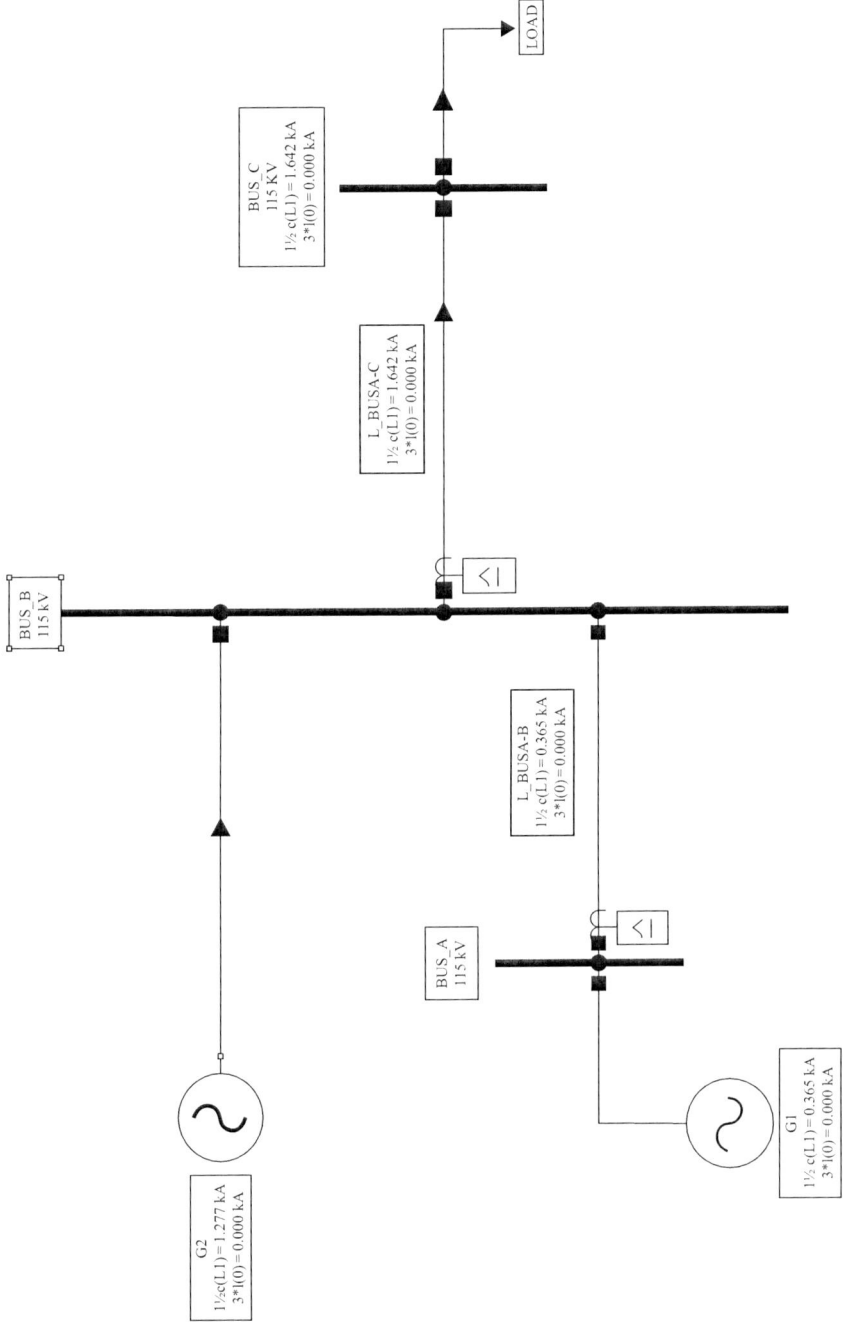

BUS_C
115 KV
1½ c(L1) = 1.642 kA
3*I(0) = 0.000 kA

LOAD

L_BUSA-C
1½ c(L1) = 1.642 kA
3*I(0) = 0.000 kA

BUS_B
115 kV

L_BUSA-B
1½ c(L1) = 0.365 kA
3*I(0) = 0.000 kA

BUS_A
115 kV

G2
1½c(L1) = 1.277 kA
3*I(0) = 0.000 kA

G1
1½ c(L1) = 0.365 kA
3*I(0) = 0.000 kA

Figure 4.21   Solution of example 4.4 by using a software package for a three-phase fault

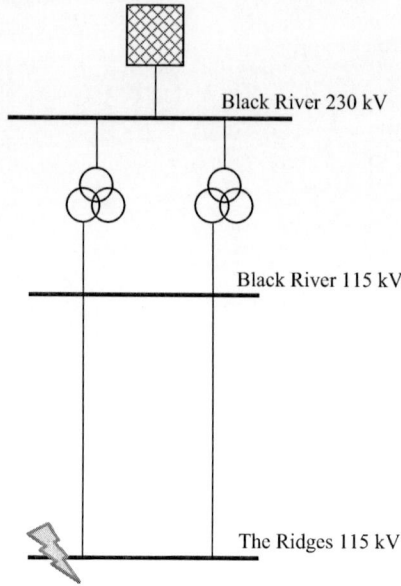

Figure 4.22   Reduced diagrams for the fault analysis in example 4.5

Considering that S_{base} = 100 kV and V_{base} = 230 kV, for Thévenin equivalent of the system:

$$S_{cc} = 2514.9 \text{ MVA}$$

$$Z_{cc} = \frac{V_{BR230}^2}{S_{cc}} = \frac{(230 \text{ kV})^2}{2514.9 \text{ MVA}} = 21.035 \ \Omega$$

$$Z_{\text{Thévenin\_pu}} = \frac{Z_{cc}}{Z_{\text{Thévenin\_base}}} = \frac{21.035 \ \Omega}{\left(\dfrac{230 \text{ kV}^2}{100 \text{ MVA}}\right)} = 0.0398 \text{ pu}$$

For the transformers:

$$Z_{Tr1} = Z_{Tr2} = Z_{Trpu} \frac{S_{new}}{S_{old}} = 0.1 \text{ pu} \frac{100 \text{ MVA}}{90 \text{ MVA}} = 0.111 \text{ pu}$$

$$Z_{Tr} = Z_{Tr1}||Z_{Tr2} = \frac{0.111 \text{ pu}}{2} = 0.0556 \text{ pu}$$

For the lines:

$$Z_{L1} = Z_{L2} = 0.4654 \ \Omega/\text{km}$$

$$L_{L1} = L_{L2} = 12 \text{ km}$$

$$Z_L = Z_{L1}||Z_{L2} = \frac{0.4654 \ \Omega/\text{km} \times 12 \text{ km}}{2} = 2.7924 \ \Omega$$

Black River 230 kV   Black River 115 kV   The Ridges 115 kV

0.0398 pu   0.0556 pu   0.0211 pu

G1
$1\angle 0°$

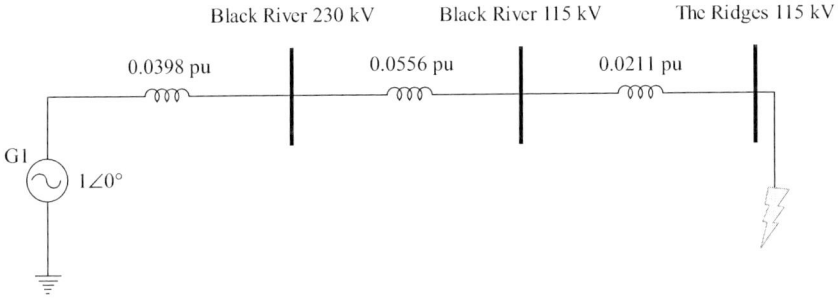

Figure 4.23 *Impedance diagram for the three-phase fault in node The Ridges 115 kV*

$$Z_{Lpu} = \frac{Z_L}{Z_{Lbase}} = \frac{2.7924\ \Omega}{\dfrac{115\ kV^2}{100\ MVA}} = 0.0211\ pu$$

Figure 4.23 shows the diagram with all the elements that affect the three-phase fault on the bus The Ridges 115 kV.

The three-phase fault current in this case is:

$$I_{3\phi pu} = \frac{V_{3\phi pu}}{Z_{Thévenin\_pu} + Z_{Tr1} + Z_{Lpu}} = \frac{1}{0.0398 + 0.0556 + 0.0211} = 8.6\ pu$$

$$I_{base} = \frac{S_{base}}{V_{base}} = \frac{100\ MVA}{\sqrt{3} \cdot 115\ kV} = 502.04\ A$$

$$I_{3\phi pu} I_{base} = 8.6 \cdot 502.04\ A = 4309.2\ A$$

The results are shown in Figure 4.24 using a software package. It can be seen the fault current distribution throughout the system.

Proposed exercises

1. A short transmission line of inductance 0.04 H and resistance 1.2 ω is suddenly short-circuited at the receiving end, while the source voltage is $V = 311.13$ sin $(377t-30°)$. Calculate the instant of the short circuit fault where the DC offset will be zero and maximum.

2. For the power system in Figure 4.25, form sequence impedance networks (including the zero-sequence network). Assume that loads do not contribute to the short circuit currents and use a common base of 120 MVA.

3. For Figure 4.26 convert the sequence impedance networks to single impedances as seen from the fault point. Use the following numerical values on a per unit basis (all on a common MVA base). Neglect resistances and ignore the load currents.

    Generators G1, G2, and G3: $Z_1 = 0.158$, $Z_2 = 0.161$, $Z_0 = 0.06$, Z_n (neutral grounding impedance) $= 0.22$

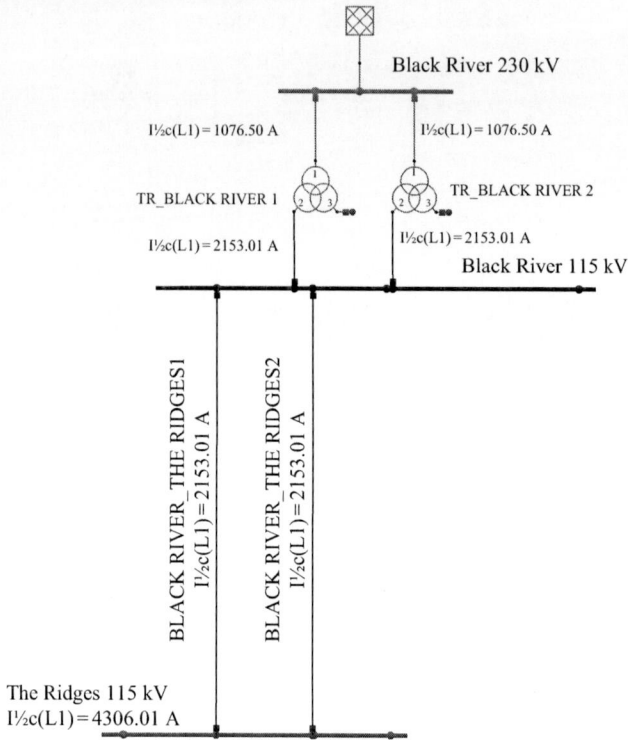

Figure 4.24   *Solution of example 4.5 by using a software package*

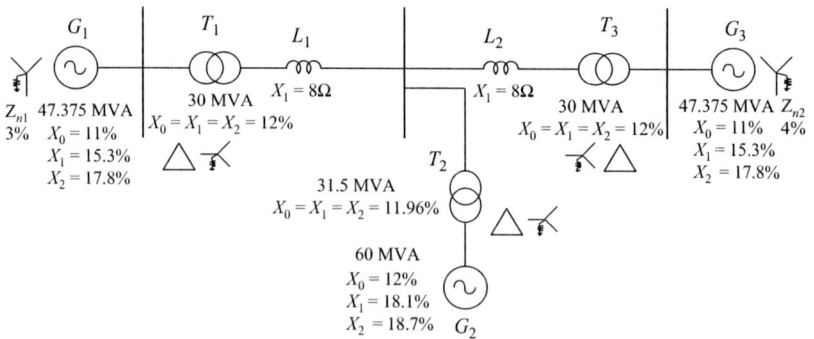

Figure 4.25   *Diagram for exercise 4.2*

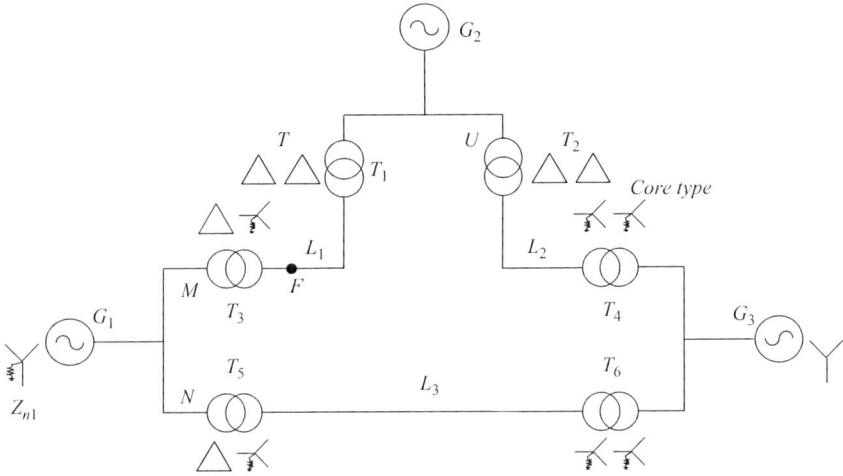

Figure 4.26   Diagram for exercise 4.3

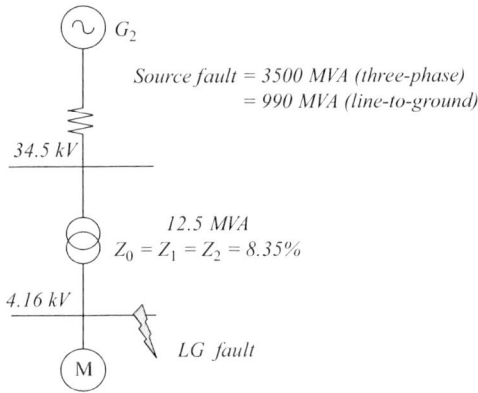

Figure 4.27   Diagram for exercise 4.4

Transmission lines L1, L2, and L3: $Z_1 = 0.15$, $Z_2 = 0.15$
Transformers T1, T2, T3, T4, T5, and T6: $Z_1 = Z_2 = 0.0563$, transformer T3: $Z_0 = 0.0563$

4. For the industrial system of Figure 4.27 specify the sequence network when a single line-to-ground fault occurs at the secondary of the transformer terminals, and calculate the fault current. Consider a load operating power factor of 0.879 and an overall efficiency of 93%.

Chapter 5

Reliability of distribution systems

Knowing that the most important benefit of a Smart Grid is the increase in the reliability indices, a proper examination of this concept is important. The term "reliability" refers to the notion that the system performs its specified task correctly for a certain duration of time. The term "availability" refers to the readiness of a system or component to be immediately ready to perform its task. Both terms have precise definitions within the reliability engineering field and have specified equations and methods to provide quantitative metrics for them.

Availability is the probability of something being energized. It is the most basic aspect of reliability and is typically measured in percent or per unit. The complement of availability is unavailability.

Unavailability can be computed directly from interruption duration information. If a customer experiences 24 h of interrupted power in a year, unavailability is equal to $24/8760 = 0.10274 = 0.274\%$. Availability is equal to $100\% - 0.2739\% = 99.7261\%$.

Each distribution system component can be described by a set of reliability parameters. Simple reliability models are based on component failure rates and component repair times, but sophisticated models make use of many other reliability parameters. Some of the most common reliability parameters are the following:

- Failure rate, indicated by the symbol λ, describes the number of times per year that a component can be expected to experience a failure. Likewise, failure rate for the whole system (indicated by the symbol f), describes the number of times per year that the system can be expected to experience a failure.
- Mean time to repair (MTTR), indicated by the letter r, represents the expected time it will take for a failure to be repaired (measured from the time that the failure occurs). A single value of MTTR is typically used for each component, but separate values can be used for different failure modes.
- Probability of operational failure (POF) is the conditional probability that a device will not operate if it is supposed to operate. For example, if an automated switch fails to function properly 5 times out of every 100 attempted operations, it has a POF of 5%. This reliability parameter is typically associated with switching devices and protection devices.

5.1 Network modeling

The analysis of reliability in physical networks is done by using the concepts of series and parallel circuits of electrical systems, which are clear and flexible tools to handle. The solutions involve simple mathematical methods that represent accurately the models under analysis.

Network modeling is a component-based technique rather than a state-based technique. Each component is described by a probability of being available, P, and a probability of not being available, Q. Since components are assumed to be either available or not available, Q and P are arithmetic complements: $Q = 1 - P$. Care has to be exercised in reliability studies since the symbol P represents the non-supplied power, especially in software packages.

If a component is described by an annual failure rate λ and a mean time to repair r in hours, the probability of being available can be computed as follows:

$$P = \frac{8760 - \lambda r}{8760} \tag{5.1}$$

Therefore:

$$Q = \frac{\lambda r}{8760} \tag{5.2}$$

As mentioned before, networks are composed of series and parallel components. Two components are in series if both must be available for the connection to be available. Two components are in parallel if only one of the components needs to be available for the connection to be available.

For a series network, the probability of an available path is equal to the product of the individual component availabilities. For a parallel network, the probability of an unavailable path is equal to the product of the individual component unavailabilities. Components in series reduce availability and components in parallel improve availability.

$$P_{\text{series}} = \prod P_{\text{component}} \tag{5.3}$$

$$Q_{\text{parallel}} = \prod Q_{\text{component}} \tag{5.4}$$

The failure rate λ is a measure of unreliability. The product λr (failure rate $\times$ average downtime per failure) is equal to the forced downtime hours per year and it can be considered as a measure of forced unavailability, since a scale factor of 8760 converts one value to the other. The average downtime per failure r could be called restorability. The procedure and formulas for calculating the reliability indices are given in IEEE Standard 493-2007 and are reproduced in (5.5)–(5.10). A schematic demonstrating these formulas is shown in Figure 5.1 for two components numbered "1" and "2" connected in series and in Figure 5.2 for two components "3" and "4" connected in parallel.

Figure 5.1   Repairable components in series – both must work for success (taken from IEEE Standard 493-2007)

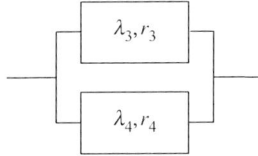

Figure 5.2   Repairable components in parallel – one or both must work for success (taken from IEEE Standard 493-2007)

In these schematics, scheduled outages are assumed to be zero and the units for f, λ, and r are, respectively, frequency of failures for the whole system, failure rate per year of each component, and hours of downtime per failure. The equations in Figure 5.1 and Figure 5.2 assume the following:

- The component failure rate is constant with age.
- The outage time after a failure has an exponential distribution.
- Each failure event is independent of any other failure event.

$$f_s = \lambda_1 + \lambda_2 \tag{5.5}$$

$$f_s r_s = \lambda_1 r_1 + \lambda_2 r_2 \tag{5.6}$$

$$r_s \cong \frac{\lambda_1 r_1 + \lambda_2 r_2}{\lambda_1 + \lambda_2} \tag{5.7}$$

$$f_p = \frac{\lambda_3 \lambda_4 (r_3 + r_4)}{8760} \tag{5.8}$$

$$f_p r_p = \frac{\lambda_3 r_3 \lambda_4 r_4}{8760} \tag{5.9}$$

$$r_p = \frac{r_3 r_4}{r_3 + r_4} \tag{5.10}$$

Example 5.1 Figure 5.3 shows a power system with two lines, each one with three failures per year and 0.75 h downtime per failure. The system load is 20 MW. Calculate the frequency of failures, the minutes per year of unreliability, and the

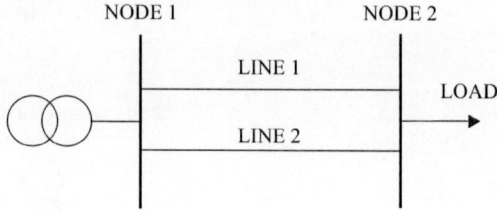

Figure 5.3   Diagram system for example 5.1

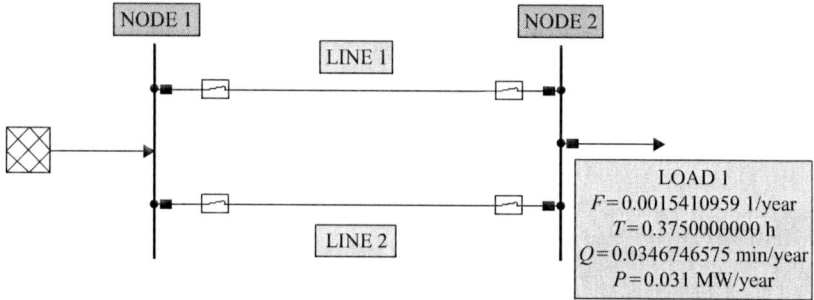

Figure 5.4   Diagram system results for the first part of example 5.1

non-supplied power. First assume that each line can support the whole load; then calculate again, assuming that each line cannot support the total load.

(a) For the first situation, where each line can support the whole load, it is valid to assume a parallel connection.

$$f_p = \frac{3 \times 3 \times (0.75 + 0.75)}{8760} = 0.0015411/\text{yr}$$

$$r_p = \frac{0.75 \times 0.75}{0.75 + 0.75} = 0.375 \text{ h}$$

$$Q_p = f_p \times r_p = 0.0015411 \times 0.375 = 0.000578 \text{ h/year} = 0.03467 \text{ min/yr}$$

or

$$Q_p = Q_1 \times Q_2 = (Q_1)^2 = (135 \text{ min/yr})^2 = (0.0002568)^2 = 0.03467 \text{ min/yr}$$

Bear in mind that:

$$Q_1 = Q_2 = 3^*0,75 \text{ h/yr} = 135 \text{ min/yr}$$

The total power interrupted due to failures is:

$$P_p = P_{\text{load}} \times f_p = 0.031 \text{ MW/yr}$$

These results were validated with a software package as shown in Figure 5.4.

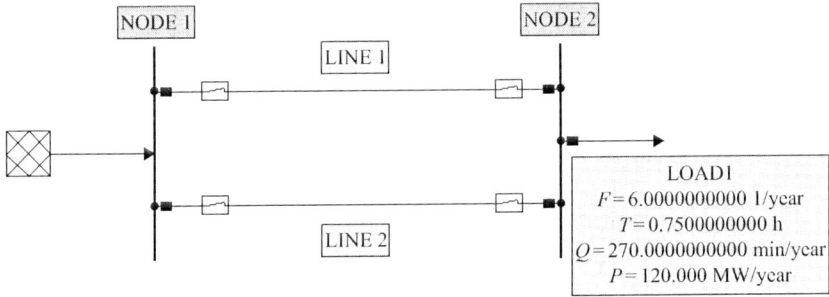

Figure 5.5 Diagram system results for the second part of example 5.1

(b) For the second case where it is assumed that each individual line can't support the total load, the reliability analysis is developed by treating the system as a series connection, although both lines are physically connected in parallel.

$$f_s = 3 + 3 = 6/\text{year}$$

$$r_s \cong \frac{3 \times 0.75 + 3 \times 0.75}{3 + 3} = 0.75 \text{ h}$$

$$Q_s = f_s \times r_s = 6 \times 0.75 = 4.5 \text{ h/year} = 270 \text{ min/yr}$$

or

$$1 - Q_s = P_s = P_1 \times P_2 = P_1^2 = (1 - Q_1)^2$$

$$Q_s = 1 - (1 - Q_1)^2 = 1 - (1 - 3 \times 0.75/8760)^2 = 1 - (1 - 0.0002568)^2$$

$$= 270 \text{ min/yr}$$

The total power interrupted due to failures is:

$$p_s = p_{\text{load}} \times f_s = 20 \times 6 = 120 \text{ MW/yr}$$

These results were validated with a software package as shown in Figure 5.5.

5.2 Network reduction

Complex systems have a high number of series and parallel connections. To process them, a network reduction is required. This is accomplished by simplifying parallel and series components into equivalent network components until a single component is obtained. The availability of the last component is equal to the availability of the original system. An example of network reduction is shown for the system shown in Figure 5.6 and the corresponding steps to reduce it.

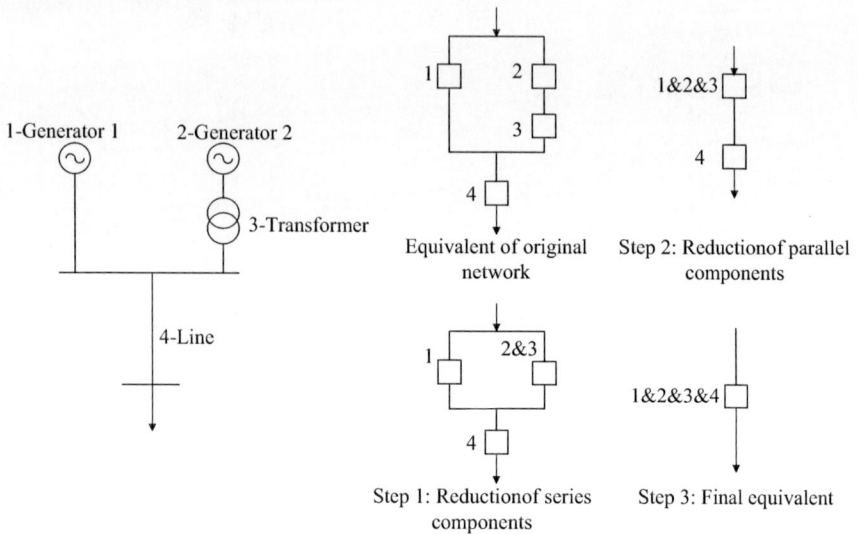

Figure 5.6   Example of network reduction

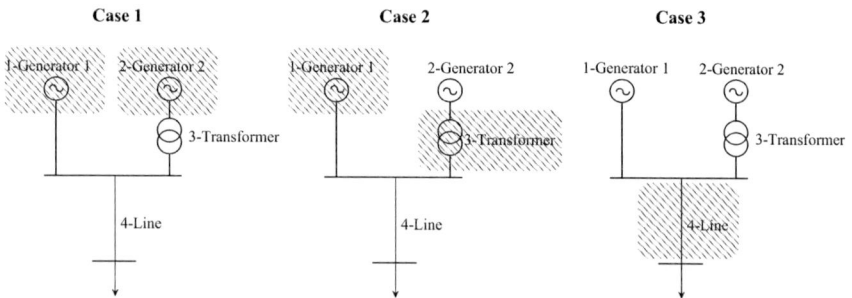

Figure 5.7   Minimal cut sets of the system of figure 5.6

In this system, a generator (component 1) and/or another generator (component 2) and a transformer (component 3) provide power to a load through a line (component 4). The three steps of the network reduction required to compute the availability of the system is well illustrated in Figure 5.6.

A good approach for network reduction is the so-called minimal cut set method whereby the minimal numbers of unavailable elements that get the system unavailable are determined. Figure 5.7 shows three scenarios each corresponding to a minimal cut set for the same system considered in Figure 5.6.

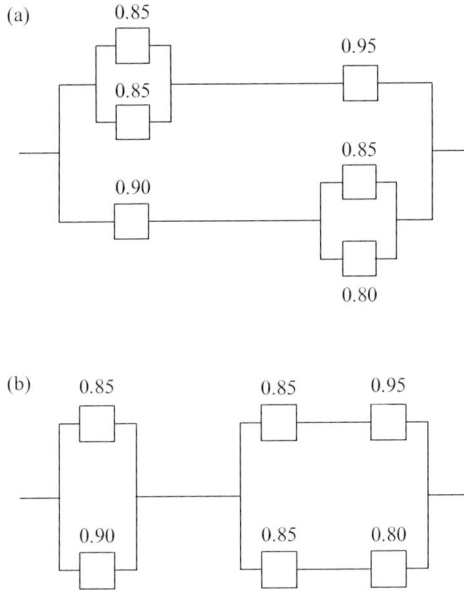

Figure 5.8 Elements of exercise 5.2

Proposed exercises

1. A system is made up by 3 elements, each one with a failure rate and MTTR given by λ_1, λ_2 and λ_3, and r_1, r_2 and r_3. Consider that in one case all elements are connected in series and in the other in parallel. For each case, determine the equivalent failure rate and the MTTR for the whole system.
2. Figure 5.8 presents the availability values of the different elements of a system that has two different arrangements. Calculate the total availability for each arrangement.

Chapter 6

Reconfiguration and restoration of distribution systems

For decades, distribution systems were designed with rigid topologies and limited possibilities for change in configuration. Improvements achieved by the manufacturers in recent years have prompted the use of feeder reconfiguration to modify the topology of distribution networks. These improvements include remotely controlled switches and breakers for pole installation, numerical protection, and appropriate communication systems

Prior to determining the location of switches to allow changes in configuration, it is highly recommended to find the best topology for a distribution system. Under normal operating conditions, feeder reconfiguration aims for a more efficient operating condition of the network. Under faulty conditions, feeder reconfiguration aims to restore the service to the maximum number of users in the shortest time.

6.1 Optimal topology

The optimal topology of a distribution system is one where operating conditions are satisfied with the lowest possible value of system losses. The operating conditions correspond to the voltage regulation, loading, and power factors. To determine the optimal topology, all the poles in overhead feeders and in particular the double dead end poles, or switching boxes in underground feeders, are potential open points. This allows to determine the best boundaries among feeders to reduce the overall losses.

Once the optimal topology is achieved, switches, breakers, or even re-closers can be installed in some of the feeders to allow for system reconfiguration.

The optimal topology can be found by running load flows, one for each case. This would result in a huge number of runs especially for large systems, thus making this approach unfeasible. That is why several methods have been developed to determine the best topology for a distribution system. These include heuristic methods, linear programming, neural networks, expert systems, fuzzy logic, simulated annealing and genetic algorithms, among others.

The heuristic method is the most commonly used since it can produce fast results with good accuracy. Linear programming is used mainly in planning applications to minimize capital investment.

There are several heuristic methods available. In one of them, first the network is considered with all tie switches closed, which converts it into a meshed network. Then one by one, the tie switch with the lowest current is opened until the network becomes radial. Another method within the heuristic techniques considers the closing of one tie switch and the opening of one sectionalizing switch at the same time, in order to transfer loads from feeders with a higher voltage drop to those with a lower voltage drop, but still keeping the network radial.

Another approach uses a feeder reconfiguration algorithm with an optimal flow pattern. In this case, unlike the first two methods, only one tie switch is closed at a time and the radial configuration is restored by opening the same or another switch, depending upon the result of the optimal flow. Although this method gives good results, its implementation can be very lengthy and therefore difficult to apply in real-time environments.

The economic benefits of optimal topology are one of the key considerations when investing not only in the hardware and software but also in switches and communication equipment, which are essential when implementing the method.

Therefore, other methods have taken into account these considerations. A very interesting formulation of the problem considers a simulated annealing approach for solving combinatorial optimization problems; other methods propose an algorithm for the purpose of reducing the operating cost in the real-time operation environment based on heuristic approaches. It is also important in optimal topology to consider variations in customer loads, and the unbalanced nature of distribution systems.

One of the major difficulties in implementing optimal topology methods in distribution systems is the large amount of data to be handled, which demands a lot of time and makes the application difficult especially in online conditions. To overcome this difficulty, some algorithms have been proposed based on partitioning the distribution network into groups of load buses, such that the line section losses between the groups of nodes are minimized. These algorithms attempt to overcome the size limitations and are recommended for online applications.

Recent papers take advantage of expert systems and neural networks to overcome the difficulties inherent in feeder reconfiguration. Fuzzy control is used to improve the programming based on heuristic information that could eventually be applied to real systems.

The best approach for determining the basic topology is by considering that all the poles in overhead lines, or in the connection boxes of underground feeders, can feasibly be opening points or feeder ends. This is illustrated in Figure 6.1 which has a simple system of two feeders.

Several specialized software packages have been developed to determine the optimal topology which some software developers refer to as "optimal separation points." The packages can have different optimization criteria, such as minimum losses, elimination of overloads, and voltage control. The most common criterion is minimum losses.

Example 6.1 Consider a distribution system that has approximately 110 feeders at 13.2 kV with a total loss figure above 15%. An optimized configuration is required

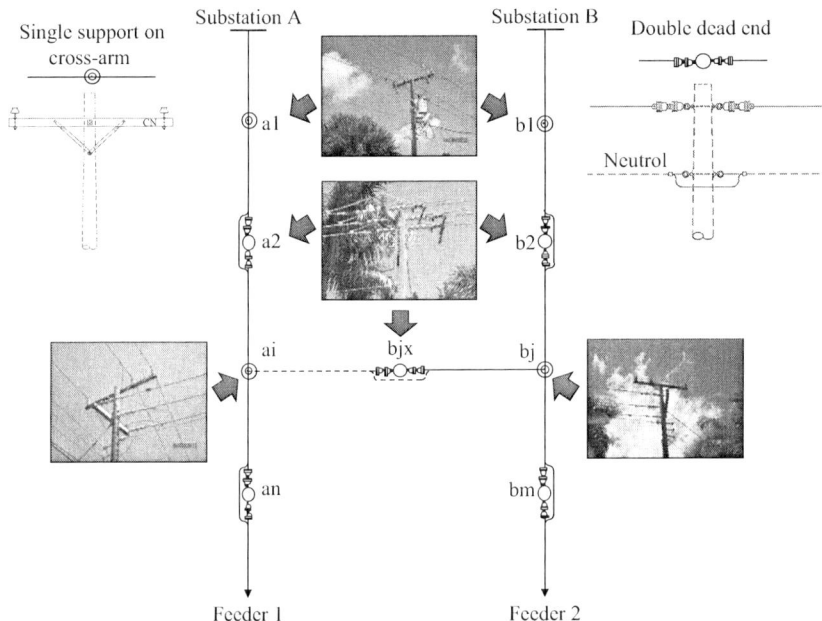

Figure 6.1   Two-feeder system showing location of potential opening points

to reduce that figure. This will require the application of several methods, among them is the feeder reconfiguration.

A pilot of around 5% of the distribution network was studied, which is a good representation of the entire system. Calculate the saving in dollars first for the pilot and then the savings projected for the entire system.

The prototype was selected such that the characteristics of the feeders were typical and the area included several social strata, hospitals, sport arenas and other key locations of the city. The analysis was carried out by running a load flow for each case and also by using an optimal topology algorithm based on a heuristic method. Table 6.1 presents the data of the feeders analyzed.

The first columns of Table 6.2 show the identification of the poles and the feeders they correspond to. The next columns present the initial and final loss levels for the circuits receiving loads (CRL) and circuits transferring loads (CTL). The loss reduction is shown in two columns. The first corresponds to the results calculated with load flow runs and the second with the optimal topology methods. It is interesting to see the excellent accuracy of the later. The last column shows the voltage profile to make sure that all nodes are within the established range. Table 6.3 is similar and presents the best three options for reconfiguring the system at the poles that offer the highest loss reduction.

Figure 6.2 shows the initial topology of the system and Figure 6.3 shows the topology of the system after reconfiguration to attain loss reduction. The numbers in both figures indicate those poles where two feeders meet and obviously correspond to double dead en arrangements.

Table 6.1 Data of feeders encompassed within the prototype

Substation	Circuit name	Circuit code	Nodes	Transformers	Installed capacity (kVA)	Load factor	Links	Total losses (kW)	(kVAR)
St. Anthony	Crystals	0106	283	152	16122	0.312	10	61.08	234.53
St. Anthony	10th Street	0109	100	47	9927	0.253	2	10.99	32.60
St. Anthony	St. Ferdinand	0110	152	81	7750	0.324	7	9.11	37.86
South	Britain	0513	469	251	27582	0.390	6	226.72	802.16
South	Lido	0517	405	222	27147	0.333	9	292.36	809.14
South	Cedar	0518	417	230	30760	0.283	8	99.98	323.07

Table 6.2 Result of reconfiguration analysis

Link no.	Current location			Recommended location		Initial losses (kW)			Final losses (kW)			Reduction (kW)		Volt min (pu)
	Link	Circuits		Link	Circuits	CRL	CTL	Total	CRL	CTL	Total	Optimal topology	Load flow	
1	100483A-100483B	0110	0517	1003445-1192621	0517	9.11	292.36	301.47	46.58	163.26	209.84	90.45	91.63	0.970302
2	105861B-105861A	0110	0106	1184784-1060406	0106	9.11	61.08	70.19	35.87	23.21	59.08	11.05	11.11	0.973206
3	121478B-121478A	0110	0106	1059041-1059050	0106	9.11	61.08	70.19	43.51	16.66	60.20	9.88	9.99	0.970410
4	105880B-105880A	0110	0106	1059084-1059092	0106	9.11	61.08	70.19	50.48	14.45	64.93	5.18	5.26	0.967839
5	105850B-105850A	0110	0106	105850A-1184695	0106	9.11	61.08	70.19	9.37	60.48	69.85	0.33	0.34	0.989122
6	105392B-105392A	0110	0106	105392A-1053914	0106	9.11	61.08	70.19	9.11	61.08	70.19	0.00	0.00	0.989240
7	100349B-100349A	0518	0517	1192540-1192531	0517	99.98	292.36	392.34	101.83	289.43	391.26	1.10	1.08	0.955976
8	100285B-100285A	0518	0517	1002741-1203495	0517	99.98	292.36	392.34	100.64	291.12	391.76	0.61	0.58	0.956378
9	100370B-100370A	0518	0517	1003976-1192621	0517	99.98	292.36	392.34	130.15	252.65	382.80	9.11	9.54	0.947586
10	119244B-119244A	0518	0517	119244A-1192434	0517	99.98	292.36	392.34	99.98	292.36	392.34	0.04	0.00	0.956510
11	100857A-100857B	0106	0518	1008544-1008528	0518	61.08	99.98	161.06	74.64	84.55	159.19	1.90	1.87	0.958511
12	106023A-106023B	0106	0518	1008536-1201573	0518	61.08	99.98	161.06	82.34	73.40	155.74	5.37	5.32	0.958584
13	100357A-100357B	0106	0518	1092499-1092481	0518	61.08	99.98	161.06	89.36	69.60	158.96	2.14	2.10	0.957062
14	100447A-100447B	0513	0517	1004484-1192426	0517	226.72	292.36	519.08	229.28	288.57	517.85	1.20	1.23	0.932889
15	106202A-106202B	0109	0513	1061321-1184474	0513	10.99	226.72	237.71	56.66	118.02	174.68	62.63	63.03	0.968333
16	106128A-106128B	0109	0513	5099684-1184482	0513	10.99	226.72	237.71	64.87	111.53	176.40	60.63	61.31	0.963412
17	100546-1192248	0106	0517	1004646-1004638	0517	61.08	292.36	353.44	127.54	189.47	317.01	35.62	36.43	0.945557
18	5105455-1062115	0110	0513	5047145-1061852	0513	9.11	226.72	235.83	45.35	138.54	183.89	51.67	51.94	0.968140
19	1008340-1060872	0518	0106			99.98	61.08	161.06						
20	100498A-100498B	0513	0517	1004930-1192337	0517	226.72	292.36	519.08	255.70	255.93	511.63	7.32	7.45	0.928727
21	100209A-100209B	0513	0517	1192370-1192353	0517	226.72	292.36	519.08	250.19	266.32	516.51	2.41	2.57	0.928589

Table 6.3 Options where the highest loss reductions are obtained

Link no.	Current location		Recommended location		Initial losses (kW)			Final losses (kW)			Reduction (kW)		Volt min (pu)
	Link	Circuits	Link	Circuits	CRL	CTL	Total	CRL	CTL	Total	Optimal topology	Load flow	
1	100483A–100483B	0110 0517	1003445–1192621	0517	9.11	292.36	301.47	46.58	163.26	209.84	90.45	91.63	0.970302
12	106023A–106023B	0106 0518	1008536–1201573	0518	61.08	99.98	161.06	82.34	73.40	155.74	5.37	5.32	0.958584
15	106202A–106202B	0109 0513	1061321–1184474	0513	10.99	226.72	237.71	56.66	118.02	174.68	62.63	63.03	0.968333
TOTAL					81.18	619.06	700.24	185.58	354.68	540.26	158.45	159.98	

A simple cost saving evaluation data considering a load factor equal to 1 gives the following results:

kWh cost: 0.10 $/kWh

Perform the calculation of the annual savings assuming that the load factor (defined as the average load divided by peak load), is equal to one.

Loss savings: 160 kW (from Table 6.3)

Annual savings: 160 kW $\times$ 8760 h $\times$ 0.10 $/kWh = $140,160

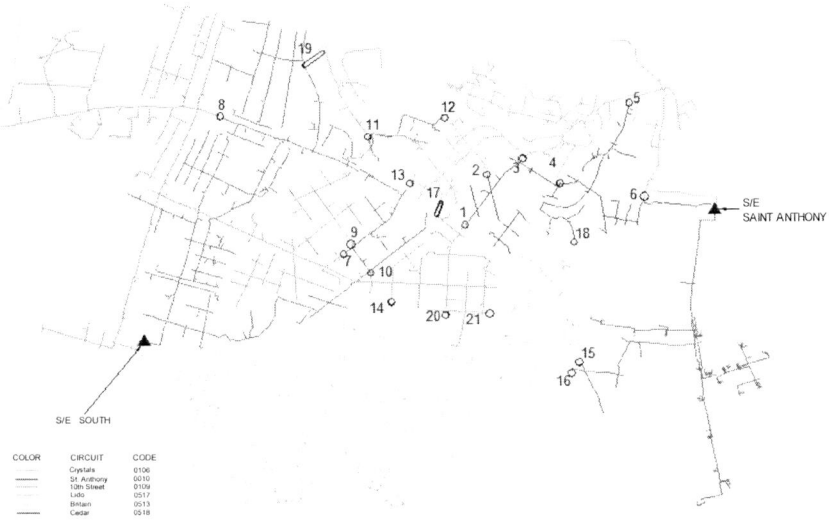

Figure 6.2   Initial topology of case study

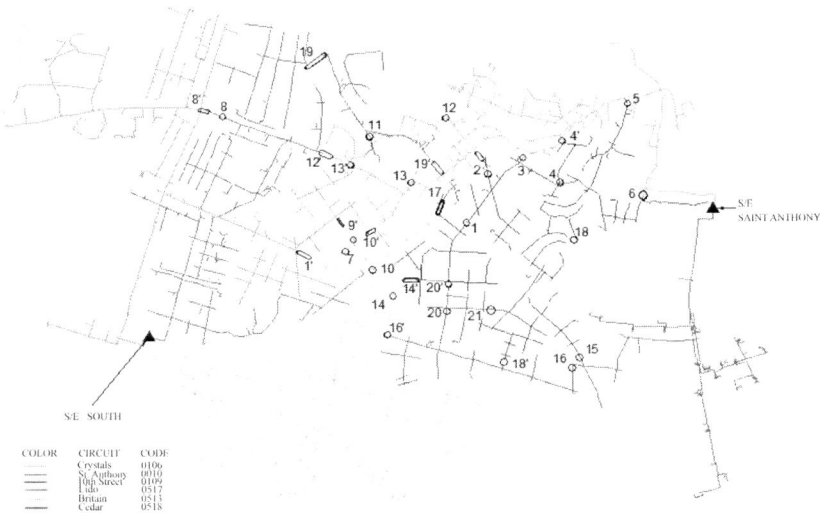

Figure 6.3   Optimal topology considering loss reduction

If the prototype represents 5% of the overall system, the total savings amount is $2,803,200 per year.

6.2 Location of switches controlled remotely

Switches can be breakers, reclosers, or sectionalizers and should have the means for remote operation to guarantee a fast reconfiguration when required.

The placement of switches is carried out in such a way that the reliability and flexibility criteria of the network are increased. This is explained in the following sections.

6.2.1 Considerations to increase reliability

The basic concepts of reliability were treated in Chapter 5. Those concepts will be now used to show how the reliability of distribution networks can be greatly improved by adding switches along the feeders. If breakers with the corresponding protective equipment are installed on sections downstream important loads, faults on those sections will not affect the loads upstream, improving then the reliability indices. The benefit of adding the switches can be measured by examining the improvement in reliability performance.

Example 6.2 Figure 6.4 shows a distribution system with three feeders. Table 6.4 contains the data for each load (active and reactive power), and Table 6.5 shows the reliability characteristics for each line. Calculate the frequency of failures, the hours per year of unavailability, the probability to being unavailable, and the non-supplied power at Node 11 for the following scenarios:

(a) Normal operation,
(b) Removing the breaker at substation N11 associated with line N11-N12,
(c) Removing the breaker at substation N10 associated with line N2-N10,
(d) Removing the circuit breaker at substation N2 associated with line N2-N8, and
(e) Removing the circuit breaker at substation N4 associated with line N4-N13.

Compare the effect from each scenario on the whole system using the reliability values.

Considering the possibility of multiple faults in this system, the calculation becomes difficult since the analysis for each node results from a combination of scenarios/combinatory selection of relationships between different distribution system elements. For that reason, power system analysis software is used.

The results for the base normal operating scenario, Case (a), are presented in Figure 6.5.

In Case (b) (see Figure 6.6), the circuit breaker just after node N11 was removed. It can be seen that the reliability decreases. This is explained because the circuit breaker removed could have cleared all the faults beyond node N11. Then faults beyond this point can only be cleared when a crew identifies, isolates, and restores the service at this point. This increases the probability that node N11 is impaired by a fault for a particular amount of time.

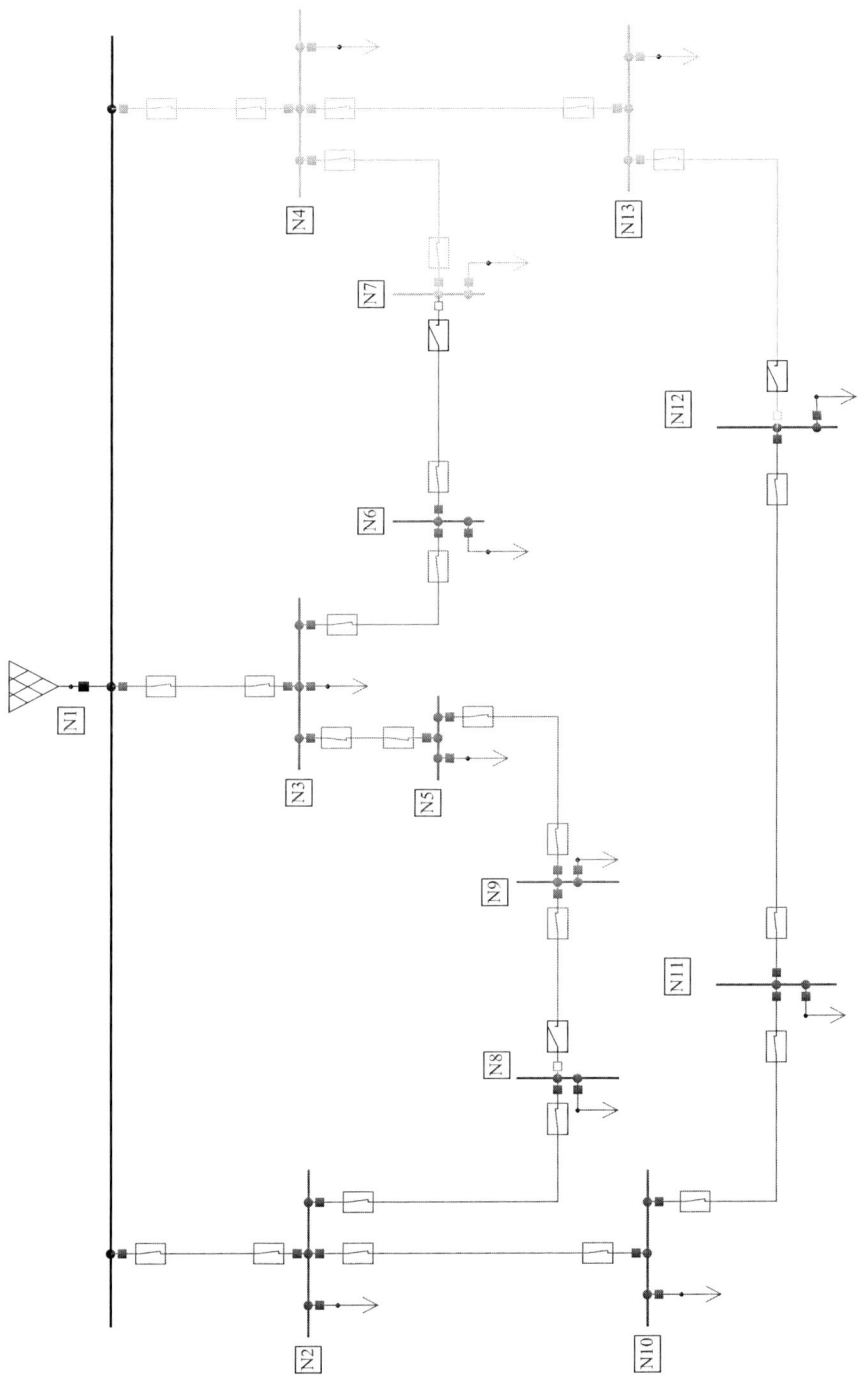

Figure 6.4   System diagram for example 6.2

Table 6.4   Load data for example 6.2

Name	P (MW)	Q (MVAR)
LN_{02}	2	1.0
LN_{03}	2	1.0
LN_{04}	2	1.0
LN_{05}	5	2.5
LN_{06}	5	2.5
LN_{07}	5	2.5
LN_{08}	5	2.5
LN_{09}	5	2.5
LN_{10}	7	3.5
LN_{11}	5	2.0
LN_{12}	2	1.0
LN_{13}	7	3.5

Table 6.5   Line data for example 6.2

Name	Length (km)	λ* (1/year-km)	r (h)
$L_{01\text{-}03}$	1.0	1.4	0.5
$L_{06\text{-}07}$	1.2	1.4	0.5
$L_{03\text{-}06}$	1.7	1.4	0.5
$L_{04\text{-}13}$	1.9	1.0	0.75
$L_{01\text{-}04}$	0.8	1.2	0.45
$L_{02\text{-}08}$	2.6	1.2	0.45
$L_{01\text{-}02}$	2.0	1.0	0.75
$L_{09\text{-}08}$	2.2	1.4	0.5
$L_{10\text{-}11}$	3.0	1.4	0.5
$L_{11\text{-}12}$	1.0	1.2	0.45
$L_{03\text{-}04}$	1.5	1.4	0.5
$L_{05\text{-}09}$	2.0	1.2	0.45
$L_{03\text{-}05}$	1.4	1.0	0.75
$L_{02\text{-}10}$	2.1	1.4	0.5
$L_{13\text{-}12}$	1.5	1.4	0.5

* Failure rate is equal to the product of columns 2 and 3 for each case.

For Case (c) (see Figure 6.7), the circuit breaker just before node N10 was removed. Here, the reliability doesn't change compared to the first scenario. Removing this element from the system doesn't change the ability to clear a fault beyond this node.

For Case (d) (see Figure 6.8), the circuit breaker removed is near node N2, between the node N2 and N8. Removing this element from the system impairs the reliability of node N11 in comparison with the first scenario.

Changes in other feeders can also impair reliability in neighboring feeders. In Case (e) (see Figure 6.9), the circuit breaker at busbar N1 connecting with busbar

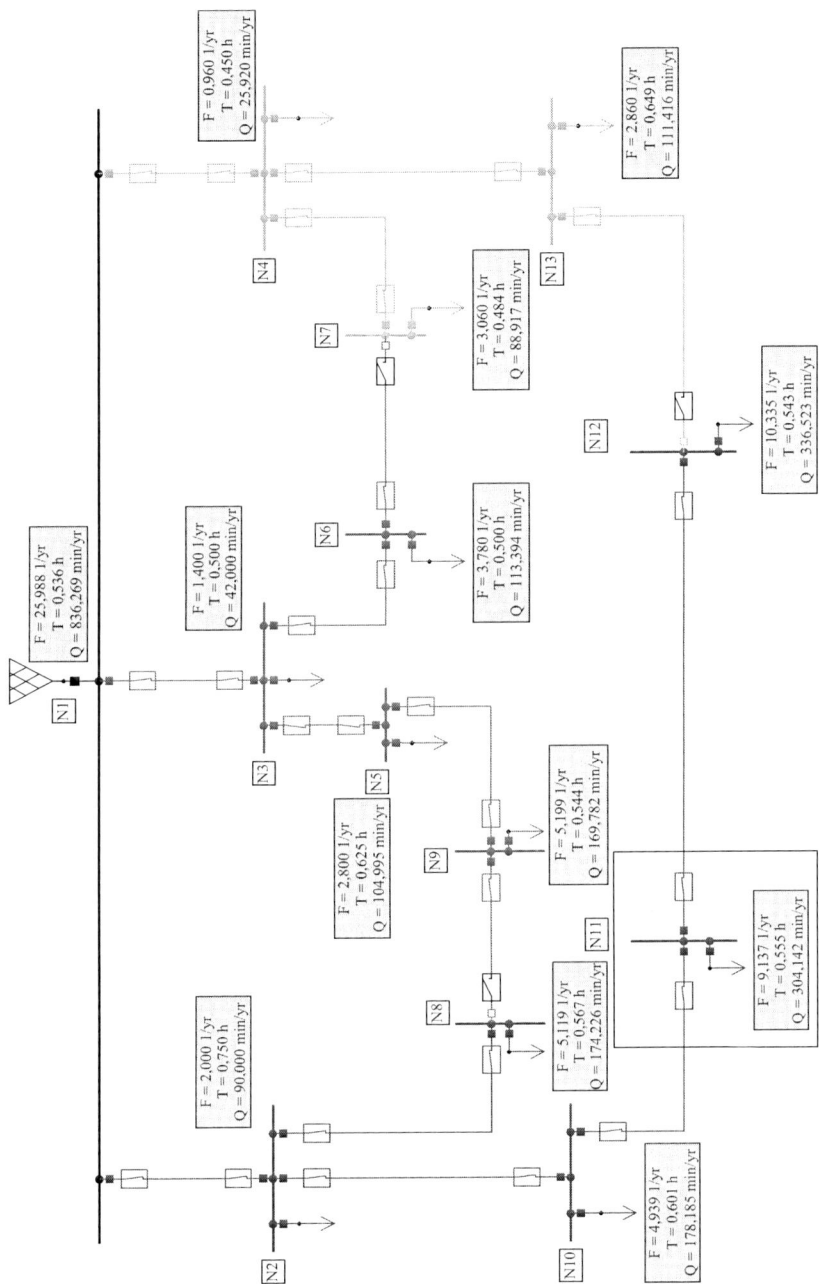

F = 0.960 1/yr
T = 0.450 h
Q = 25,920 min/yr

F = 2.860 1/yr
T = 0.649 h
Q = 111,416 min/yr

F = 25.988 1/yr
T = 0.536 h
Q = 836,269 min/yr

F = 1.400 1/yr
T = 0.500 h
Q = 42,000 min/yr

F = 3.060 1/yr
T = 0.484 h
Q = 88,917 min/yr

F = 10.335 1/yr
T = 0.543 h
Q = 336,523 min/yr

F = 3.780 1/yr
T = 0.500 h
Q = 113,394 min/yr

F = 2.800 1/yr
T = 0.625 h
Q = 104,995 min/yr

F = 5.199 1/yr
T = 0.544 h
Q = 169,782 min/yr

F = 9.137 1/yr
T = 0.555 h
Q = 304,142 min/yr

F = 2.000 1/yr
T = 0.750 h
Q = 90,000 min/yr

F = 5.119 1/yr
T = 0.567 h
Q = 174,226 min/yr

F = 4.939 1/yr
T = 0.601 h
Q = 178,185 min/yr

N1 N2 N3 N4 N5 N6 N7 N8 N9 N10 N11 N12 N13

Figure 6.5 Results for scenario (a) of example 6.2

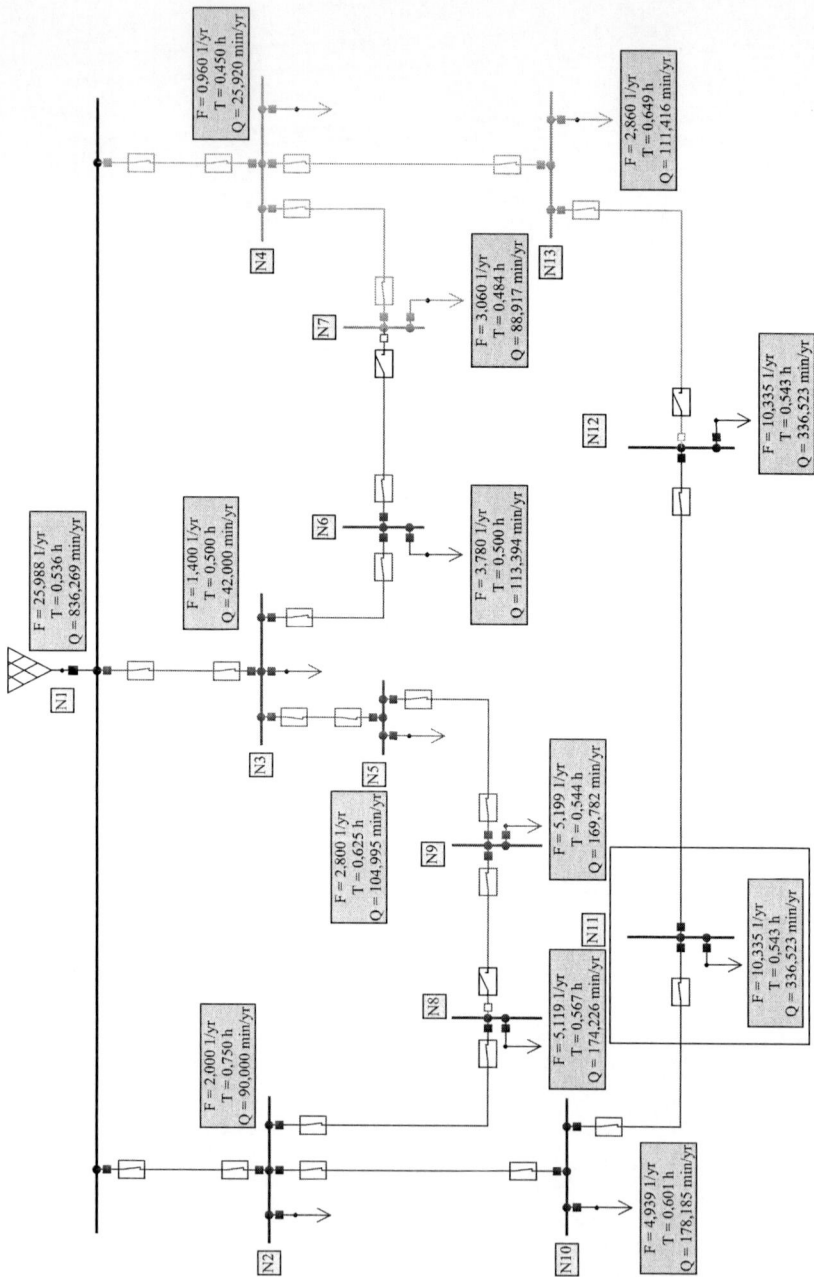

Figure 6.6 Results for scenario (b) of example 6.2

F = 0.960 l/yr
T = 0.450 h
Q = 25,920 min/yr

F = 2,860 l/yr
T = 0.649 h
Q = 111,416 min/yr

N4

N13

F = 3,060 l/yr
T = 0.484 h
Q = 88,917 min/yr

N7

F = 10,335 l/yr
T = 0.543 h
Q = 336,523 min/yr

N12

F = 25,988 l/yr
T = 0.536 h
Q = 836,269 min/yr

N1

F = 1,400 l/yr
T = 0.500 h
Q = 42,000 min/yr

N6

F = 3,780 l/yr
T = 0.500 h
Q = 113,394 min/yr

F = 2,800 l/yr
T = 0.625 h
Q = 104,995 min/yr

N3

N5

F = 5,199 l/yr
T = 0.544 h
Q = 169,782 min/yr

N9

F = 10,335 l/yr
T = 0.543 h
Q = 336,523 min/yr

N11

N8

F = 2,000 l/yr
T = 0.750 h
Q = 90,000 min/yr

F = 5,119 l/yr
T = 0.567 h
Q = 174,226 min/yr

N2

N10

F = 4,939 l/yr
T = 0.601 h
Q = 178,185 min/yr

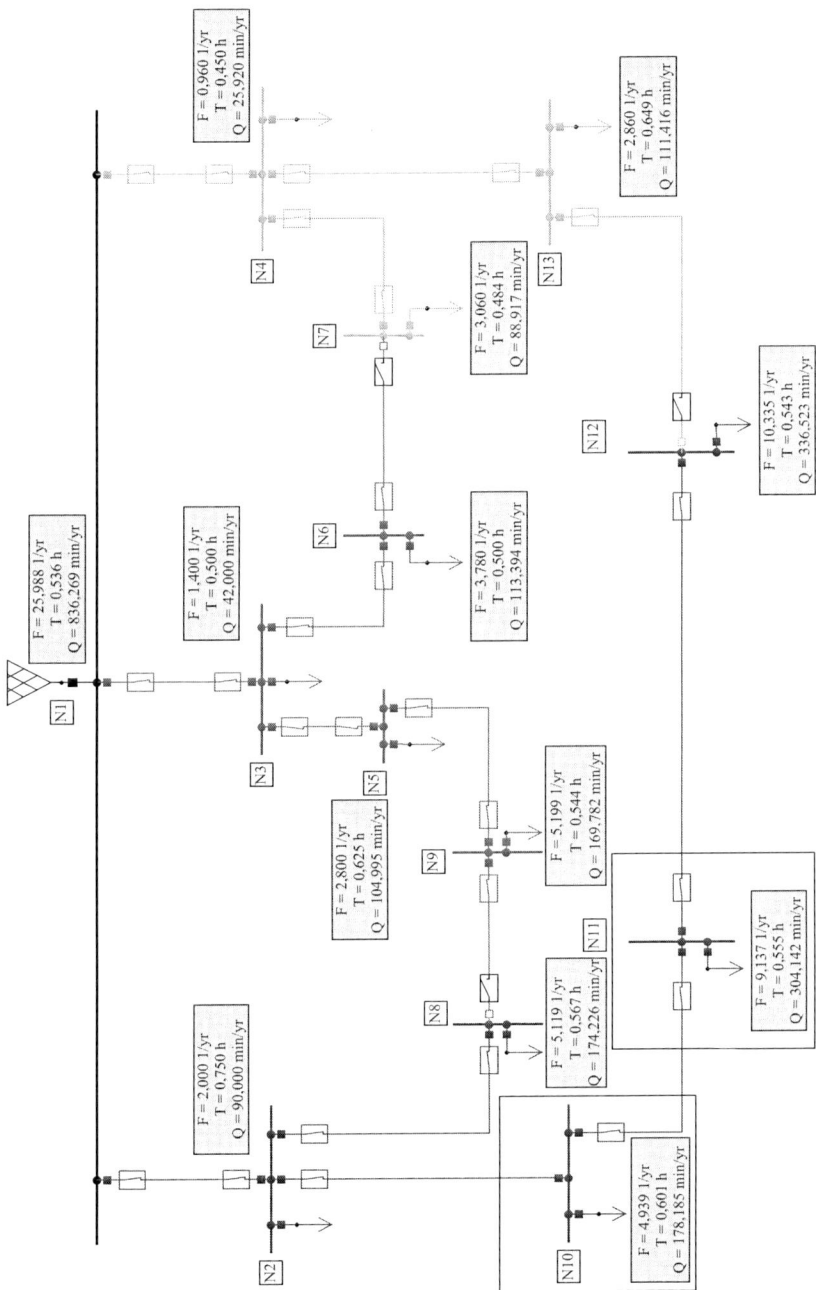

Figure 6.7 Results for scenario (c) of example 6.2

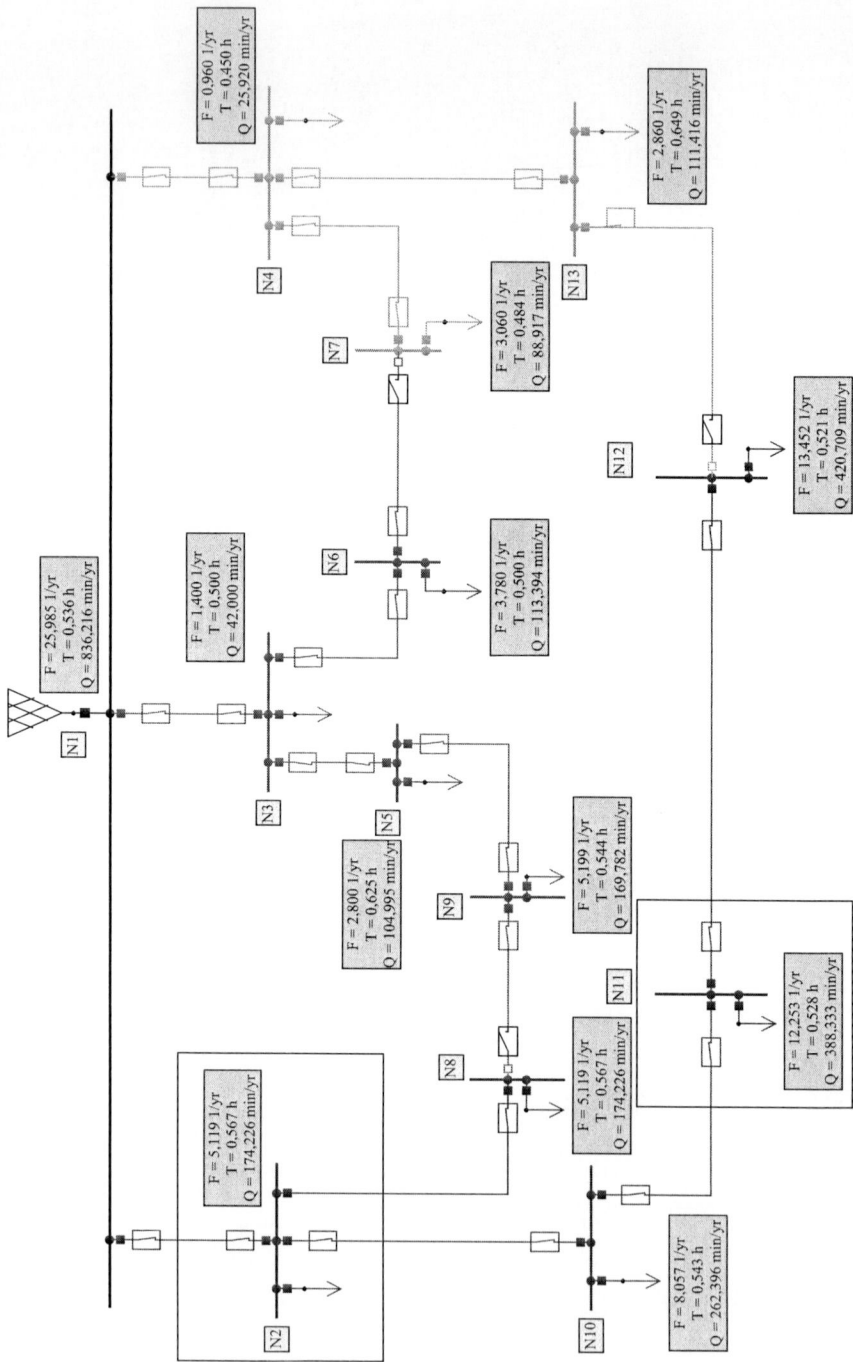

F = 25,985 l/yr
T = 0,536 h
Q = 836,216 min/yr

F = 0,960 l/yr
T = 0,450 h
Q = 25,920 min/yr

F = 2,860 l/yr
T = 0,649 h
Q = 111,416 min/yr

F = 1,400 l/yr
T = 0,500 h
Q = 42,000 min/yr

F = 3,060 l/yr
T = 0,484 h
Q = 88,917 min/yr

F = 3,780 l/yr
T = 0,500 h
Q = 113,394 min/yr

F = 13,452 l/yr
T = 0,521 h
Q = 420,709 min/yr

F = 2,800 l/yr
T = 0,625 h
Q = 104,995 min/yr

F = 5,199 l/yr
T = 0,544 h
Q = 169,782 min/yr

F = 12,253 l/yr
T = 0,528 h
Q = 388,333 min/yr

F = 5,119 l/yr
T = 0,567 h
Q = 174,226 min/yr

F = 5,119 l/yr
T = 0,567 h
Q = 174,226 min/yr

F = 8,057 l/yr
T = 0,543 h
Q = 262,396 min/yr

N1
N2
N3
N4
N5
N6
N7
N8
N9
N10
N11
N12
N13

Figure 6.8 Results for scenario (d) of example 6.2

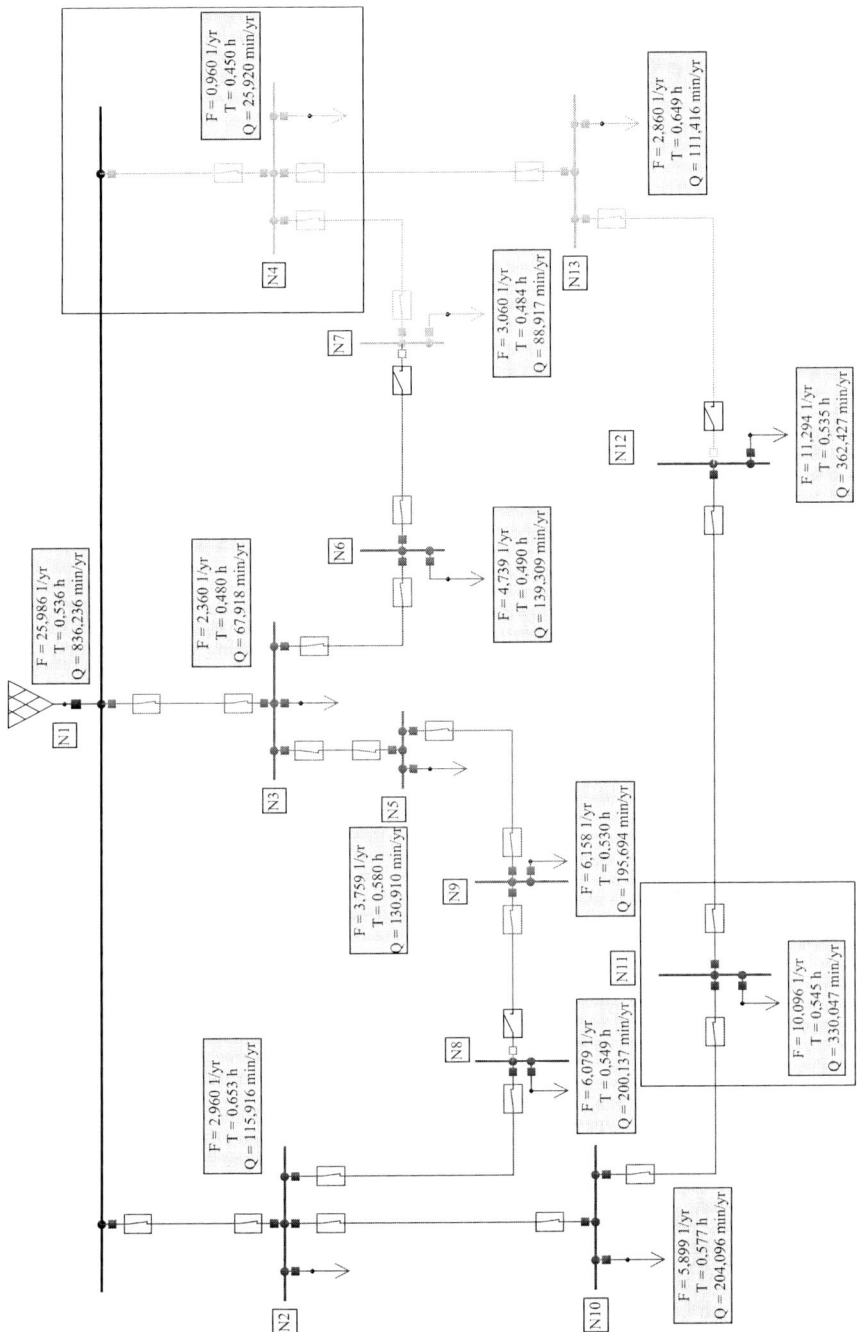

Figure 6.9 Results for scenario (e) of example 6.2

F = 0,960 1/yr
T = 0.450 h
Q = 25,920 min/yr

F = 2,860 1/yr
T = 0.649 h
Q = 111,416 min/yr

N4

F = 3,060 1/yr
T = 0.484 h
Q = 88,917 min/yr

N13

N7

F = 25,986 1/yr
T = 0.536 h
Q = 836,236 min/yr

F = 2,360 1/yr
T = 0.480 h
Q = 67,918 min/yr

F = 11,294 1/yr
T = 0.535 h
Q = 362,427 min/yr

N12

N6

F = 4,739 1/yr
T = 0.490 h
Q = 139,309 min/yr

N1

N3

N5

F = 3,759 1/yr
T = 0.580 h
Q = 130,910 min/yr

F = 6,158 1/yr
T = 0.530 h
Q = 195,694 min/yr

N9

N11

F = 10,096 1/yr
T = 0.545 h
Q = 330,047 min/yr

N8

F = 6,079 1/yr
T = 0.549 h
Q = 200,137 min/yr

F = 2,960 1/yr
T = 0.653 h
Q = 115,916 min/yr

N2

N10

F = 5,899 1/yr
T = 0.577 h
Q = 204,096 min/yr

Table 6.6   Comparison of reliability data for node N11 under different scenarios

	F (1/year)	r (h)	Q (min/year)	P (MW/year)	W (MWh/year)
Scenario a	9.137	0.555	304.142	45.683	25.345
Scenario b	10.335	0.543	336.523	51.676	28.044
Scenario c	9.137	0.555	304.142	45.683	25.345
Scenario d	12.253	0.528	388.333	61.266	32.361
Scenario e	10.096	0.545	330.047	50.478	27.504

N4 was removed. The results are summarized in Table 6.6, which shows for each scenario the impact on the reliability indices. The table has six columns. The first corresponds to the scenario; the second to the failure rate; the third to the MTTR; the fourth to the unavailability; the fifth to the total power interrupted during the events, and the sixth to the energy not supplied due to the faults.

6.2.2   Considerations to increase flexibility

Distribution networks should not be considered to have rigid configurations any more since the location of remotely controlled switches allows changing topologies. From this perspective, the switches can be located to serve two purposes:

- The feeders are sectionalized into equally loaded portions as far as it is possible.
- Flexibility switches are installed at the boundaries so that one or more load section may be routed through them.

Figure 6.10 illustrates the flexibility of each normally open (NO) link. Those links located in the first section of the feeders, such as number 5, do not offer the possibility of partial load transfers through the operation of the normally closed (NC) switches, but only allow a full load transfer. This is an invalid option since the radiality of the feeder would be given up.

The switch located in the second section of the feeder, number 6, offers the possibility of transferring only two-thirds of the load to keep the radiality.

The switch located in the last section of the feeders, identified by number 7, allows the transfer of one- or two-thirds of the feeder load. Therefore, the switches located in this section have the highest flexibility to carry out remotely controlled transfers.

Normally there should be at least one NO switch between two adjacent feeders. NC switches are placed so that they are in the main path of the feeders to carry out load transfers. Ideally, there should be at least one NC switch per installed MW, or a sufficient number to give appropriate flexibility to the feeder. Most NO and NC switches can nowadays be operated remotely from a control center thanks to the availability of motorized devices and the good communication systems available. Figure 6.11 illustrates an example of this type of switches.

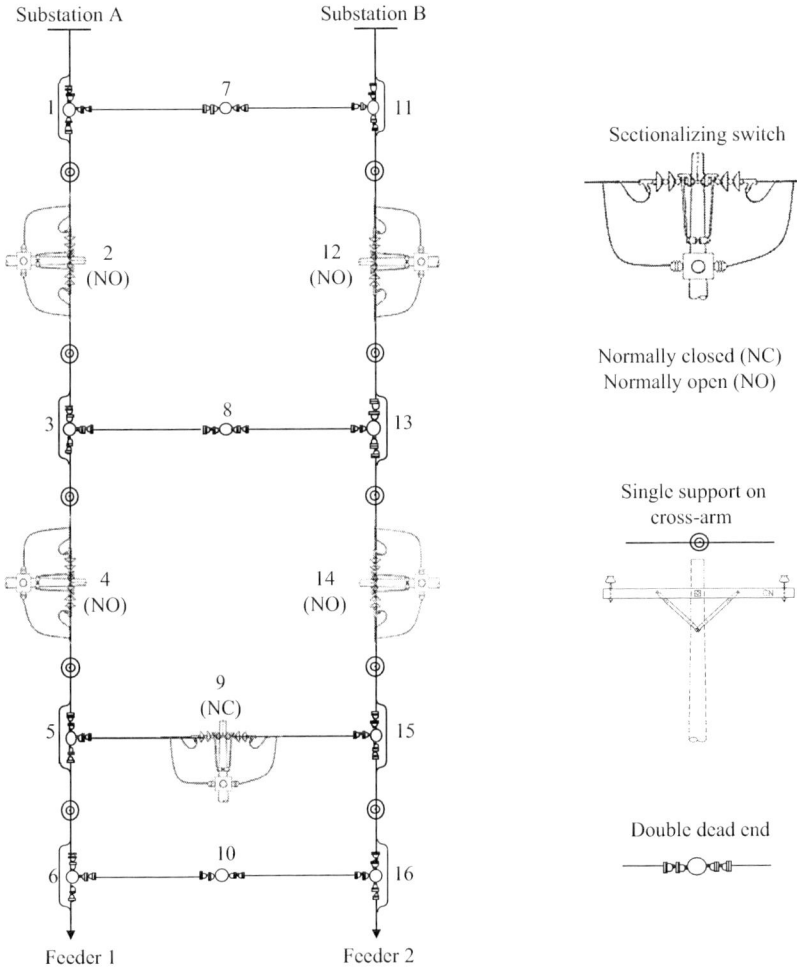

Figure 6.10 Two-feeder system showing location of switches controlled remotely

6.3   Feeder reconfiguration for improving operating conditions

Feeder reconfiguration correspond to the modification of the topology of a network through the closing of a switch that links two feeders and the opening of another switch so as to maintain the radial condition of the feeders.

The reconfiguration is carried out for better operation of the network and specifically to reduce the losses due to the Joule effect. Distribution systems should be operated at minimum cost subject to a number of constraints: all loads are served, overcurrent protective devices are coordinated, voltage drops are within limits, radial configuration is maintained, and lines, transformers, and other equipment operate within current capacity.

Figure 6.11   Location of switches in distribution systems

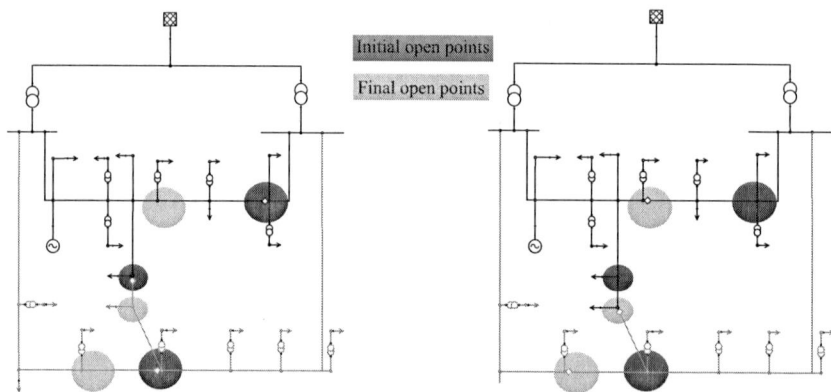

Figure 6.12   Distribution system illustrating loss reduction

The exchange process starts by assuming a radial configuration and therefore with the distribution network operating in a radial configuration. One of the tie switches is closed, and then another switch is opened in the loop created, which restores a radial configuration. The switch pairs are chosen through heuristics and approximate formulas to create the desired change in loss reduction. The branch exchange process is stopped when no more loss reductions are possible.

Figure 6.12 shows a system with four feeders. Here the reconfiguration brings about savings on significant potential losses.

6.4 Feeder reconfiguration for service restoration

One of the most important functions in distribution automation (DA) is service restoration in case of a fault in the primary feeder. The main goal of automatic service restoration is the execution of a series of operations of the tie (NO) and section (NC) switches, aimed at restoring the power supply to the maximum number of areas that have been impaired after a fault in a primary feeder or substation. Usually it would reconfigure the network by transferring loads from the healthy portions of faulty feeders to neighboring feeders that are operating normally.

There are multiple possibilities for reconfiguring the network for service restoration but the group solution is finite; it is defined by the combination of binary states of the section switches and tie switches.

This group grows exponentially with an increase in equipment. It can become a problem of scale. The solution consists in choosing a combination of states for the equipment in the areas not affected, which satisfies some functional objective.

A system reconfiguration changes the topology, line flows, and short circuit values. Thus, different solutions may be feasible for the same fault.

In these circumstances, equipment, voltage profiles, and equipment loading have to be considered. Relay pickup values and time dials have to be rechecked to avoid nuisance tripping or unwanted pickups. This would require the use of different group settings.

6.4.1 Fault location, isolation, and service restoration (FLISR)

The steps to restore service when a fault happens can be summarized as follows:

1. The corresponding relay operates and trips the breaker. If reclosing units operate and the fault remains, the feeder is open.
2. The fault is located and the associated section switches open to isolate it.
3. The feeder is re-energized up to the location of the first section switch that was open – upstream restoration.
4. The healthy sections are transferred to one or more neighboring feeders by operating NO switches – downstream restoration.
5. The faulty section is repaired by the crew and the system is taken back to the normal configuration.

The first four steps should be completed as far as possible within a minute to avoid affecting SAIFI and SAIDI indices. In most countries these indices are affected with outages that last one or more minutes.

These steps are illustrated by using the system shown in Figure 6.13 which corresponds to the normal configuration of a distribution system with four substations, each with one feeder, labeled 1, 2, 3 and 4, associated to breakers B1, B2, B3 and B4 respectively. The system has 14 loading sections labeled Z1 to Z14 and 13 switches in total labeled S1 to S13. Switches 4, 9, and 12 are NO and the rest are NC. Figures 6.14–6.17 illustrate the first four steps already mentioned, following a fault at section Z2 in order to restore the system as far as possible. The fifth step of the process would be to take the system back to normal.

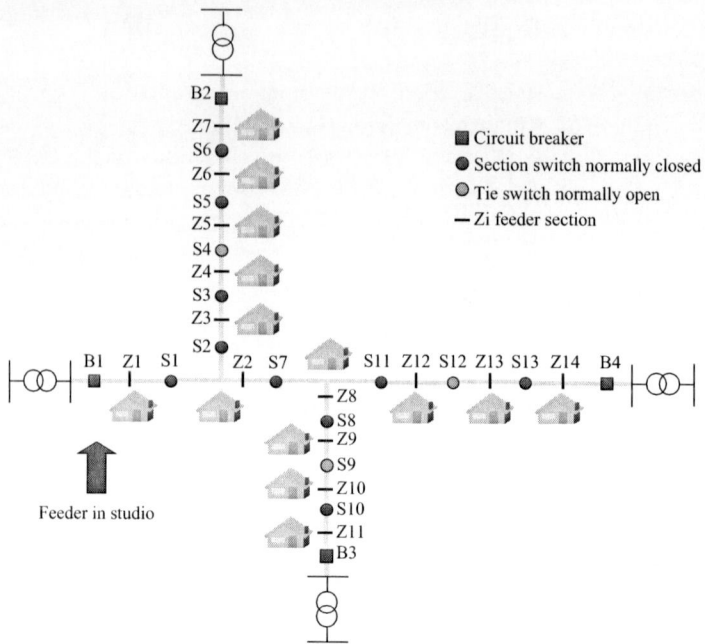

Figure 6.13   Normal configuration

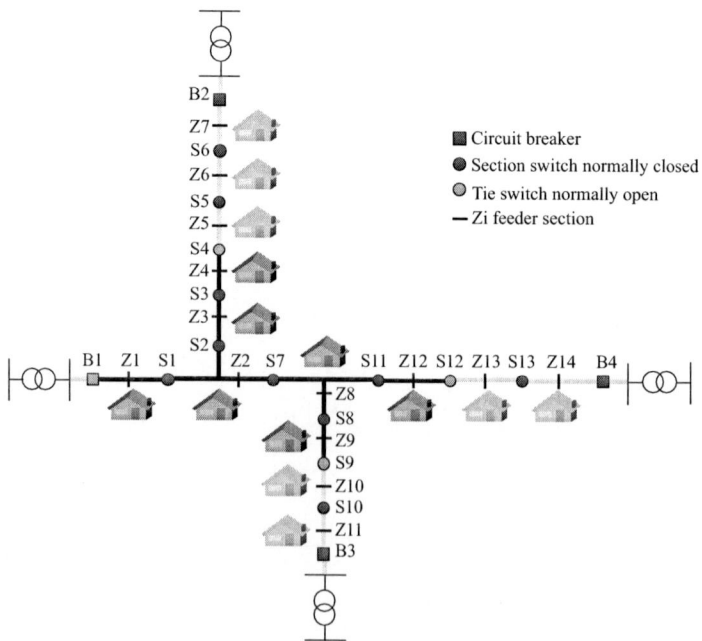

Figure 6.14   Fault clearing by relay at B1 – step 1

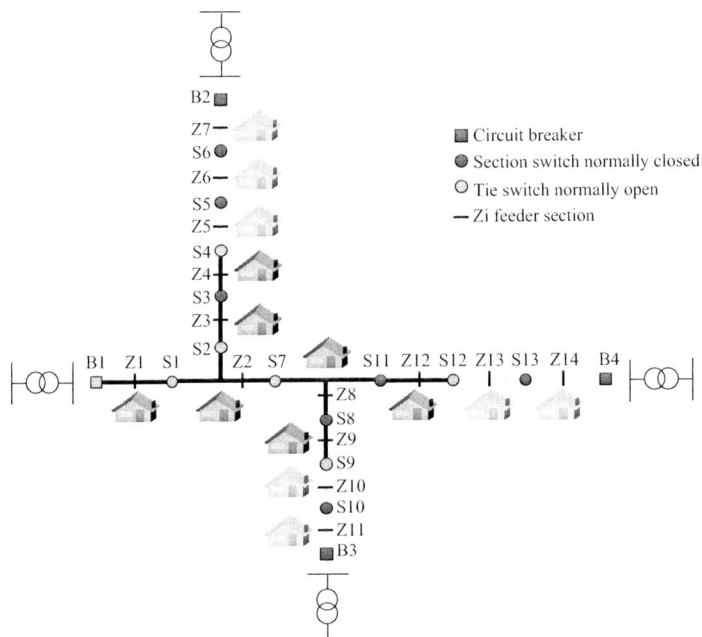

Figure 6.15 Fault isolation by opening corresponding switches – step 2

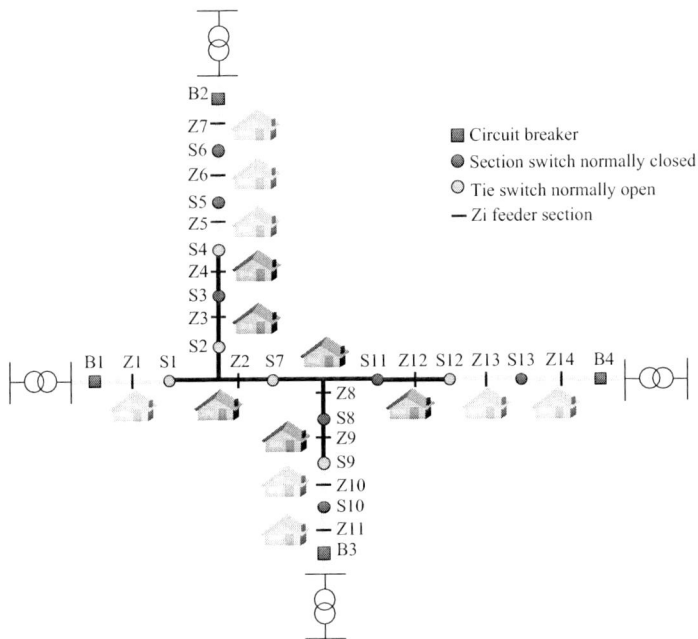

Figure 6.16 Reclosing B1 – step 3 (upstream restoration)

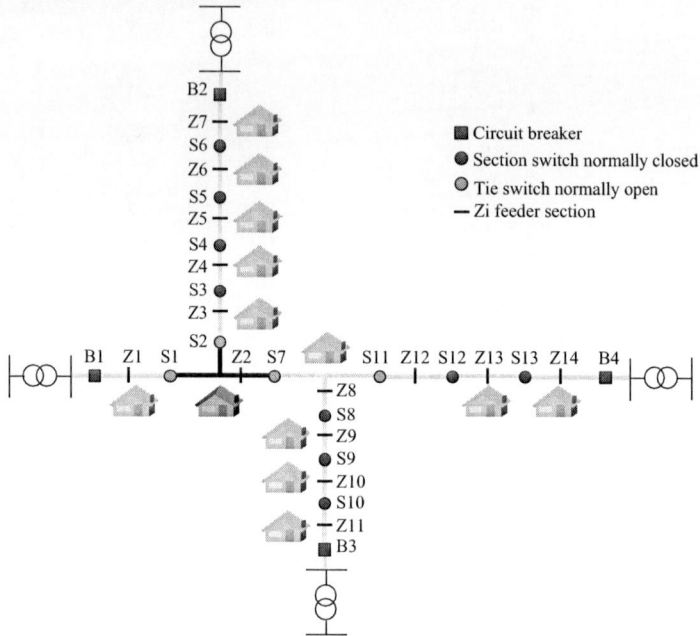

Figure 6.17 *Reconfiguration by operating switches NO and NC – step 4 (downstream restoration)*

Table 6.7 *Reconfiguration options for system of example 6.3*

Fault	Open	Close
F_1	1 and 2	T1 or T3
F_2	8 and 9	T1 or T2
F_3	12 and 13	T2 or T3

Example 6.3 The following example shows a distribution system with three feeders coming out from three different substations where switches have been located to increase the flexibility of system operation. Indicate the five steps for the system reconfiguration after fault F1 happens.

The system has in total 14 section switches (NC) and three tie switches (NO) labeled T1, T2, and T3. It can be noted that the tie switches are chosen after the section switches. The five steps for system reconfiguration after fault F_1 happens are the following:

1. Relay clears fault at feeder I.
2. Switches 1 and 2 open to isolate the fault.
3. Relay recloses feeder I to re-energize up to the location of switch 1 (upstream restoration)

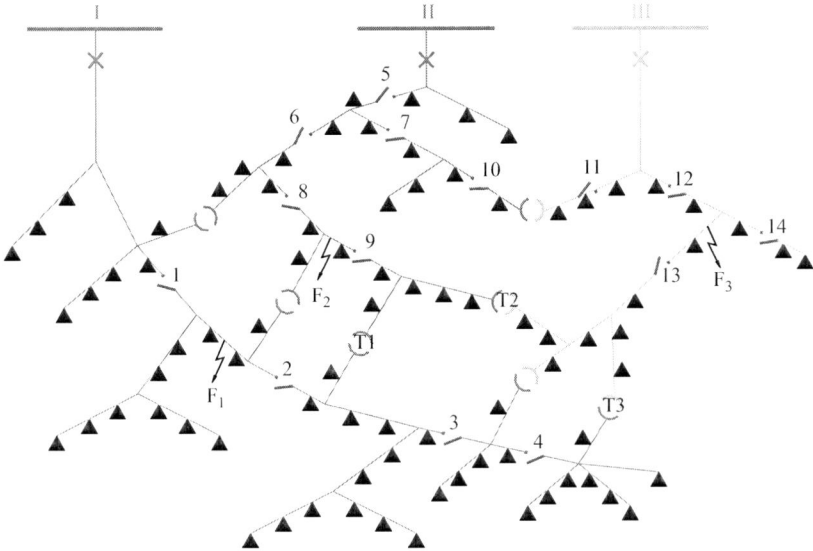

Figure 6.18   *Three-feeder system illustrating switch location for service*
restoration

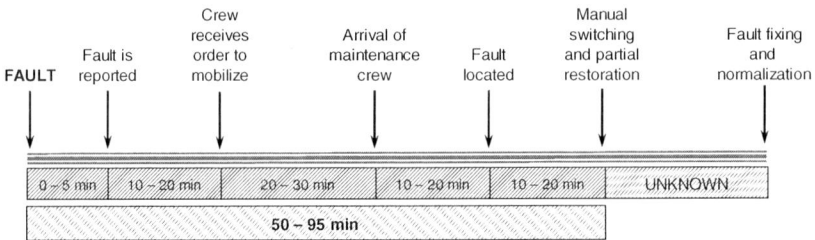

Figure 6.19   *Time to fix faults for typical systems*

4. System reconfigures by choosing the best option from the possibilities indicated in Table 6.7 for fault F_1 (downstream restoration).
5. Fault is fixed and system is restored to normal configuration.

The same analysis can be conducted for the other two faults F_2 and F_3. Table 6.7 shows the operation of the switches for the three different faults indicated in Figure 6.18. Note that for fault F_1, once the section switches 1 and 2 are open, the restoration can be achieved by closing either T1 or T3 or even a combination involving the opening of switches 3 or 4 and the closing of T1 and T3.

6.4.2   *Manual restoration vs. FLISR*

When a permanent fault occurs, customers on healthy sections of the feeder may experience a lengthy outage if FLISR is not available. Figure 6.19 is an illustration of the time taken at a typical distribution system to fix a fault and restore service to the users associated with these healthy sections when manual switches have been deployed without any automation. In this case the maintenance crew has to get the faulty section and perform the required switching to allow both the upstream and downstream restoration as practicable.

6.4.3 Restrictions on restoration

When carrying out the reestablishment the operations that are executed should allow the system to satisfy these restrictions:

- The current-carrying capacity of the transformers and lines should be within specified limits.
- The voltage drop should stay inside an established margin.
- The system should continue being radial.
- The number of operations of the equipment stays within limits.
- Important customers have priority.
- System must be balanced as much as possible.
- The coordination of the protection must be maintained.

Example 6.4 To illustrate the nature of service restoration, the same system of Figure 6.13 is used. The total loading of each feeder is shown in Figure 6.20 assuming that each section (Z1 to Z14) has a loading of 1 pu. Indicate how the operation is affected when the fault happens and the system is reconfigured after step 4 that was mentioned previously.

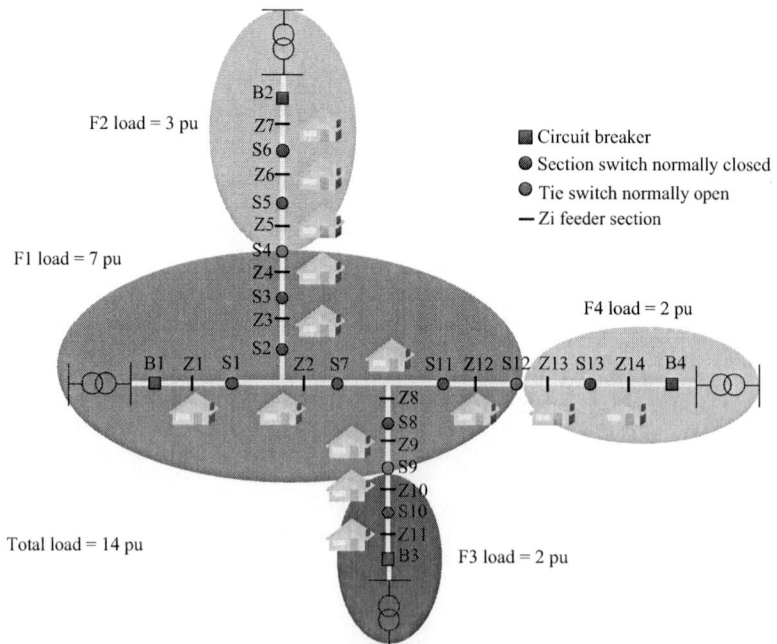

Figure 6.20   Loading for normal condition

In the scenario of a fault in the feeder associated with breaker 1 (B1), the system should clear and isolate the fault, and then energize the healthy sections making use of neighboring circuits, provided that they have sufficient capacity. The options for solutions grow exponentially with an increase in equipment such as section and tie switches. The restoration algorithm should find a configuration that fulfills the operative restrictions. The reconfigured system is shown in Figure 6.21 where all sections are re-energized except section Z2 that needs to be fixed by the maintenance crew. The reconfigured system of Figure 6.21 clearly has different loadings in the feeders 1–4. Table 6.8 illustrates the section loadings prior to the fault and after the system is reconfigured and also the ratio of currents after the reconfiguration and before the fault.

It is clear that the situation after reconfiguration does not offer any risk to trip feeder 1. However, it certainly is risky for feeders 2, 3, and 4 which experience an

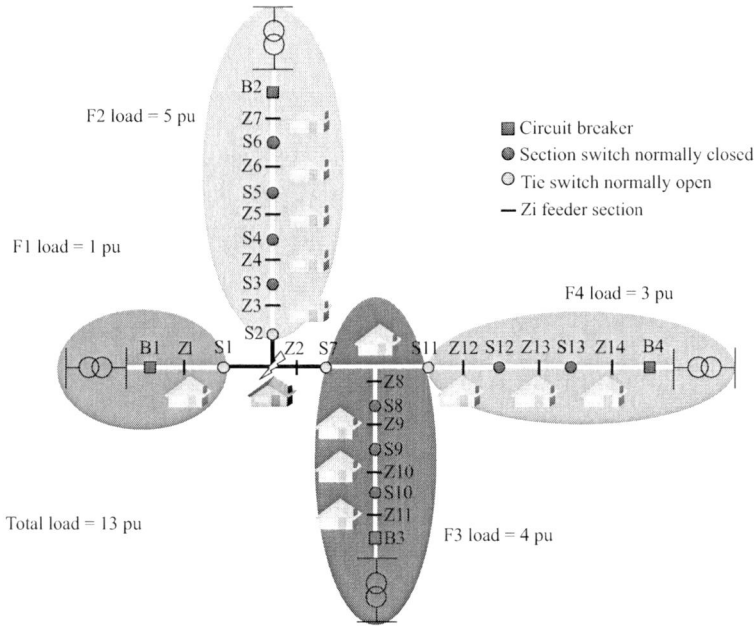

Figure 6.21   Loading after reconfiguration

Table 6.8   Loading values of example 6.4

Feeder	Sections before	Sections after	I_{after}/I_{before}
B1	7	1	1/7
B2	3	5	5/3
B3	2	4	4/2
B4	2	3	3/2

important increase in loading that might trip the corresponding breakers if the pickup value of the corresponding relays is exceeded. Measures including adaptive relaying have to be taken to avoid any nuisance tripping that could worsen the operation of the system.

6.4.4   *FLISR central intelligence*

Central intelligence is still the commonest method of automation to control a power system that includes substations. Central intelligence, of course, relies on powerful and fast SCADA systems and therefore requires a very good communication infrastructure. In the case of distribution systems, central intelligence should have proper control not only of the substation elements but also of all the devices deployed on the feeders. These include breakers, switches, voltage regulators, and capacitors. All of them should be monitored and controlled directly from the control system. This method is very efficient, but it requires a lot of communication.

The central intelligence in FLISR allows to:

- analyze faults based upon input from real-time SCADA (fault detectors and fault currents) and crew reports;
- isolate the fault based upon current state of distribution network;
- generate recommended switching for isolation of a nominated fault and restoration of unaffected customers; and
- propose multiple restoration plans based upon network status, topology, equipment limits, and desired objectives.

An illustration of feeder reconfiguration using central intelligence taken from an ALSTOM case study is given in Figures 6.22–6.27.

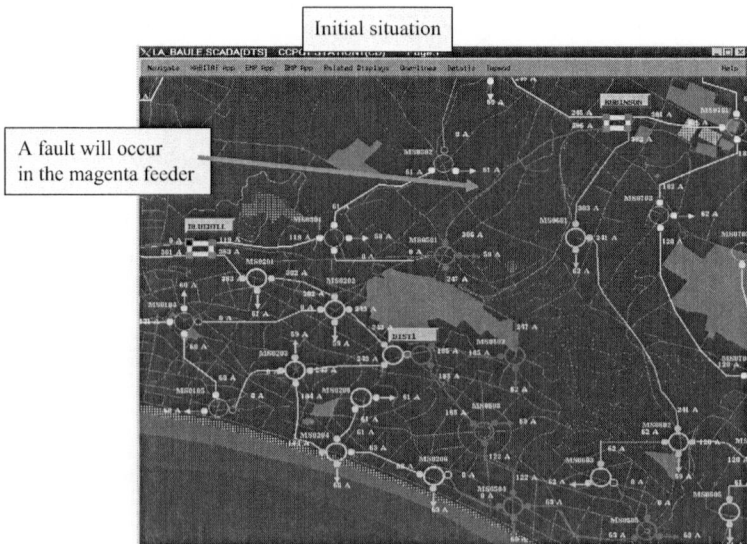

Figure 6.22   Detection of a fault in a FLISR with central intelligence (reproduced by permission of ALSTOM)

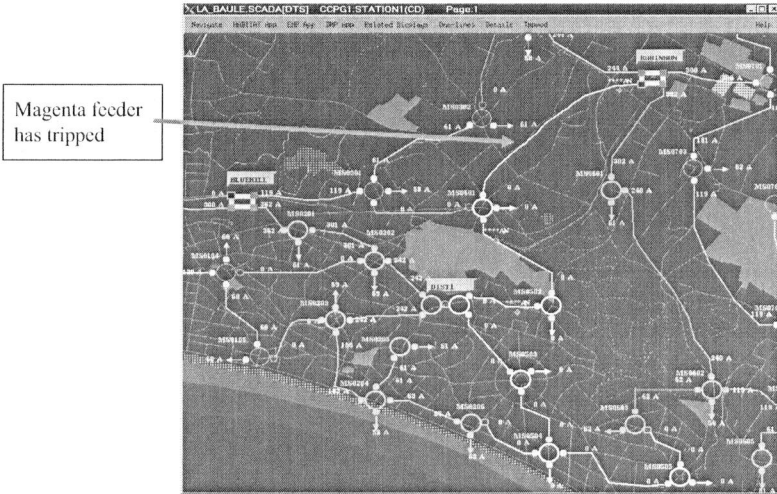

Figure 6.23   Fault isolation by opening corresponding feeder (reproduced by permission of ALSTOM)

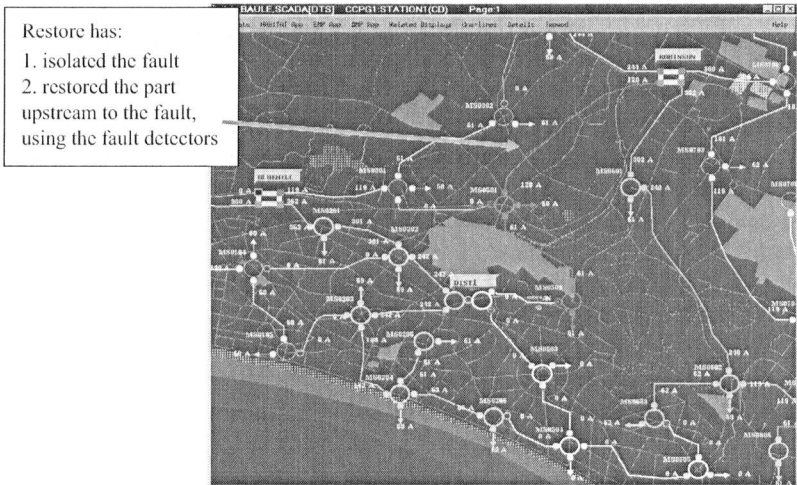

Figure 6.24   Upstream restoration (reproduced by permission of ALSTOM)

6.4.5   FLISR distributed intelligence

Distributed intelligence automatically isolates faulted distribution segments using a "team" of peer devices. It requires proper communication and protective relay coordination. The objective is to reconfigure adjacent feeders to restore power to customers beyond the fault (to minimize the number of outages, their size, and duration).

All possible restoration plans are then proposed to the operator

Fault Detection Isolation & Restoration　　Application is: Ready

Restoration Plans

Done　　Application Log / Operator Log

Isolate & restore non faulty zones　　on ROBINSON.CB.LS-3　　processing done

Clear Request　　Isolation Plan　　Restoration Plans

Weighting Factors ...

	Global	Restored Load (MW)	Non Restored Load (MW)	Level	Total Switchings	Manual Switchings	Margin (MVA)	Lowest Voltage (p.u.)	Losses Added (MW)	SAIFI (interup. /year)	R...
RS002 Detail..	204	3.320	0.000	1	1	0	2.160	0.971	0.130	0.668	
RES003 Detail..	192	3.320	0.000	1	1	0	1.034	0.947	0.281	0.670	
RES004 Detail..	4	3.320	0.000	1	3	0	2.218	0.958	0.194	0.689	
RES005 Detail..	4	3.320	0.000	1	3	0	2.218	0.958	0.194	0.689	
RES006 Detail..	-222	2.195	1.125	1	2	0	3.344	0.979	0.071	0.871	
RES007 Detail..	-233	2.195	1.125	1	2	0	2.218	0.958	0.163	0.872	
RES008 Detail..	-233	2.195	1.125	1	2	0	2.218	0.958	0.163	0.872	
RES009 Detail..	-233	2.195	1.125	1	2	0	2.218	0.958	0.163	0.872	
RES010 Detail..	-233	2.195	1.125	1	2	0	2.218	0.958	0.163	0.872	
RES011 Detail..	-421	2.195	1.125	1	4	0	3.402	0.970	0.099	0.871	

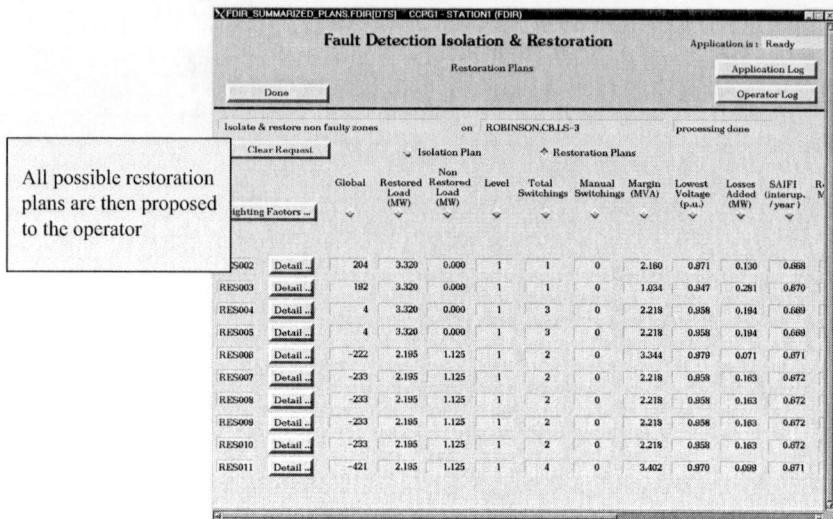

Figure 6.25　Possible downstream restoration plan with FLISR (reproduced by permission of ALSTOM)

One restoration plan is selected

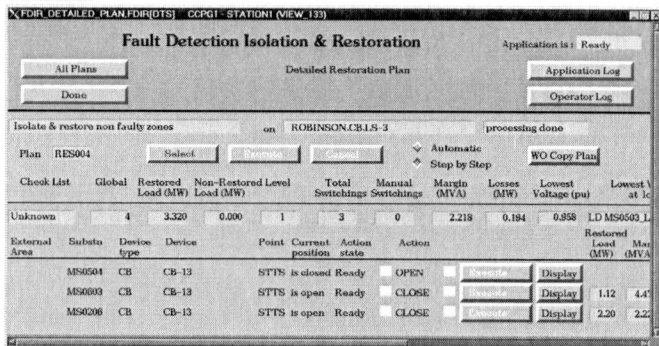

Fault Detection Isolation & Restoration　　Application is: Ready

All Plans　　Detailed Restoration Plan　　Application Log

Done　　Operator Log

Isolate & restore non faulty zones　　on ROBINSON.CB.LS-3　　processing done

Plan　RES004　　Select　　Automatic / Step by Step　　WO Copy Plan

Check List	Global	Restored Load (MW)	Non-Restored Load (MW)	Level	Total Switchings	Manual Switchings	Margin (MVA)	Losses (MW)	Lowest Voltage (pu)	Lowest V at l...
Unknown	4	3.320	0.000	1	3	0	2.218	0.194	0.958	LD MS0503_L

External Area	Substn	Device	Device type	Point	Current position	Action state	Action			Restored Load (MW)	Mar (MVA)
	MS0504	CB	CB-13	STTS is closed	Ready	OPEN			Display		
	MS0603	CB	CB-13	STTS is open	Ready	CLOSE			Display	1.12	4.4?
	MS0206	CB	CB-13	STTS is open	Ready	CLOSE			Display	2.20	2.2?

Figure 6.26　Selection of the best option for restoration (reproduced by permission of ALSTOM)

Some software applications have been developed to find solutions using FLISR distributed intelligence. Figure 6.28 shows an application to a faulted segment. Figure 6.28(a) shows the normal configuration of the groups in the distribution system. Figure 6.28(b) presents a fault in group 6 between the switches F, G, and J. All switches in the affected group experience an overcurrent followed by a loss of voltage. Therefore, switch J opens to isolate the fault. Figure 6.28(c) shows how group 6 detects the fault and initiates opening of switches F, G, and I based on voltage loss. Finally, Figure 6.28(d) shows how the service is restored to

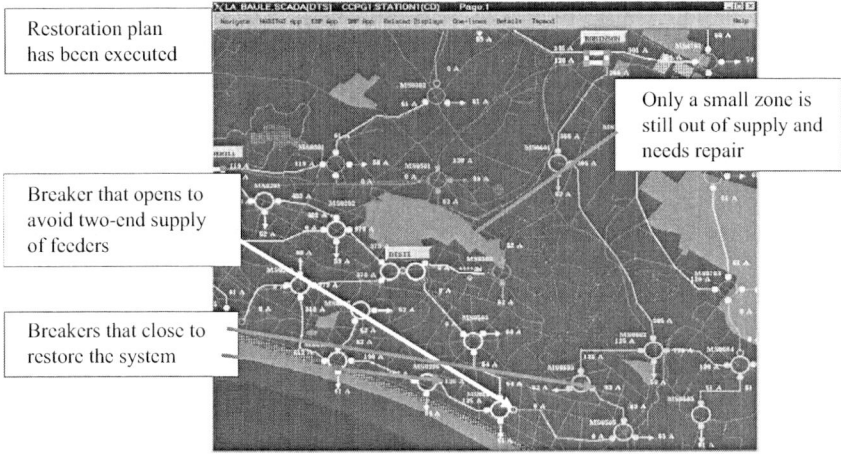

Figure 6.27 Downstream restoration of healthy distributed system section (reproduced by permission of ALSTOM)

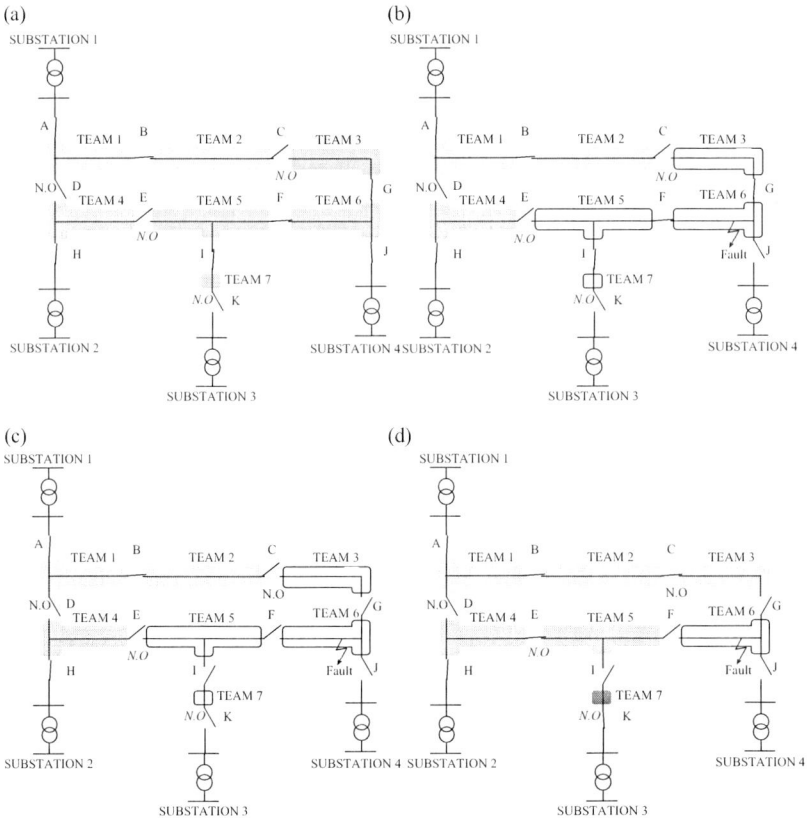

Figure 6.28 Example of a FLISR distributed intelligence solution

all the un-faulted segments in a short time: team 7 closes switch K; teams 4 and 5 close switch E; and teams 2 and 3 close switch C.

It is worth noting that other, more advanced algorithms can locate the fault more accurately. If this were the case, switch I would not need to have been opened at step (c) since the opening of switches F, G, and J isolate the fault.

Distributed intelligence is growing and its application is very popular for systems with long feeders such as those in the United States. Distributed intelligence avoids congesting control centers since the communication is among elements associated only within groups. However, this method has the disadvantage of not considering the whole system. This means that the optimal topology for the whole system might not be found.

6.4.6 FLISR local intelligence

Local intelligence has been used for many years even before SCADA systems were developed. Local intelligence is extremely simple to use, very reliable, and does not require a special communication system, although in some cases a good communication system helps to obtain better results. Local intelligence performs actions like opening or closing devices in order to satisfy previously established conditions.

To illustrate the concept of local intelligence, a system with two feeders is presented in Figure 6.29. Each feeder has two switches. The first feeder, shown to the left, is fed from substation A and the second feeder to the right is fed from substation B. Figure 6.30 shows that a fault between switches R03 and R04 has occurred. Switch R04 operates first. Since the fault remains, additional operations are attempted until

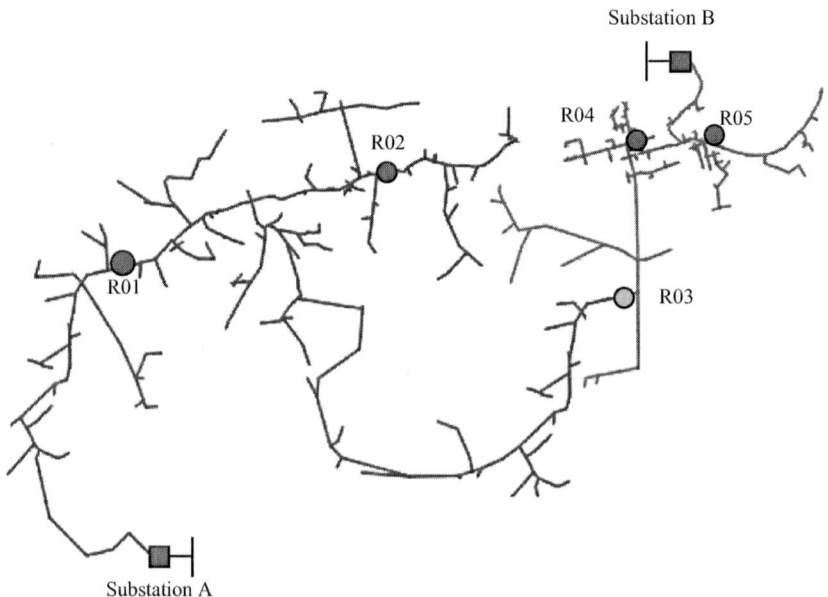

Figure 6.29 Example of a distribution system with FLISR local intelligence

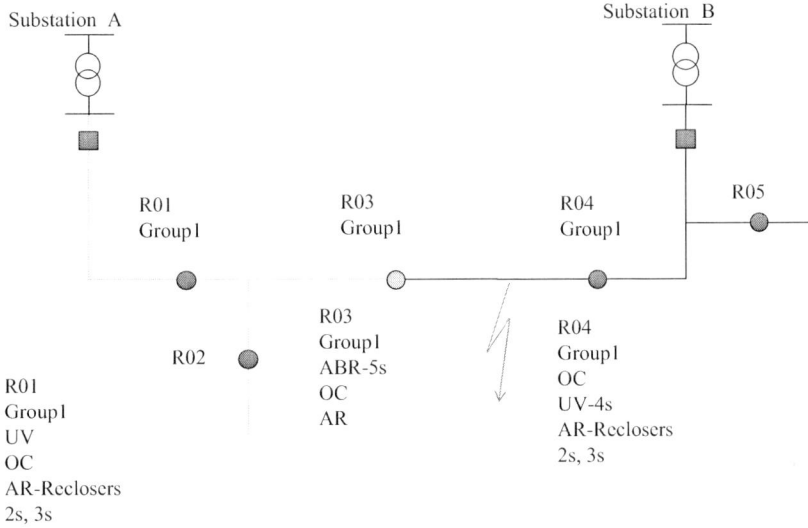

Figure 6.30   Fault between R03 and R04

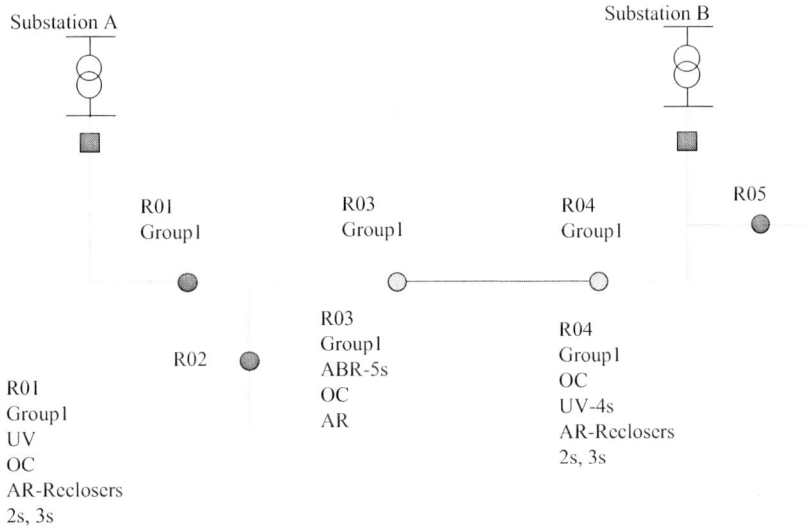

Figure 6.31   Final topology after local intelligence operation

the switch locks out. Then the other switch operates in connection with the automatic backfeed restoration (ABR) function, attempting to close. The ABR function generates a close action if supply is lost from the load side of a switch that is normally open. In this particular case the fault is still present and the switch eventually locks out. In the end the fault is cleared. This is shown in the Figure 6.31.

Chapter 7

Volt/VAR control

Keeping voltages within pre-defined ranges is one of the main targets in the operation of power systems. To achieve this goal, reactive flow has to be monitored since the two variables are closely related, as mentioned in Chapter 3. Therefore, it is common to treat both topics simultaneously.

In recent years and in particular with the development of Smart Grids the term "VVC" was introduced. The term "VVC" stands for Volt/VAR control and refers to the technique of using voltage regulating devices and reactive power controls to maintain voltage levels within the accepted ranges at all points of the distribution system under all loading conditions. Modern software techniques and the progress on communication technologies allow nowadays achieving these goals and therefore the service quality has improved remarkably in most utilities throughout the world.

Voltage regulation and reactive power control are performed by switching capacitor banks installed along the lines, substation transformer load tap changers (LTCs), shunt capacitor banks installed at substations, and voltage regulators. The control of the devices is discussed later in this chapter.

The main objectives of implementing VVC can be summarized as follows:

- Maintain acceptable voltages at all points along the feeder under all loading conditions.
- Increase the efficiency of distribution systems without violating any loading and voltage constrains.
- Support the reactive power needs of the bulk power system during system emergencies.
- Keep power factor within the accepted ranges which normally are higher than 0.9 inductive.

Voltages are the nominal single-phase supply voltages. All voltages are rms and therefore the peak AC voltage is greater by a factor of $\sqrt{2}$. In most countries if not all, they are in the range of 100–240 V.

For the case of the United States the normal rated voltage is 120 V according to ANSI C84.1-2006. Figure 7.1 shows a maximum of 126 V and a minimum of 114 V since the voltage regulation is 5%. In the United Kingdom the statutory limits of the LV supply are 230 V minus 6% and 230 V plus10%. This represents a range from 216.2 to 253 V.

Desirable service voltage range 126–114 VAC

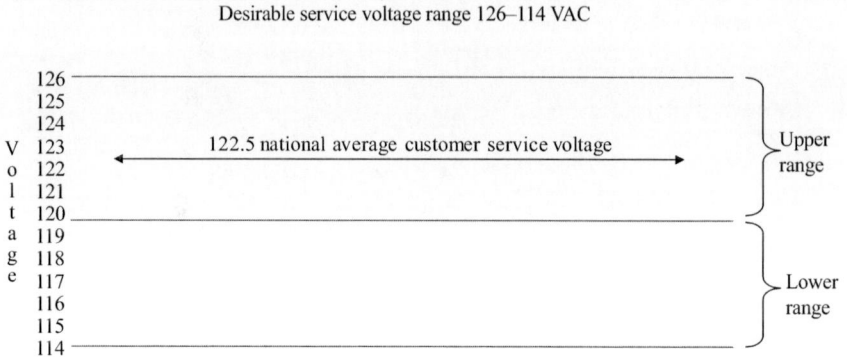

Figure 7.1   Rated voltages specified for the United States

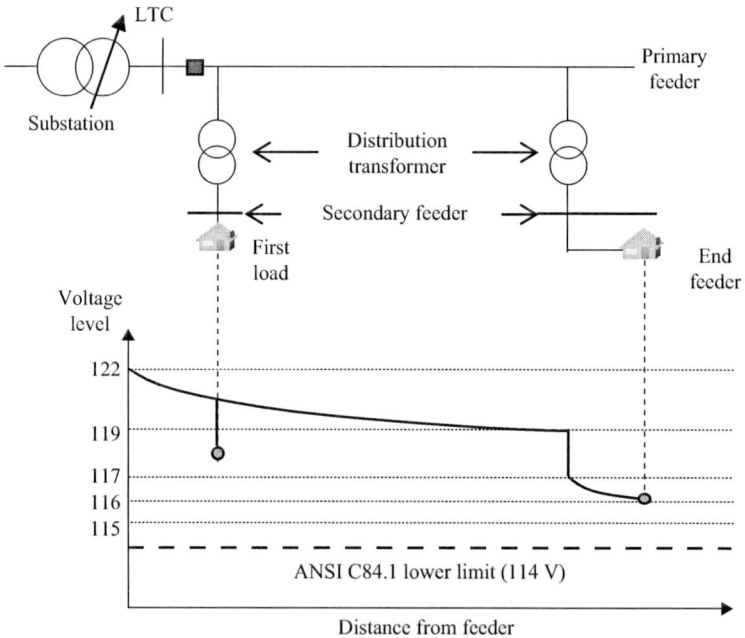

Figure 7.2   Voltage regulation limits in the United States

Figure 7.2 illustrates a portion of a typical system that illustrates the effect of the loading of a feeder on the voltage profile which risks the fulfillment of the voltage ranges specified by the standard.

Different devices are required to achieve the VVC including on-load tap changers at the main substation transformers, voltage regulators, and capacitor banks. Distributed energy resources (DER) can also help the VVC on the feeders where they are deployed. FACTS equipment (like STATCOM) and distributed generation (DG) if available should be included under the VVC analysis and be part of it.

7.1   Definition of voltage regulation

Voltage regulation can be defined as the percentage of voltage drop of a line (e.g., a feeder) with respect to the receiving-end voltage.

Therefore:

$$\% \text{ regulation} = \frac{|V_\text{s}| - |V_\text{r}|}{|V_\text{r}|} \times 100 \qquad (7.1)$$

where:

$|V_\text{s}|$ Sending voltage
$|V_\text{r}|$ Receiving-end voltage

7.2   Options to improve voltage regulation

There are numerous ways to improve the distribution system's overall voltage regulation:

- use of generator voltage regulators,
- application of voltage-regulating equipment in the distribution substations,
- application of capacitors in the distribution substation,
- balancing of the loads on the primary feeders,
- increasing feeder conductor size,
- changing feeder sections from single-phase to multiphase,
- transferring loads to new feeders,
- installing new substations and primary feeders,
- increasing or primary voltage level,
- application of voltage regulators out on the primary feeders,
- application of shunt capacitors on the primary feeders,
- application of series capacitors on the primary feeders, and
- use of DG or DER.

The commonest way to regulate voltage in power systems is by means of tap changers in transformers. Normally they are used to control the voltage level at the LV side of transformers associated to transmission or distribution substations. They could control voltage at the HV side in generator plants. However, this is not usual since in this case the voltage is controlled by the excitation of the generators.

Tap changers can be of the no-load operating type NLTC or on-load operation type OLTC. The first one is used mainly on small transformers or non-important transformers. OLTC tap changers could be mechanical type or thyristor assisted. Mechanical tap changers physically make a connection before releasing the previous connection point employing various tap selectors. They avoid the presence of elevated circulating currents through a diverter switch that provides a temporary high impedance value in series with the short circuited winding section.

Tap changers assisted by thyristors are used to take the load current while the main contacts switch from one source to the other. This prevents arcs in the main contacts and can result in an extended life cycle in the maintenance activities. They require an additional low voltage supply for the thyristor circuit. This type of tap changers is not very popular due to their complexity and the elevated cost.

Calculations of voltage regulation with tap changes are rather straightforward and were already presented in Chapter 3 and when the basics of load flow was discussed.

The most economical way of regulating the voltage along the feeders within the required limits is to apply both step voltage regulators and shunt capacitors. Of course, a fixed capacitor is not a voltage regulator and cannot be directly compared to regulators; however, in some cases, automatically switched capacitors can replace conventional step-type voltage regulators for voltage control on distribution feeders. The following sections make specific reference to voltage regulators and capacitors.

7.3   Voltage regulators

A step-type voltage regulator is fundamentally an autotransformer with many taps (or steps) in the series winding that adjusts itself automatically by changing the taps until the desired voltage is obtained. It can be either station-type or distribution-type.

Voltage regulators manufactured by Cooper Power Systems are very popular. They are designed to correct the line voltage from 100% boost to 10% buck (i.e., 10%) in 32 steps, with a 5% or 8% voltage change per step as shown in Figure 7.3. The effect of each step is shown in Figure 7.4.

The voltage regulators can be located at the substation busbar or on the feeders. The settings are calculated such as to keep the voltage within certain ranges at a regulating point along the feeders. The procedure to determine the settings has been very well treated in the book *Electric Power Distribution System Engineering* by Turan Gönen, from which (7.2)–(7.7) and the corresponding explanations have been taken.

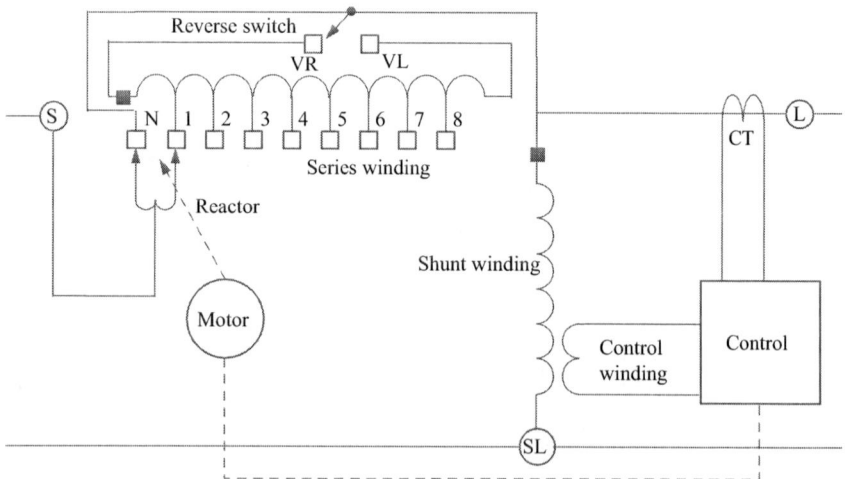

Figure 7.3   Schematic of a single phase 32-step voltage regulator

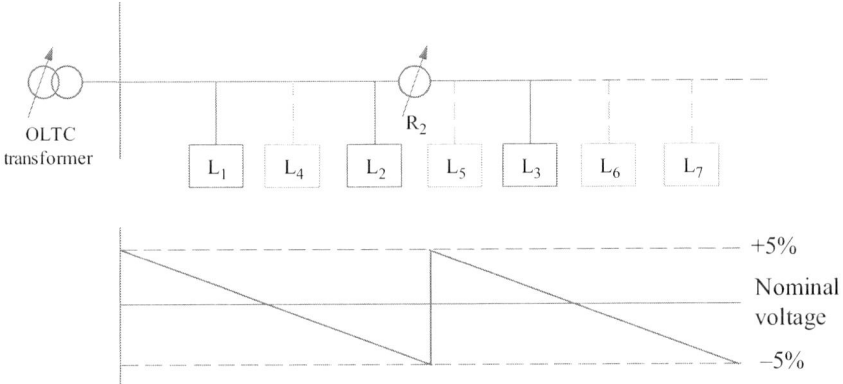

Figure 7.4 Voltage profile with step-type voltage regulators

The settings correspond to the dial of the resistance and reactance elements built in the so-called line-drop compensator (LCD) which is in the control panel of the device.

In case there is no load tapped off between the regulator and the controlled point, then the R dial setting of the LCD is determined from:

$$R_{set} = \frac{CT_P}{PT_N} \times R_{eff} \tag{7.2}$$

where:

CT$_P$ Rating of the current transformer's primary
PT$_N$ Potential transformer's turns ratio (V_{pri}/V_{sec})
R$_{eff}$ Effective resistance of a feeder conductor from regulator station to regulation point, Ω

$$R_{eff} = r_a \times \frac{l - s_1}{2} \tag{7.3}$$

where:

r_a Resistance of a feeder conductor from regulator station to regulation point, Ω/mi per conductor
s_1 Length of three-phase feeder between regulator station and substation, mi (multiply length by 2 if feeder is in single phase)
l Primary feeder length, mi

Also, the Z dial setting of the LCD can be determined from:

$$X_{set} = \frac{CT_P}{PT_N} \times X_{eff} \tag{7.4}$$

where:

X_{eff} Effective reactance of a feeder conductor from regulator to regulation point, Ω

$$X_{\text{eff}} = x_{\text{L}} \frac{l - s_1}{2} \tag{7.5}$$

and

$$x_{\text{L}} = x_{\text{a}} + x_{\text{d}} \ (\Omega/\text{mi}) \tag{7.6}$$

where:

x_{a} Inductive reactance of individual phase conductor of feeder at 12-in. spacing
x_{d} Inductive reactance spacing
x_{L} Inductive reactance of feeder conductor

The difference between the two voltage values is the total voltage drop between the regulator and the regulation point, which can also be defined as:

$$VD = |I_{\text{L}}|R_{\text{eff}} \cos \theta + |I_{\text{L}}|X_{\text{eff}} \sin \theta \tag{7.7}$$

From which R_{eff} and X_{eff} values can be determined easily if the load power factor of the feeder and the average R/X ratio of the feeder conductors between the regulator and the regulation point are known.

Automatic voltage regulation is provided by bus regulation at the substation, or individual feeder regulation in the substation, or supplementary regulation in the main by regulators mounted on poles, or a combination of the above.

Many utilities have experienced that the most economical way of regulating the voltage within the required limits is to apply both step voltage regulators and shunt capacitors.

7.4   Capacitor application in distribution systems

Capacitor location in distribution networks is a very important option to reduce electrical losses and therefore it has been widely used. Capacitors not only help to save losses but also play an important role in power factor correction and in the improvement of the voltage profile especially in long feeders. Even with the benefits of capacitor location in distribution systems, their location has to be analyzed carefully not only for the high costs involved, but also for the overvoltages which capacitors may produce in networks with harmonic circulation when resonance conditions may occur.

A capacitor is a device consisting essentially of two electrodes separated by a dielectric insulating material that is capable of supplying magnetizing kVAR to the system. The capacitive reactance has the nature of a negative inductive reactance. This property is utilized in electrical circuits to compensate the effects of inductive reactance and the lagging reactive kVA of inductive loads.

Source of supply

(a) System diagram

(b) Equivalent circuit

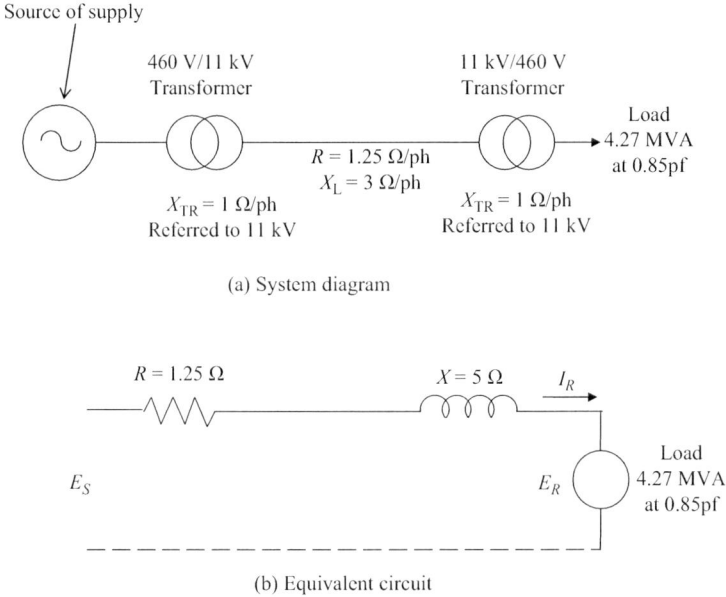

Figure 7.5   Diagrams for example 7.1

Capacitors can be classified as series or parallel according to the type of connection they have. Series capacitors are connected in series with lines to compensate for inductive reactance. Shunt capacitors are connected in parallel with lines to compensate the reactive power or current required by an inductive load.

Example 7.1 Consider a simple 11 kV radial line transmitting power by an overhead system to a lagging power factor load. Figure 7.5 presents the system diagram and the equivalent circuit.

Analyze the sending and receiving-end conditions for each of the following three cases:

(a) Without capacitors: The line-to-line receiving-end voltage is assumed constant at 11 kV. All calculations are referred to 11 kV base voltage.
(b) With shunt capacitors: Three single phase capacitor units, each having a reactance of 45.7 Ω, are connected between phase and neutral adjacent to the load. Calculate the current taken by each capacitor, I_C, I_R, and E_S.
(c) With series capacitors: In this case a capacitor unit is connected in series with each phase, the reactance of each unit being 5 Ω. Calculate I_R and E_S.

For the case (a):

The line-to-line receiving-end voltage is assumed constant at 11 kV. All calculations are referred to 11 kV base voltage.

$$I_R = \frac{4.27 \text{ MVA}}{\sqrt{3} \cdot 11 \text{ kV} \times 0.85} \angle\left(-\cos^{-1}(0.85)\right) = 264\angle(-31.78°) \text{ A per phase}$$

Taking E_R as reference vector:

$$E_S = E_R + \left[\sqrt{3}\left(I_R \angle(\phi_R)\right)(R + jX_L)10^{-3}\right] \text{ kV}$$

$$= 11 + \left[\sqrt{3}\left(264\angle(-31.78°)\right)(1.25 + j5)10^{-3}\right] \text{ kV}$$

$$= 11 + \left[2.36\angle(17.86°)\right] \text{ kV}$$

$$= 13.25\angle1.33° \text{ kV}$$

Figures 7.6 and 7.7 show the vector diagram for the current case. That is the base case for the other points.

For the case (b):

Three single phase capacitor units, each having a reactance of 45.7 Ω, are connected between phase and neutral adjacent to the load.

Then, current taken by each capacitor:

$$I_C = \frac{11,000}{\sqrt{3} \times 45.7}\angle(90°) \text{ A}$$

$$= 139\angle(90°) \text{ A, leading the applied voltage by } 90°$$

The inductive component of the load current

$$= 264\sin(31.78°)\angle(-90°)$$

$$= 139\angle(-90°) \text{ lagging the applied voltage by } 90°$$

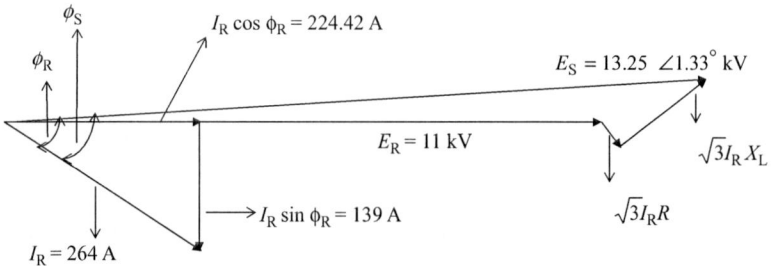

Figure 7.6   Voltage and current vector diagram for the base case

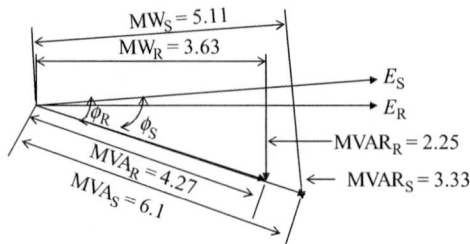

Figure 7.7   Power diagram for the base case

Therefore, the current taken by the capacitor will neutralize the inductive component of the load current, and the actual current taken from the line will be only 10.5 A, in phase with the applied voltage:

$$I_R = 224.42\angle(0°)$$
$$E_S = 11 + \left[\sqrt{3} \times 224.42(1.25 + j5)10^{-3}\right] \text{kV}$$
$$E_S = 11.65\angle(9.6°) \text{kV}$$

For the case (c):

In this case a capacitor unit is connected in series with each phase, the reactance of each unit being 5 Ω.

$$I_R = 264\angle(-31.78°)$$

$$E_S = E_R + \left[\sqrt{3}\left(I_R\angle(\varphi_R)\right)\left(R + j(X_L - X_C)\right)10^{-3}\right] \text{kV}$$
$$= 11 + \left[\sqrt{3}\left(264\angle(-31.78°)\right)\left(1.25 + j(5 - 5)\right)10^{-3}\right] \text{kV}$$
$$= 11 + \left[0.57\angle(-31.78°)\right] \text{kV}$$
$$= 11.48\angle(-1.5°) \text{kV}$$

For the purpose of comparison the relevant sending- and receiving-end conditions for the three cases are tabulated in Table 7.1.

The following can be seen:

1. Both shunt and series capacitors reduce the voltage drop and also the MVAR demand at source of supply.
2. The MVAR rating or the shunt capacitor bank for the three phases

$$= 3I_C^2 X_C 10^{-6} \text{ MVAR}$$
$$= 3(224.42)^2 45.7 \times 10^{-6} \text{ kVAR}$$
$$= 6.9 \text{ MVAR}$$

The MVAR rating of the series capacitor bank

$$= 3(264)^2 5 \times 10^{-6} \text{ MVAR}$$
$$= 1.05 \text{ VAR}$$

Thus, for the same improvement in voltage regulation, the series capacitor bank is much smaller than the shunt capacitor bank. At the same time, it should be noted that series capacitors reduce MVAR demand at source far less than shunt capacitors do.

3. Shunt capacitors reduce the line power losses by reducing the receiving-end current.

Table 7.2 shows a comparison between the shunt capacitor and the series capacitor.

7.4.1   Feeder model

A general model is considered here, which corresponds to a distributed load along the feeder with the possibility of having a concentrated load at the end whose value

Table 7.1 Comparison of results for example 7.1

Type of system	Line voltage		Voltage drop	Power factor		Sending-end power			Receiving-end power			Active power loss
	Sending end (kV)	Receiving end (kV)		Sending end	Receiving end	Active power (MW)	Reactive power (MVAR)	Apparent power (MVA)	Active power (MW)	Reactive power (MVAR)	Apparent power (MVA)	
Case A, without capacitors	13.25	11	2.35 kV (20.4%)	0.84 lag	0.85 lag	5.1	3.3 lag	6.1	4.3	2.7 lag	5.1	0.8 MW (18.6%)
Case B, with shunt capacitors	11.65	11	1.65 kV (5.9%)	0.986 lag	1.0	4.47	0.76 lag	4.53	4.3	Nil	4.3	0.17 MW (3.9%)
Case C, with series capacitors	11.48	11	0.48 kV (4.4%)	0.864 lag	0.85 lag	4.53	2.65 lag	5.3	4.3	2.7 lag	5.1	0.23 MW (5.35%)

Table 7.2 Comparison between shunt capacitor and series capacitor

	Shunt capacitor	**Series capacitor**
Size	45.7 Ω	5 Ω
Voltage regulation	Good	Good
Loss reduction	Very good	Nothing
Power factor	Very good	Fair
Stability	Good	Very good
Cost	Reasonable	High

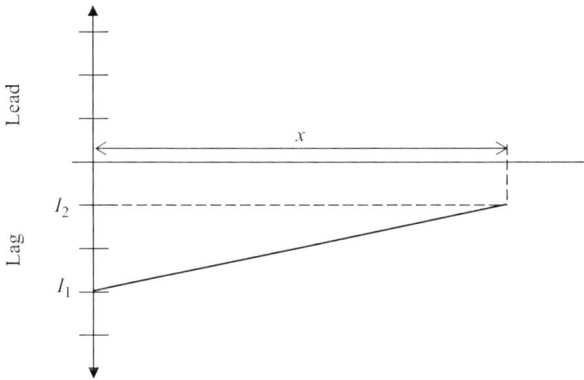

Figure 7.8 Current profiles for feeders with uniformly distributed loads

can also be set to zero, as proposed by Neagle and Samson in reference. If the total feeder current at the substation end is I_1 and the concentrated load at the other end takes a current I_2, the expression for the current along the feeder depends on the distance from the substation according to the following equation:

$$i(x) = I_1 - (I_1 - I_2)x \qquad (7.8)$$

If a factor $p = \frac{I_2}{I_1}$ is introduced, the equation becomes:

$$i(x) = I_1[(p-1)x + 1] \qquad (7.9)$$

The profile corresponding to that expression is shown in Figure 7.8. If the load is uniformly distributed along the feeder and there is no concentrated load at the end, I_2, and therefore p, is equal to zero, and (7.9) becomes:

$$i(x) = I_1(1 - x) \qquad (7.10)$$

Likewise, if the load is concentrated at the end of the feeder, $p = 1$, the expression for the current is:

$$i(x) = I_1 = I_2 \qquad (7.11)$$

7.4.2   Capacitor location and sizing

The reduction in losses by using network reconfiguration can be further enhanced by placing capacitors along the feeders. The following sections will develop theories for such applications which have been proposed in several works. The different magnitudes, i.e., power, energy, impedance, and current, will be dealt with in pu values.

The power loss dissipated in a circuit with a total current I, a resistance value R, and a power factor angle ϕ can be expressed as:

$$P_1 = I^2 R = (I\cos\phi)^2 R + (I\sin\phi)^2 R \tag{7.12}$$

If a capacitor with a current I_c is installed, the reactive part of the current will be compensated and therefore the new total losses (P_2) can be expressed as:

$$P_2 = I^2 R = (I\cos\phi)^2 R + (I\sin\phi - I_C)^2 R \tag{7.13}$$

Therefore, the loss reduction can be found as:

$$\Delta P = P_1 - P_2 = 2IRI_C\sin\phi - I_C^2 R \tag{7.14}$$

Equation (7.14) shows that only the reactive component of the load current is required for power loss reduction studies. Therefore in the rest of this section, that component will be referred to as I_{1R}.

If a feeder has a uniformly distributed loading whose current is that given by (7.9), the total losses due to the *reactive* component without capacitors will again be referred to as P_1 and can be found with the following equation:

$$P_1 = 3\int_0^1 \left[I_{1R}(1 + (p-1)x)\right]^2 R\,dx \tag{7.15}$$

where:

$p = I_{2R}/I_{1R}$
I_{1R} Reactive current at the substation end
I_{2R} Reactive current at the feeder end
x Distance along the feeder ranging from 0 to 1

The result is given in the following equation:

$$P_1 = RI_{1R}^2(p^2 + p + 1) \tag{7.16}$$

7.4.3   Reduction in power losses with one capacitor bank

Figure 7.9 shows a distribution feeder with one capacitor bank installed at a distance a from the source.

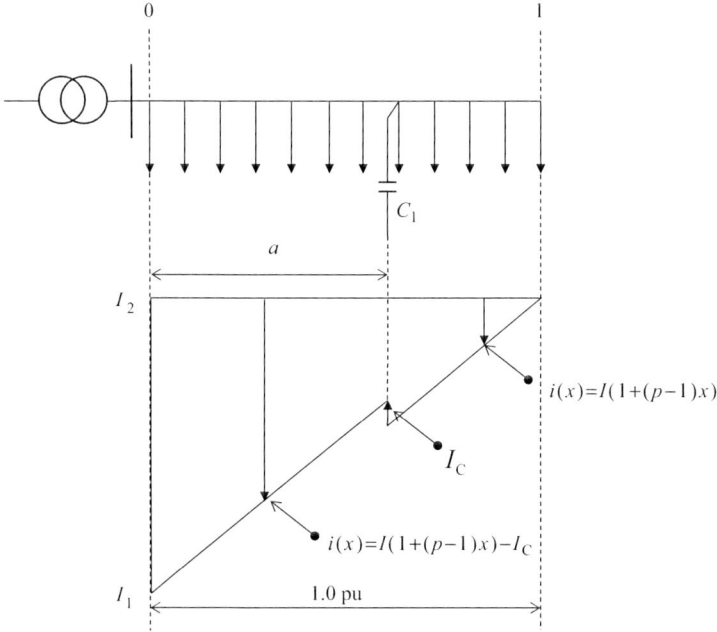

Figure 7.9   Distribution feeder with one capacitor bank

If the capacitor bank takes a current I_C as shown in Figure 7.9, the new total power losses P_2, due to the reactive current including the capacitor bank, are calculated as:

$$P_2 = 3 \int_0^a \left[I_{1R}(1 + (p-1)x) - I_C \right]^2 Rdx + 3 \int_a^1 \left[I_{1R}(1 + (p-1)x) \right]^2 Rdx$$

(7.17)

The result of this expression is:

$$P_2 = R I_{1R}{}^2 (p^2 + p + 1) - 3R I_{1R}{}^2 \left[(p-1)\frac{I_C}{I_{1R}}a^2 + 2\frac{I_C}{I_{1R}}a - \left(\frac{I_C}{I_{1R}}\right)^2 a \right] \quad (7.18)$$

The power loss reduction due to the capacitor installation is found by subtracting (7.18) from (7.16) as follows:

$$\Delta P = P_1 - P_2 = 3R I_{1R}{}^2 \left[(p-1)\frac{I_C}{I_{1R}}a^2 + 2\frac{I_C}{I_{1R}}a - \left(\frac{I_C}{I_{1R}}\right)^2 a \right] \quad (7.19)$$

The optimum sizing and location of one capacitor bank to reduce power losses is found by taking partial derivatives of (7.19) with respect to I_C and the distance a as follows:

$$\frac{\partial \Delta P}{\partial I_C} = 3R I_{1R}{}^2 \left[(p-1)\frac{1}{I_{1R}}a^2 + 2\frac{1}{I_{1R}}a - 2\frac{I_C}{I_{1R}{}^2}a \right] \quad (7.20)$$

$$\frac{\partial \Delta P}{\partial a} = 3RI_{1R}^2 \left[2(p-1)\frac{I_C}{I_{1R}}a + 2\frac{I_C}{I_{1R}} - \left(\frac{I_C}{I_{1R}}\right)^2 \right] \tag{7.21}$$

By equating (7.20) and (7.21) to zero and solving simultaneously for I_C and a, the following values are obtained:

$$I_C = \frac{2}{3}I_{1R} \tag{7.22}$$

$$a = \frac{2}{3(1-p)} \tag{7.23}$$

The maximum power loss reduction with one capacitor is obtained by replacing (7.22) and (7.23) into (7.19). The corresponding expression is:

$$\Delta P_{MAX} = R\,I_{1R}^2 \frac{8}{9(1-p)} \tag{7.24}$$

Equations (7.23) and (7.24) are valid if $p \le 1/3$. For larger values of p, the magnitude of a becomes greater than 1 or even negative, which does not have physical meaning. Therefore for p greater than 1/3, the capacitor bank should be located at the end of the feeder which corresponds to a value of a equal to 1. If this value of a is introduced in (7.19), the value of the capacitor current is:

$$I_C = \frac{1}{2}\,I_{1R}(p+1) \quad \text{for } p > 1/3 \tag{7.25}$$

Likewise, if the value of I_C given in (7.23) and $a = 1$ are introduced in (7.19) the reduction in losses when a capacitor bank is installed and $p > 1/3$ is given by the following equation:

$$\Delta P = \frac{3}{4}\,R\,I_{1R}^2(p+1)^2 \quad \text{for } p > 1/3 \tag{7.26}$$

7.4.4   Reduction in power losses with two capacitor banks

Figure 7.10 shows a distribution feeder with two capacitor banks, one installed at a distance a and the other at a distance b from the source.

Initial losses:

$$P_1 = 3\int_0^1 [I(1-x)]^2 R\,dx = I^2 R \tag{7.27}$$

Losses if capacitors are connected:

$$P_2 = 3\int_0^a R[I(1-x) - 2I_C]^2 dx + 3\int_a^b R[I(1-x) - I_C]^2 dx + 3\int_b^1 R[I(1-x)]^2 dx$$

$$= 3R\left(II_C a^2 - 2II_C a + 3I_C^2 a - 2II_C b + II_C b^2 + I_C^2 b + \frac{I^2}{3} \right)$$

$$\tag{7.28}$$

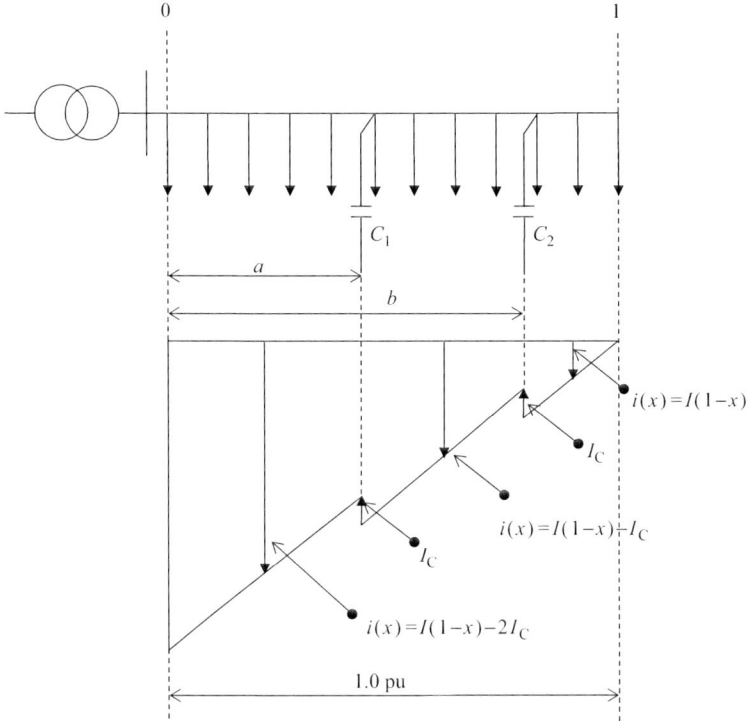

Figure 7.10 Distribution feeder with two capacitor banks

Therefore the loss reductions are expressed as:

$$\Delta P = 3R(a(2II_C - II_Ca - 3I_C^2) + b(2II_C - II_Cb - I_C^2)) \tag{7.29}$$

It can be demonstrated that the optimal values of a and b are:

$$a = 1 - \frac{3I_C}{2I} \tag{7.30}$$

$$b = 1 - \frac{I_C}{2I} \tag{7.31}$$

7.4.5 *Loss reductions with three capacitor banks*

$$\Delta P = 3R(a(2II_C - II_Ca - 5I_C^2) + b(2II_C - II_Cb - 3I_C^2) + c(2II_C - II_Cc - I_C^2)) \tag{7.32}$$

The optimum localization of capacitors:

$$a = 1 - \frac{5I_C}{2I} \tag{7.33}$$

$$b = 1 - \frac{3I_C}{2I} \tag{7.34}$$

$$c = 1 - \frac{I_C}{2I} \tag{7.35}$$

7.4.6   Consideration of several capacitor banks

The expressions presented in (7.19), (7.29), and (7.32) can be generalized for any feeder with n sections to find the power and energy loss reductions when capacitors are located in different places.

For more than one segment on the feeder, expressions (7.19), (7.29), and (7.32) are still valid, although new techniques that consider heuristic strategies have been proposed. Under these circumstances the expressions can be used by putting $a = 1$, which is then directly applicable to one segment at a time. Therefore when a capacitor is installed downstream at a segment n, the reduction in the power losses are given as a function of that capacitor current, and of currents at the beginning and at the end of the segment, respectively, as follows:

$$\Delta P_m = P_{1m} - P_{2m} = 3 R_m \left[(I_{1Rm} + I_{2Rm}) I_{Cn} - I_{Cn}^2 \right] \tag{7.36}$$

For the segment where a capacitor is located, the expressions for power and energy have already been developed. General expressions are obtained by considering the effect of one capacitor at a time for every section toward the source as follows:

$$\Delta P = \Delta P_n + \sum_{m=1}^{n-1} \Delta P_m \tag{7.37}$$

That equation means that for a capacitor installed in a segment n, the power loss reduction is:

$$\Delta P = 3 R_n I_{1Rn}^2 \left[(p-1) \frac{I_{Cn}}{I_{1Rn}} a^2 + 2 \frac{I_{Cn}}{I_{1Rn}} a - \left(\frac{I_{Cn}}{I_{1Rn}} \right)^2 a \right]$$

$$+ \sum_{m=1}^{n-1} 3 \left[(I_{1Rm} + I_{2Rm}) I_{Cn} - I_{Cn}^2 \right] R_m \tag{7.38}$$

Likewise if a number of k banks are considered:

$$\Delta P = \sum_{n=1}^{k} \left\{ 3 R_n I_{1Rn}^2 \left[(p-1) \frac{I_{Cn}}{I_{1Rn}} a^2 + 2 \frac{I_{Cn}}{I_{1Rn}} a - \left(\frac{I_{Cn}}{I_{1Rn}} \right)^2 a \right] \right.$$

$$\left. + \sum_{m=1}^{n-1} 3 \left[(I_{1Rm} + I_{2Rm}) I_{Cn} - I_{Cn}^2 \right] R_m \right\} \tag{7.39}$$

7.4.7   *Capacitor sizing and location using software*

It is impractical to perform the calculation of the optimal number and size of capacitor banks by hand. A number of software packages are available which normally request the following information:

- maximum size allowed (kVAR) for each installation,
- available places (poles),
- maximum number of banks,
- step in kVAR of the bank,
- load factor, and
- existing banks.

Figure 7.11 corresponds to a real distribution system with 12 feeders. The losses of the system are high and the power factor low. Therefore possibilities to improve the system are considered by installing capacitor banks in the most critical feeders.

One of the feeders is considered to determine the number and size of capacitor banks which is shown in Figure 7.12. Two conditions of load factors are analyzed: 0.6 and 0.9. Two locations are allowed of banks having steps of 50 kVAR. A maximum of 2 MVAR has been established.

After running the analysis with a software package, two banks are considered for each loading factor as shown in Table 7.3. The bold letters in the Loss column correspond to the losses of the system before the capacitors were used. An

Figure 7.11   Distribution system to illustrate application of capacitors

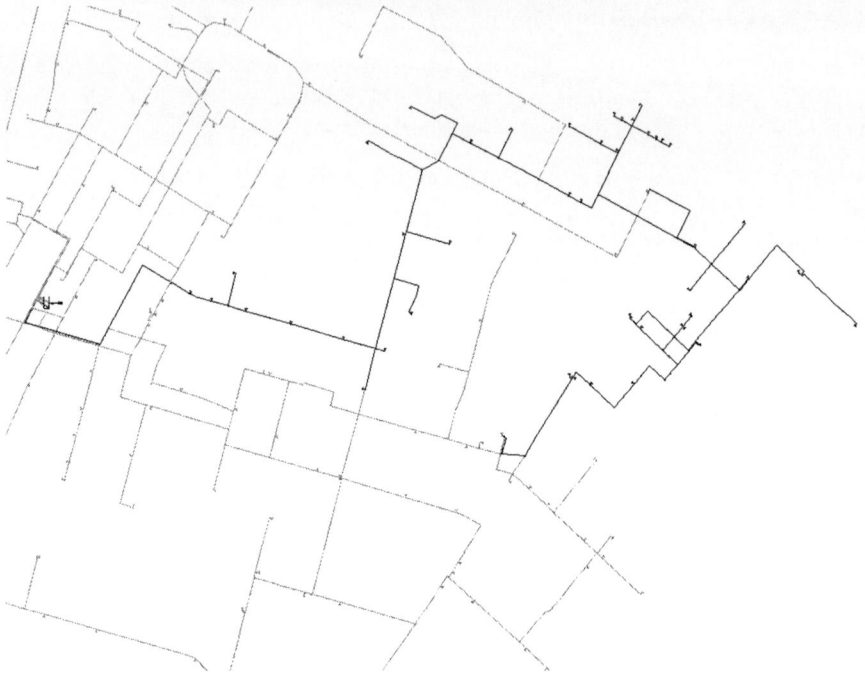

Figure 7.12   Feeder used for the illustration

Table 7.3   Results of software simulation

Load factor	Bus name	Size (kVAR)	Loss (MW)	Additional loss reduction (%)
0.6			**0.2502**	
	306	1150	0.1972	21.1
	280	850	0.1821	6.03
0.9			**0.6208**	
	306	1750	0.4839	22
	301	250	0.4709	2.09

economic analysis not shown in this example is recommended to define the option to select from the two presented in the results.

7.5   Modeling of distribution feeders including VVC equipment

Modeling of distribution feeders involving the elements mentioned above can be a comprehensive task due in particular to the non-symmetry that distribution systems have. All the details of modeling distribution lines are beyond the scope of this book.

In order to illustrate the modeling, a small system taken from IEEE will be considered. The data from the model are introduced in the illustration presented below. The system has the following characteristics:

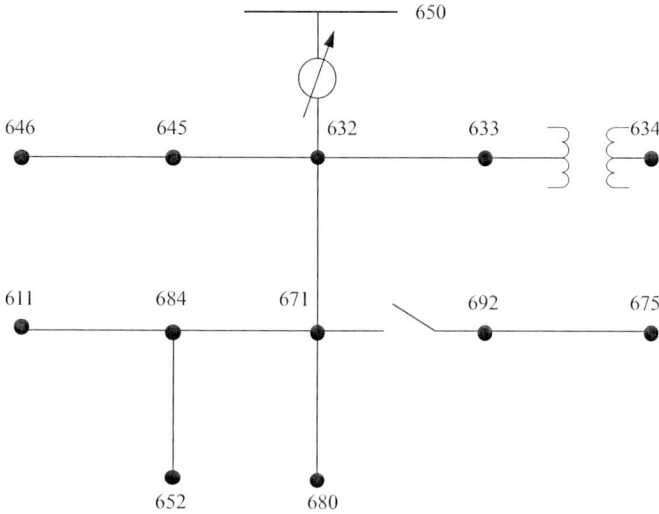

Figure 7.13   Layout of the feeder used in the modeling

1. short and relatively highly loaded for a 4.16 kV feeder,
2. one substation voltage regulator consisting of three single phase units con-
    nected in wye,
3. overhead and underground lines with variety of phasing,
4. shunt capacitor banks,
5. in-line transformer, and
6. unbalanced spot and distributed loads.

Figure 7.13 presents a radial feeder taken from IEEE that will be used to
illustrate the modeling and the type of analysis that can be performed.

Figure 7.14 shows the modeling of the voltage regulator installed in the system
considered. The modeling requires a previous analysis of the device used and the
setting of the taps, based on the system requirements.

Figure 7.15 shows the load flow results for the system modeled, including voltage
levels at the different nodes and active and reactive flow through all the elements.

7.6   Volt/Var control considering SCADA

Years ago the information to monitor and command equipment in charge of the
control of the voltage and reactive power came from the current and voltage transformers feeding them. These days much better results are achieved by employing
software packages interacting with SCADA systems. Figure 7.16 illustrates various
elements of a VVC which are controlled and monitored by a SCADA system.

Distributed resources could present a challenge in the implementation of VVC.
The impact on feeders can be remarkable based on their position in the system and
the generator size; consequently, corresponding coordination with the VVC devices
must be accomplished.

Voltage Regulator

Voltage Regulator
Reliability
Harmonic Analysis
Other Analysis
Info
User Data

Voltage Regulator

Name: REGULATOR-1
Type:

Ur1 .. kV: 0   Sr .. MVA: 0   URr .. %:   0
Ir1 .. A: 0   Ukr .. %:   0

Regulator connection   Open Delta   Properties
  Y-connected   L2L3-L1L3   Automatic regulation active
  Closed Delta   L3L1-L2L1   Line compensation active
  Open Delta   L1L2-L3L2   Regulate on fixed Uset
                                                Same regulators

Phase L1 / L1L2   Phase L2 / L2L3   Phase L3 / L3L1
Tap min: 16   Tap min: 0   Tap min: 0
Tap max: 16   Tap max: 0   Tap max: 0
Tap act: 10   Tap act: 8   Tap act: 11
Delta U .. %: 0.625   Delta U .. %: 0   Delta U .. %: 0
Line Compensation Ub=120V   Line Compensation Ub=120V   Line Compensation Ub=120V
Bandwidth.V: 2   Bandwidth.V: 0   Bandwidth.V: 0
Irp CT .. A: 700   Irp CT .. A: 0   Irp CT .. A: 0
PT Ratio: 20   PT Ratio: 0   PT Ratio: 0
Volt Level..V: 122   Volt Level..V: 0   Volt Level..V: 0
R setting .V: 3   R setting .V: 0   R setting .V: 0
X setting .V: 9   X setting .V: 0   X setting .V: 0
Fixed voltage   Fixed voltage   Fixed voltage
U set .. %: 100   U set .. %: 100   U set .. %: 100

Copy   Paste   Library   Export   OK   Cancel   Color   Help

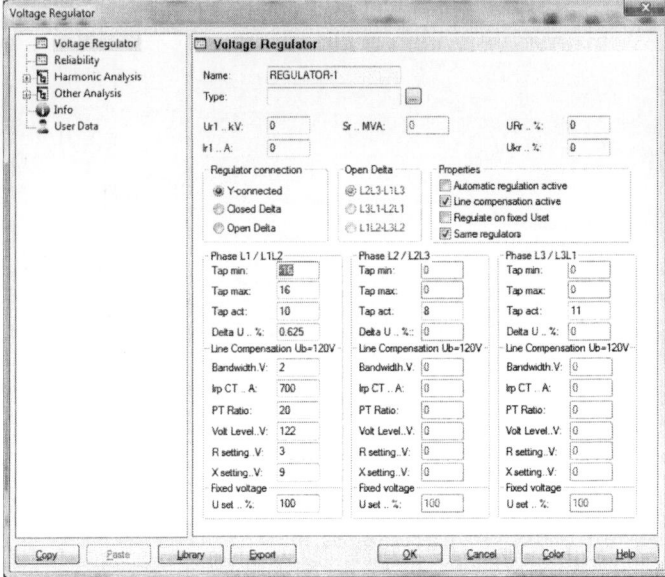

Figure 7.14   Voltage regulator modeling

7.7   Requirements for Volt/VAR control

VCC implementation is getting more popular due to the favorable experiences in service quality and increased efficiency. The system should be able to maintain the range of voltages specified by the user on feeders and branches. Normally power factors above 0.9 are required, although some utilities request 0.95 and above.

It is important to enable the operator to override in the event of topology changes as a result of a FLISR sequence or reconfiguration to reduce losses under normal conditions. On the other hand, various setting conditions must be developed for the different operational scenarios particular to the system. It helps the operator overview the actions, especially in the event of emergencies where time is crucial.

7.8   Integrated Volt/VAR control

The integrated VCC (IVVC) is considered an exceptional function of distributed automation that helps identify the most suitable control procedures for the devices involved in voltage regulation and VAR control to help achieve the specified operational goals of utilities avoiding violations to the fundamental operational constraints like high/low voltage limits and load limits.

The established VVC operating goals consider conditions like minimal electrical losses, minimal electrical demand, and reduced energy consumption. The decision criterion of IVVC requires the employment of an online power flow (OLPF).

The IVVC offers improvements compared to VVC based on the fact that it can accommodate to scenarios resulting from FLISR or feeder reconfiguration. It could

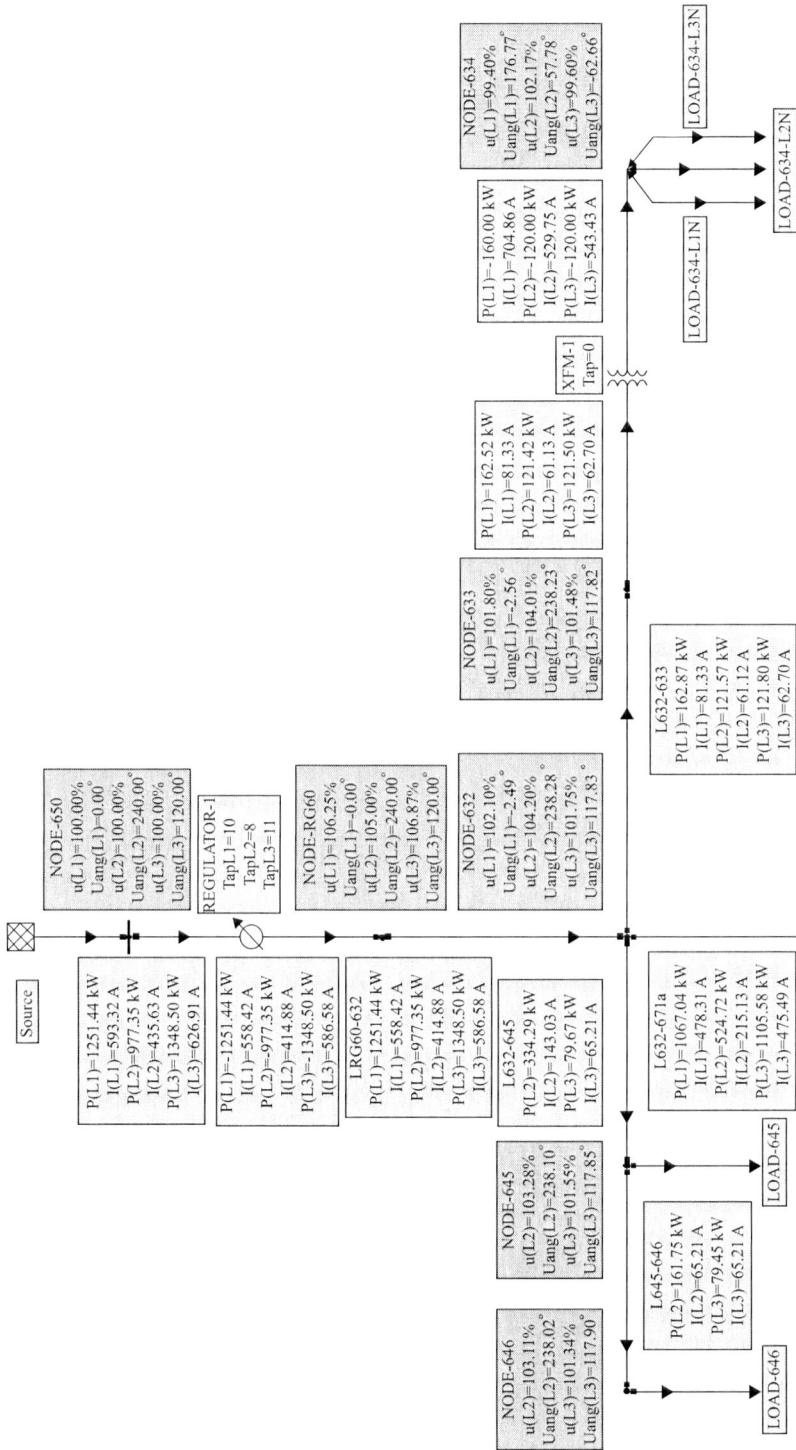

Figure 7.15 Load flow run for the system used with VVC

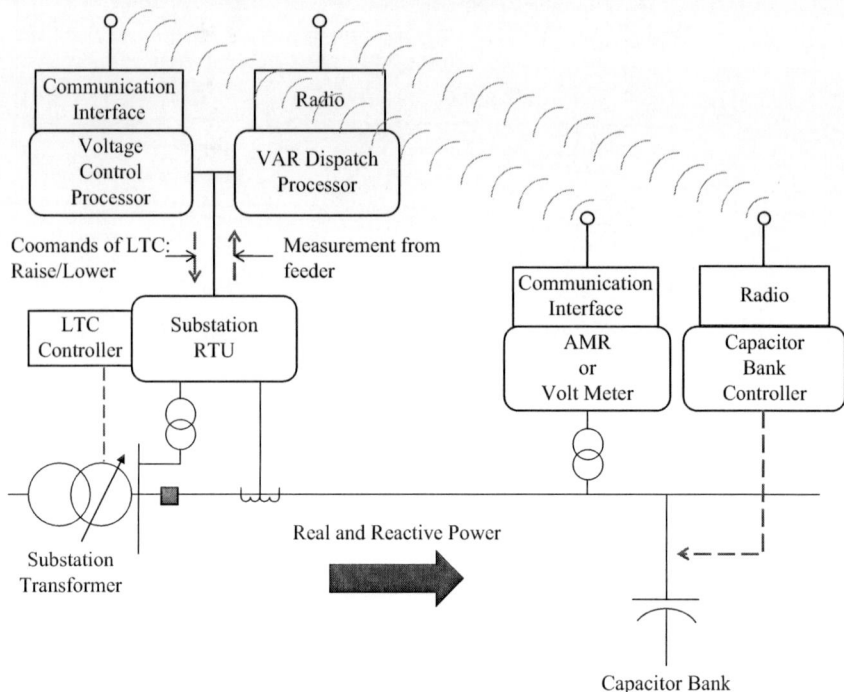

Figure 7.16   Components of a VVC assisted with SCADA

include modeling the dynamic effects of DER. Figure 7.17 illustrates the system employing IVVC along with DER.

IVVC control can be achieved similar to feeders, i.e. in a centralized or decentralized controller way.

For the case of centralized Volt/VAR controlled system, the distribution system SCADA integrates the IVVC which determines how to proceed and provides information for the recording system. Figure 7.18 presents the aforementioned case. The solution is considered very reliable and offers an overall illustration of the system. However, it is more expensive and could possibly congest the SCADA system.

On the other hand, the decentralized Volt/VAR controller is a standalone system that does not involve the overall system SCADA. It relies on local interaction with various devices associated with IVVC as observed in Figure 7.19. The recording takes place once all operations have been completed.

Proposed exercises

1. A 13.2 kV feeder has a total load of 5.5 MW and 3.2 MVAR. The load is uniformly distributed. Perform the following calculations:
    (a) Location and size of the optimal capacitor bank to achieve the maximum loss reductions for load factors of 0.9 and 0.6, respectively.
    (b) Consider the same feeder if two capacitors banks were to be installed.

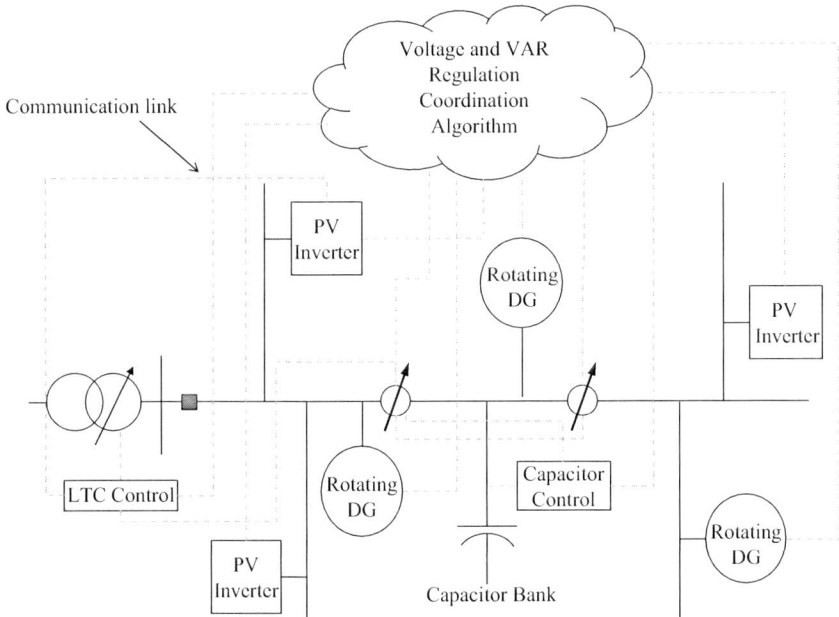

Figure 7.17   Integrated Volt/VAR control including distributed generation

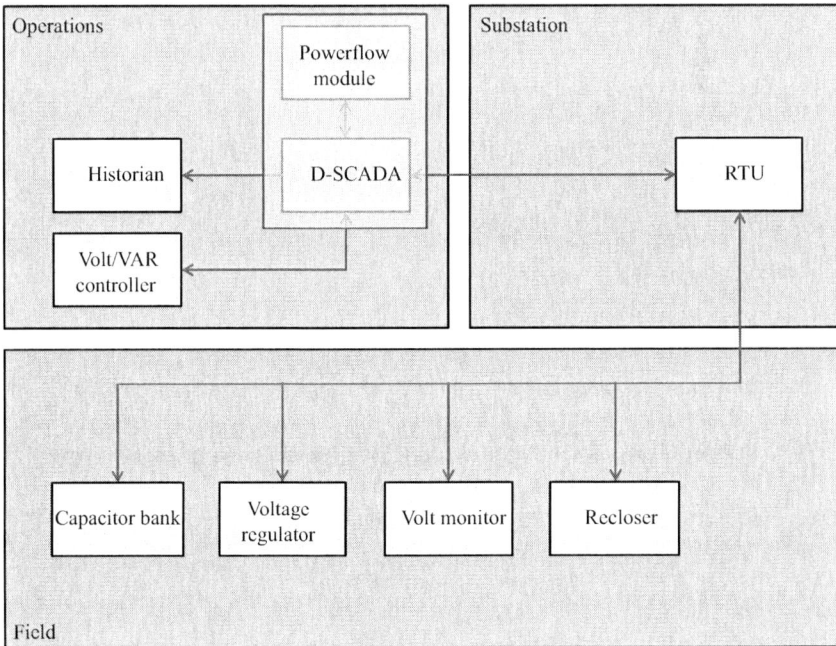

Figure 7.18   Context diagram for centralized integrated Volt/VAR control

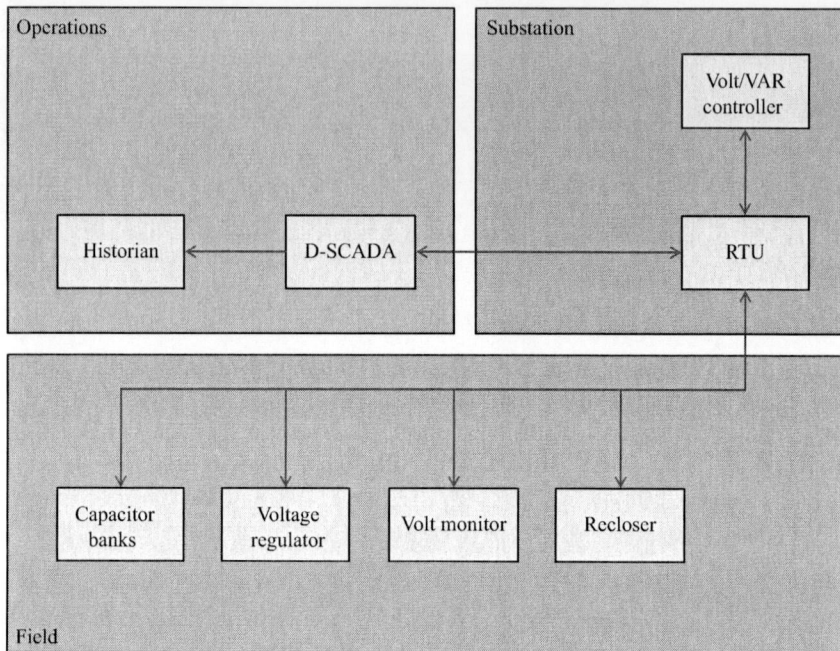

Figure 7.19   Context diagram for decentralized integrated Volt/VAR control

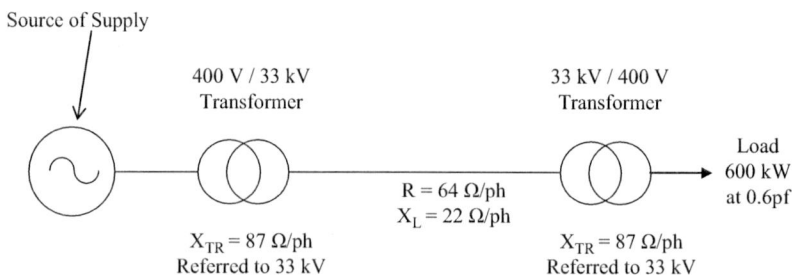

Figure 7.20   Diagrams for exercise 7.2

2. Consider a simple 11 kV radial line transmitting power by an overhead system
    to a lagging power factor load. Figure 7.20 presents the system diagram and the
    equivalent circuit.

    Calculate shunt and series capacitors (size and impedance) to improve the
power factor to 0.85 for each case. The line-to-line receiving-end voltage is
assumed constant at 33 kV. All calculations are referred to 33 kV base voltage.

Chapter 8

Harmonic analysis

Harmonics have become one of the most critical issues in power systems due to the high penetration of them and the magnitudes. This is particularly critical in distribution systems. Harmonics create problems that affect considerably the power quality of a system and therefore are treated in that field.

Power quality can be defined as the goodness of the electric power quality supply in terms of its voltage waveshape, its current waveshape, its frequency, its voltage regulation, as well as level of impulses and noise, and the absence of momentary outages.

Other definition defines power quality as the measure, analysis, and improvement of bus voltage, usually a load bus voltage, to maintain that voltage to be a pure sinusoid and at rated levels of voltage and frequency. A proper quality level is ultimately a compromise of the utility, the end user, and the equipment manufacturer.

Power quality involves several categories of phenomena, which are better presented in Table 8.1.

The categories and characteristics are presented in Table 8.2 taken from the IEEE Standard 1159-2009. Category 5 deals with waveform distortion that includes the harmonic handling, which is the topic developed in this chapter.

8.1 General considerations about harmonics

Harmonics are voltage or current sinusoidal signals with frequencies that are multiples of the power frequency which can produce harmful conditions in power systems.

The harmonic signals in power and distribution systems are produced by different equipment like arc furnaces, nonlinear loads, AC/DC rectifiers, adjustable speed drives, and static UPS systems among others.

Nonlinear loads cause harmonic currents to change from a sinusoidal current to a nonsinusoidal current by drawing short bursts of current each cycle or interrupting the current during a cycle.

Harmonic distortion is a growing concern for many customers and for the overall power system due to increasing application of power electronics equipment.

In low voltage there are other harmonic sources like electric welders, devices with saturated cores, electronic ballasts for fluorescent lighting, and UPS for computers. The latter although have lower values, their high number of units has increased the effect on distribution systems.

Table 8.1 Summary of power quality variation categories (reproduced by permission of DRANETZ)

Example waveshape or rms variation	Power quality variation and category	Method of characterizing	Typical causes	Example power condition solutions
	Impulsive transients Transients Disturbance	• Peak magnitude • Rise time • Duration	• Lightning • Electrostatic discharge • Load switching • Capacitor switching	• Surge arresters • Filters • Isolation transformers
	Oscillatory transients Transients Disturbance	• Waveforms • Peak magnitude • Frequency components	• Line/Cable switching • Capacitor switching • Load switching	• Surge arresters • Filters • Isolation transformers
	Sags/Swells rms disturbance	• rms vs. time • Magnitude • Duration	• Remote system • Faults	• Ferroresonant transformers • Energy storage technologies • UPS
	Interruptions rms disturbance	• Duration	• System protection • Breakers • Fuses • Maintenance	• Energy/storage technologies • UPS • Backup generators

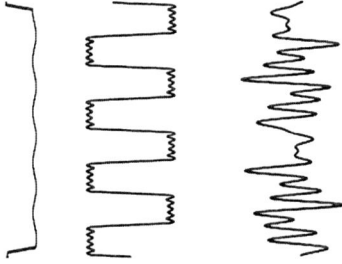

	Undervoltages/Over-voltages Steady-state variation	• rms vs. time • Statistics	• Motor starting • Load variations • Load dropping
	Harmonic distortion Steady-state variation	• Harmonic spectrum • Total harmonic distortion • Statistics	• Nonlinear loads • System resonance	• Voltage regulators • Ferroresonant transformers • Active or passive filters • Transformers with cancellation or zero sequence components • Static VAR systems
	Voltage flicker Steady-state variation	• Variation magnitude • Frequency of occurrence • Modulation frequency	• Intermittent loads • Motor starting • Arc furnaces	

Table 8.2   Categories and typical characteristics of power system electromagnetic phenomena (taken from IEEE Standard 1159-2009)

Categories	Typical spectral content	Typical duration	Typical voltage magnitude
1.0 Transients			
1.1 Impulsive			
1.1.1 Nanosecond	5 nsec rise	<50 nsec	
1.1.2 Microsecond	1 μsec rise	50 ns – 1 msec	
1.1.3 Millisecond	0.1 msec rise	>1 msec	
1.2 Oscillatory			
1.2.1 Low frequency	<5 kHz	0.3–50 msec	0–4 pua
1.2.2 Medium frequency	5–500 kHz	20 μsec	0–8 pu
1.2.3 High frequency	0.5–5 MHz	5 μsec	0–4 pu
2.0 Short-duration root-mean-square (rms) variations			
2.1 Instantaneous			
2.1.1 Sag		0.5–30 cycles	0.1–0.9 pu
2.1.2 Swell		0.5–30 cycles	1.1–1.8 pu
2.2 Momentary			
2.2.1 Interruption		0.5 cycles – 3 sec	<0.1 pu
2.2.2 Sag		30 cycles – 3 sec	0.1–0.9 pu
2.2.3 Swell		30 cycles – 3 sec	1.1–1.4 pu
2.3 Temporary			
2.3.1 Interruption		>3 sec – 1 min	<0.1 pu
2.3.2 Sag		>3 sec – 1 min	0.1–0.9 pu
2.3.3 Swell		>3 sec – 1 min	1.1–1.2 pu
3.0 Long duration rms variations			
3.1 Interruption, sustained		>1 min	0.0 pu
3.2 Undervoltages		>1 min	0.8–0.9 pu
3.3 Overvoltages		>1 min	1.1–1.2 pu
3.4 Current overload		>1 min	
4.0 Imbalance			
4.1 Voltage		Steady state	0.5–2%
4.2 Current		Steady state	1.0–30%
5.0 Waveform distortion			
5.1 DC offset		Steady state	0–0.1%
5.2 Harmonics	0–9 kHz	Steady state	0–9 kHz
5.3 Interharmonics	0–9 kHz	Steady state	0–2%
5.4 Notching		Steady state	
5.5 Noise	Broadband	Steady state	0–1%
6.0 Voltage fluctuations	<25 Hz	Intermittent	0.1–7% 0.2–2 Pstb
7.0 Power frequency variations		<10 sec	± 0.10 Hz

The harmonics produce in general negative effects such us overheating of cables, transformers, and rotate machines; mal-operation of protection, control, and measurement devices especially if they are electronic made, and overvoltages especially when resonance conditions appear due to capacitors.

8.2 Mathematical background

Fourier series allows to represent any continuous periodic function by a sinusoidal component plus a series of sinusoidal harmonics with frequencies multiple of the fundamental one.

Any periodical signal can be expanded by Fourier series if it satisfies the Dirichlet's conditions:

- should have a definite number of discontinuities in a period,
- should have a finite number of maximums and minimum in a period, and
- the integral of the function in a period should have a finite value.

Under these conditions a function $f(t)$ with a period 2π is represented by Fourier series as follows:

$$f(t) = \frac{A_0}{2} + \sum_{n=1}^{\infty} \left(A_n \cos(n\,\omega t) + B_n \sin(n\,\omega t) \right) \tag{8.1}$$

where:

$$A_0 = \frac{2}{T} \int_{-T/2}^{T/2} f(t)\,dt$$

$$A_n = \frac{2}{T} \int_{-T/2}^{T/2} f(t)\cos(n\,\omega t)\,dt$$

$$B_n = \frac{2}{T} \int_{-T/2}^{T/2} f(t)\sin(n\,\omega t)\,dt$$

$$n = 1, 2, 3, \ldots$$

$$\omega = \frac{2\pi}{T} = \text{Angular Frequency}$$

This equation can be written as follows:

$$f(t) = \frac{A_0}{2} + \sum_{n=1}^{\infty} C_n \cos(n\,\omega t) \tag{8.2}$$

where:

$$C_n \sqrt{(A_n^2 + B_n^2)}$$

$$\alpha_n = \text{Arctg}\left(\frac{B_n}{A_n}\right)$$

8.3   Verification of harmonic values

With the results obtained from the load flow at power frequency and the harmonic load flow, the rms voltages of the nodes and currents at the branches and capacitors are calculated, as well as the total harmonic distortion (THD) or total demand distortion (TDD) as defined in the IEEE Standard 519-1992.

The following expressions are applied:

$$V_{\text{rms}} = \sqrt{V_1^2 + V_2^2 + \ldots + V_n^2} \tag{8.3}$$

$$I_{\text{rms}} = \sqrt{I_1^2 + I_2^2 + \ldots + I_n^2} \tag{8.4}$$

$$\text{THD}_V = \frac{\sqrt{V_2^2 + V_3^2 + \ldots + V_n^2}}{V_1} \times 100\,\% \tag{8.5}$$

$$\text{THD}_I = \frac{\sqrt{I_2^2 + I_3^2 + \ldots + I_n^2}}{I_1} \times 100\,\% \tag{8.6}$$

$$\text{TDD} = \frac{\sqrt{I_2^2 + I_3^2 + \ldots + I_n^2}}{I_{\text{D}}} \times 100\% \tag{8.7}$$

where:

V_{rms} rms voltage of a node
I_{rms} rms current through a branch or element of the system
THD_V Total harmonic voltage distortion in a node
THD_I Total harmonic current distortion of the current through a branch or element of the system
TDD Total harmonic current distortion of the current through a branch or element of the system based on the maximum demand current
V_1 Fundamental component of the voltage in a node
I_1 Fundamental component of the current through a branch or element of the system
V_n Harmonic component of order n of the voltage in a node
I_n Harmonic component of order n of the current through a branch or element of the system
I_{D} Calculated maximum demand current through a branch or element of the system

8.4   Parallel resonance

A parallel resonance can occur when a capacitor is installed at the same busbar that has the harmonic source. Parallel resonance produces very high impedance to the harmonic current of the resonant frequency. Since most loads that generate harmonics can be treated as current sources, very high voltages and currents occur in the elements connected in parallel. This can be explained with the help of Figure 8.1.

If the equivalent impedance behind a busbar is X_{th} and a capacitor of impedance X_c is installed, the equivalent impedance is:

$$Z_{eq} = \frac{X_{th} X_c}{X_{th} + X_c} \tag{8.8}$$

The parallel resonant condition occurs when the denominator of the above expression becomes zero:

$$X_{th} + X_c = 0$$
$$X_{th} = -X_c$$

The reactances to the angular frequency of resonance (ω_n) are calculated as follows:

$$X_{th} = \omega_n L \quad \text{and} \quad X_c = \frac{1}{\omega_n C}$$

where:

L Inductance of the system equivalent
C Capacitance of the capacitor bank

Besides, with the impedance corresponding to the fundamental frequency (ω), the following expressions apply:

$$\omega L = \frac{V^2}{MVA_{sc}}$$

$$\frac{1}{\omega C} = \frac{V^2}{MVA_{cap}}$$

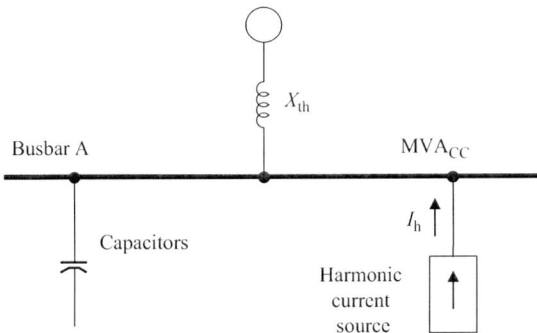

Figure 8.1   Connection of equipment producing parallel resonance

Where:

V nominal line voltage in kV
MVA_{sc} three-phase short circuit level at busbar in MVA
MVA_{cap} Total reactive power of the capacitor bank at fundamental frequency in MVA

Solving for the values of L and C and replacing for the resonant condition, the following expression is obtained:

$$\frac{\omega_n V^2}{\omega MVA_{sc}} = \frac{\omega V^2}{\omega_n MVA_{cap}}$$

Solving for the parallel resonance $f_p (\omega_n = 2\pi f_p)$, it is obtained:

$$f_p = f\sqrt{\frac{MVA_{sc}}{MVA_{cap}}} \tag{8.9}$$

where:

f_p Parallel resonance frequency (Hz)
f Fundamental frequency (Hz)

8.5   Series resonance

A series resonance can occur when a capacitor is installed in series with an inductive element and the impedances are the same for a specific harmonic. In this case the equivalent impedance is zero. Under series resonance conditions, the system offers very low impedance to harmonic voltages at resonance frequency. Harmonic filters then are designed to introduce a series resonance in the shunt element connected to the busbar of the system. They are made up normally by a reactance and a capacitor with equal reactance values but opposite in sign at the resonance frequency. This allows ground harmonic currents injected by harmonic sources. Figure 8.2 shows the connection of elements when series resonance

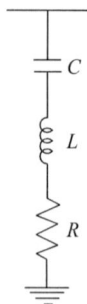

Figure 8.2   Connection of equipment producing series resonance

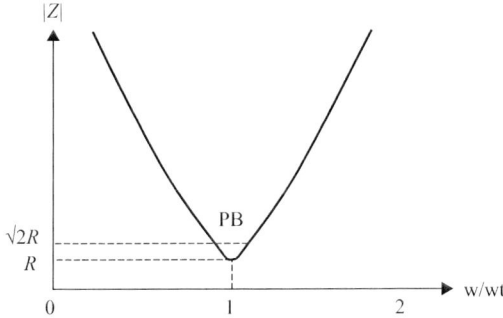

Figure 8.3 Z vs. W *in series resonance filters*

occurs, and Figure 8.3 illustrates the curve of impedance versus frequency, which shows the resonance frequency value for a general case.

8.6   Validation of harmonic values

8.6.1   *Harmonic limits*

The most important concepts to be illustrated involve the evaluation of harmonic current limits at individual customers and harmonic voltage limits on the overall system. These limits are typically evaluated at the point of common coupling (PCC) between the supplier and the customer.

8.6.2   *Voltage distortion limits*

The utility is responsible for maintaining the quality of voltage on the overall system. Table 8.3 summarizes the voltage distortion guidelines for different system voltage levels given by the IEEE Standard 519-1992.

$$\text{THD}_{V_n} = \frac{\sqrt{\sum_{h=2}^{\infty} V_h^2}}{V_n} \times 100\% \tag{8.10}$$

Table 8.3   Harmonic voltage distortion limits in percent of nominal fundamental frequency voltage (taken from IEEE Standard 519-1992)

Bus voltage at PCC (V_n)	Individual harmonic voltage distortion (%)	Total voltage distortion THD (V_n) (%)
$V_n \leq 69$ kV	3.0	5.0
69 kV $< V_n \leq 161$ kV	1.5	2.5
$V_n > 161$ kV	1.0	1.5

Table 8.4   Current distortion limits (taken from IEEE Standard 519-1992)

	I_{sc}/I_L	$h < 11$	$11 \leq h < 17$	$17 \leq h < 23$	$23 \leq h < 35$	$35 \leq h$	TDD
$V_n \leq 69$ kV	<20	4.0	2.0	1.5	0.6	0.3	5.0
	20–50	7.0	3.5	2.5	1.0	0.5	8.0
	50–100	10.0	4.5	4.0	1.5	0.7	12.0
	100–1000	12.0	5.5	5.0	2.0	1.0	15.0
	>1000	15.0	7.0	6.0	2.5	1.4	20.0
69 kV $< V_n \leq$	<20*	2.0	1.0	0.75	0.3	0.15	2.5
161 kV	20–50	3.5	1.75	1.25	0.5	0.25	4.0
	50–100	5.0	2.25	2.0	1.25	0.35	6.0
	100–1000	6.0	2.75	2.5	1.0	0.5	7.5
	>1000	7.5	3.5	3.0	1.25	0.7	10.0
$V_n > 161$ kV	<50	2.0	1.0	0.75	0.3	0.15	2.5
	≥50	3.5	1.75	1.25	0.5	0.25	4.0

*All power generation equipment is limited to these values of current distortion, regardless of actual I_{sc}/I_L.

where:

V_h Magnitude of individual harmonic components (rms volts)
n Harmonic order
V_n Nominal system rms voltage (rms volts)

8.6.3   Current distortion limits

The harmonic currents from an individual customer are evaluated at the PCC where the utility can supply other customers. Table 8.4 summarizes the current distortion limits given by the IEEE Standard 519-1992. The limits are dependent on the customer load in relation to the system short circuit capacity at the PCC. Note that all current limits are expressed as a percentage of the customer's average maximum demand load current.

8.7   Verification of harmonic values

For capacitor application, IEEE Standard 18-1992 should be applied. The following conditions are taken from this standard:

- The maximum permanent voltage including harmonics of any capacitor bank should not be higher than 110% of its nominal voltage.
- The crest voltage should not exceed $1 \times \sqrt{2}$ of rated rms voltage, including harmonics but excluding transients.
- The maximum permanent current including harmonics through any capacitor bank should not be higher than 180% of its nominal current including fundamental and harmonic currents.
- The maximum permanent power including harmonics of any capacitor bank should not be higher than 135% of its nominal power.

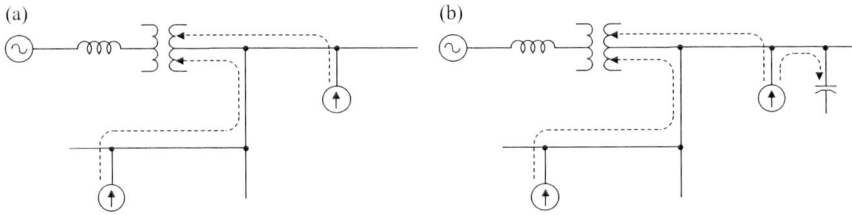

Figure 8.4   General flow of harmonic currents in radial power system: (a) without power capacitor; (b) with power capacitor

If any of the above limits is overcome, changes in the capacitor size and location should be considered to avoid exposing the system to serious difficulties. Figure 8.4 shows the effect on the harmonic currents when a capacitor is placed in the power system.

8.8   Resizing and relocation of capacitor banks

If the feeders that are going to give up load in a reconfiguration process have harmonic sources, especial care has to be considered for resonance with the capacitors installed in the feeders that are going to receive the loads.

Whenever a resonance situation occurs in a distribution system due to a capacitor bank, the solutions to be considered first are the relocation and/or resizing of the banks in order to vary the resonance frequency to a value that does not correspond to any generated by the harmonic sources. On the other hand, the short circuit level varies too, which also changes the resonance frequency.

A novel algorithm is proposed here for the analysis of distribution systems which can be applied in conjunction with the reconfiguration software or as a stand-alone analysis tool. The sources and loads are connected to earth for every harmonic under consideration, except for the fundamental frequency. The algorithm is applied for every harmonic source existing in the system and can be summarized in the following steps.

First a harmonic value is selected from one of the existing sources in the system and the feeder is organized from the node that has the harmonic source up to last node that in this case will be the power source.

Once the feeder is ordered, the load impedances of each node are calculated with the active and reactive powers of the loads and the same for the capacitors connected to the same node. For this, it is important to consider the impact of the frequencies of the respective harmonics into the impedance calculations.

Afterwards a value of 1.0 V is given to every node for the harmonic frequency under analysis. With this voltage value, the load currents are calculated in all elements, starting from the farthest away node to the harmonic source and then the current values are accumulated, taking into account the connectivity of the nodes. This routine could be called the "upstream iteration."

With the currents so calculated and starting from the node with the harmonic source, the voltages in every node are updated by calculating the voltage drop in all branches for the harmonic frequency under analysis. This routine could be called the "downstream iteration."

The new harmonic voltages so obtained in each node are used to recalculate the load currents. The cycle is repeated until convergence is obtained, i.e. when the variation in the voltage nodes is less than a pre-determined value, 0.0001 V for this case.

With the accumulated current of last iteration, the current supplied by the harmonic source is calculated. The actual value of the current source is divided by the calculated value and a factor is so obtained. The solution of the load flow is then obtained by multiplying the voltages and currents of the last iteration by this factor. The whole process is repeated for each harmonic frequency of every harmonic source.

The above process is illustrated in Figure 8.5(a) that shows a feeder with 17 nodes with current harmonic sources in 3 of them. For the sake of the illustration it

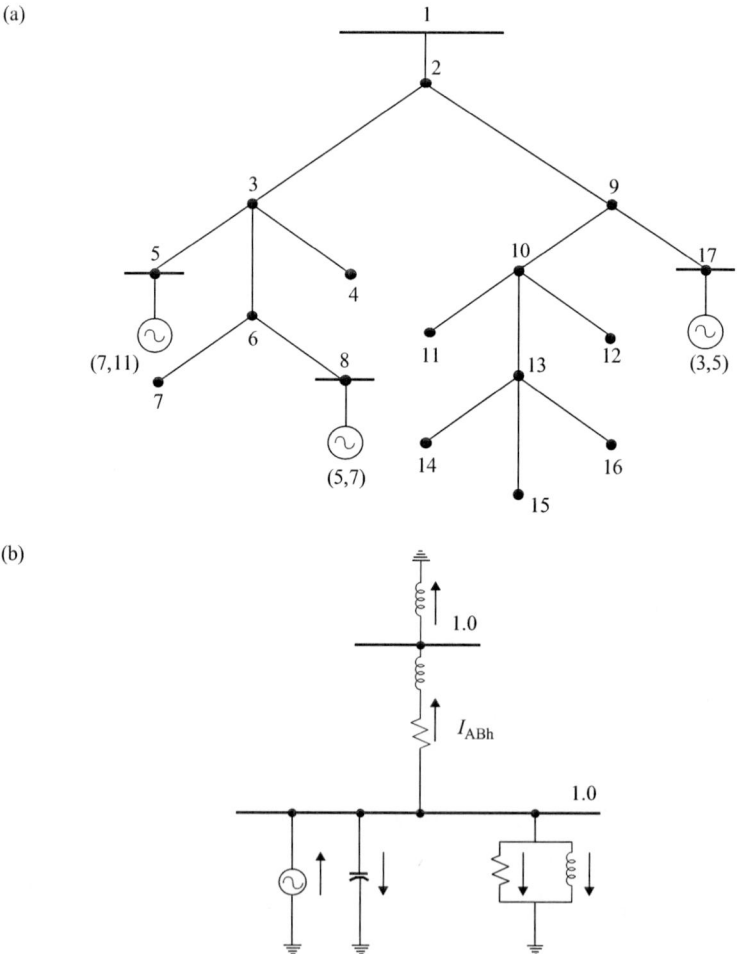

Figure 8.5   *System example to illustrate harmonic evaluation procedure*

is supposed that the nodes with harmonic sources have also a capacitor and a load made up by active and reactive elements.

According to the process mentioned, one of the harmonic sources is chosen first and then the feeder is reordered from it. Then the impedances are calculated as shown in the two nodes simplified diagram of Figure 8.5(b) for the harmonic under consideration.

Afterwards a voltage of 1.0 V is given to both nodes as shown again in Figure 8.5(b). With this voltages, the current at node B is first calculated since it is the farthest away node to the harmonic source. This current is the same to that circulating through branch A–B. The current of the capacitor and the reactive and resistive loads of node A can also be calculated knowing the voltage at A. With the branch current a voltage drop can be calculated, and therefore a new voltage for node B. This allows the calculation of a new current for node B which is the same for branch A–B. Then a new voltage drop through the branch is calculated and the voltage at node B updated.

The previous process is repeated until no further variation of voltage at node B is found. With the final current of the branch and the currents of the elements of node A the total current of the harmonic source is calculated. The real value of the current of the source is compared with the calculated value and the factor so obtained is used to calculate the real current values of the branch A–B and the currents of the elements connected to nodes A and B. The voltage of nodes A and B are also calculated with the same factor.

8.9   Models

Harmonics are analyzed by means of harmonic load flows which require a proper model for the harmonic sources, for the system equivalent, for the loads, and for the branches that consider the frequency of the respective harmonic.

The harmonic sources considered in the harmonic load flow are represented as independent current sources, each with a determined number of harmonic frequency. For each harmonic frequency the source has a definite magnitude in amperes.

In order to obtain a result that combines all the harmonic sources, the superposition theorem is applied.

Depending on the source location a branch can have different currents and therefore it is essential to consider the phasor angles of voltages and currents. The harmonic load flow requires a proper model for the harmonic sources, for the system equivalent, for the loads, and for the branches that consider the frequency of the respective harmonic. The modeling for each case is explained in this section.

8.9.1   Harmonic sources

The harmonic sources considered in the harmonic load flow are represented as independent current sources, each with a determined number of harmonic frequency. For each harmonic frequency the source has a definite magnitude in amperes.

In order to obtain a result that combines all the harmonic sources, the superposition theorem is applied. Depending on the source location a branch can have different currents and therefore it is essential to consider the phasor angles of voltages and currents.

8.9.2   System model

The system behind the substation busbar where the feeder is connected is represented by a Thévenin equivalent reactance represented as follows:

$$X_{Th}(f) = \frac{hV^2}{S_{sc}} \tag{8.11}$$

where:

$X_{th}(f)$ Thévenin equivalent reactance of the system as a frequency function
h Harmonic order
V Nominal system voltage
S_{sc} Three-phase short circuit power at the substation busbar

8.9.3   Load model

The load model as proposed in reference considers a parallel of a resistance and an inductive reactance. For each one of the components of the load model, the following expressions are used:

$$R = \frac{V^2}{P_{60}} \tag{8.12}$$

$$X(f) = \frac{hV^2}{Q_{60}} \tag{8.13}$$

where:

R Resistive load component
$X(f)$ Reactive load component as a frequency function
V Nominal load voltage
P_{60} Nominal active power load at the power frequency
Q_{60} Nominal reactive power load at the power frequency
h Harmonic order

8.9.4   Branch model

The branches connecting the nodes in distribution systems are represented by an impedance made up by a resistance and an inductive reactance connected in series. The resistance is considered as independent from the frequency, but not the reactance whose expression is as follows:

$$X_L(f) = 2\pi f L = hX_{L60} \tag{8.14}$$

where:

$X_L(f)$ Branch reactive component as a frequency function
f Frequency
L Branch inductance

h Harmonic order

X_{L60} Branch inductive reactance at power frequency

It can be noted that the parallel admittance of the branch that has the capacitive effect has been neglected, which is a valid consideration in distribution systems, especially those made up by overhead lines, rather than underground cables.

Example 8.1 A distribution feeder at 13.2 kV, shown schematically in Figure 8.6, will be considered in which a bank of 1300 kVAR will be installed. The three-phase short circuit level in the respective node is 7 kA and the load current has 20% of the 5th harmonic, 14.3% of the 7th harmonic, 9.1% of 11th harmonic, and 7.7% of the 13th harmonic. The nominal current in the node is 151.6 A at a pf = 0.7 which is assumed to be the same for all harmonics.

Perform the following actions:

- Calculate the possible resonance frequency.
- For the fundamental and harmonics 5th, 7th, 11th, and 13th, calculate the Thévenin and capacitance impedances, harmonic currents, and voltages across the capacitances.
- Calculate the THD for voltages and currents.
- Analyze the size of the capacitor bank with the results obtained.

For the calculations of the currents, bear in mind that the Thévenin equivalent is used only with the fundamental frequency. For the harmonic equivalents, there is not source and the load becomes a current source. The equivalent circuits then are as in Figure 8.7.

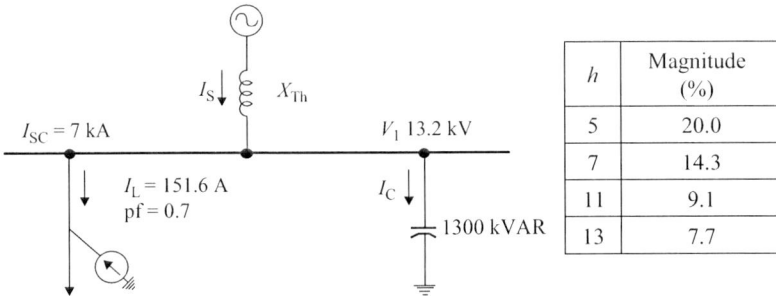

h	Magnitude (%)
5	20.0
7	14.3
11	9.1
13	7.7

Figure 8.6 Simplified diagram of the system

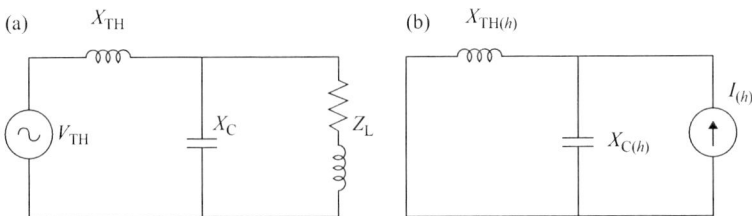

Figure 8.7 (a) Model for fundamental frequency; (b) model for harmonic h

The possible resonance frequency for this node is:

$$f_r = 60 \text{ Hz} \sqrt{\frac{\sqrt{3} \times 13.2 \text{ kV} \times 7 \text{ kA}}{1.3 \text{ MVAR}}} = 665.73 \text{ Hz}$$

$I_L = 151.6 \text{ A}$

$I_L(5°) = 20\% I_L = 0.2(151.6 \text{ A}) = 30.32 \text{ A}$

$I_L(7°) = 14.3\% I_L = 0.143(151.6 \text{ A}) = 21.67 \text{ A}$

$I_L(11°) = 9.1\% I_L = 0.091(151.6 \text{ A}) = 13.79 \text{ A}$

$I_L(13°) = 7.7\% I_L = 0.077(151.6 \text{ A}) = 11.67 \text{ A}$

$PF = 0.7 \rightarrow \cos^{-1}(0.7) = 45.57 \text{ lagging}$

Calculation of reactance impedances

$$X_{tH(60)} = \frac{V_{LN}}{I} = \frac{13.2 \text{ kV}}{\sqrt{3} \times 7 \text{ kA}} = j1.088 \text{ } \Omega$$

$X_{tH(5 \times 60)} = 5X_{TH(60)} = 5(j1.088) = j5.44 \text{ } \Omega$

$X_{tH(7 \times 60)} = 7X_{TH(60)} = 7(j1.088) = j7.62 \text{ } \Omega$

$X_{tH(11 \times 60)} = 11X_{TH(60)} = 11(j1.088) = j11.97 \text{ } \Omega$

$X_{tH(13 \times 60)} = 13X_{TH(60)} = 13(j1.088) = j14.15 \text{ } \Omega$

Calculation of capacitance impedances

$$X_{C(60)} = \frac{V^2}{Q} = \frac{(13.2 \text{ kV})^2}{1.3 \text{ MVAR}} = -j134.03 \text{ } \Omega$$

$$X_{C(\frac{60}{5})} = \frac{X_{C(60)}}{5} = \frac{-j134.03}{5} = -j26.80 \text{ } \Omega$$

$$X_{C(\frac{60}{7})} = \frac{X_{C(60)}}{7} = \frac{-j134.03}{7} = -j19.15 \text{ } \Omega$$

$$X_{C(\frac{60}{11})} = \frac{X_{C(60)}}{11} = \frac{-j134.03}{11} = -j12.18 \text{ } \Omega$$

$$X_{C(\frac{60}{13})} = \frac{X_{C(60)}}{13} = \frac{-j134.03}{13} = -j10.31 \text{ } \Omega$$

$$I_S = I_C + I_L$$

$$I_{C(60)} = \frac{V}{X_C} = \frac{13.2 \text{ kV}}{\sqrt{3}(-j134.03 \text{ } \Omega)} = j56.86A$$

$I_{S(60)} = I_{C(60)} + I_{L(60)} = j56.86 + 151.6\angle{-45.57°} = 117.92\angle{-25.84°} \text{ A}$

Figure 8.8 shows the equivalent circuit for fundamental frequency.

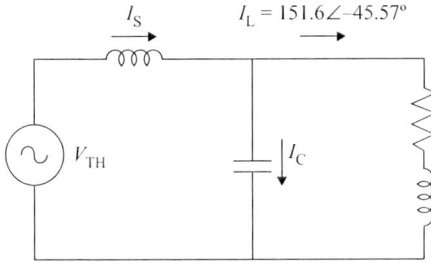

Figure 8.8 Circuit for fundamental frequency

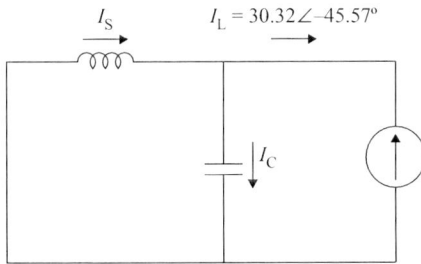

Figure 8.9 Circuit for 5th harmonic

For the 5th harmonic (300 Hz), the equivalent circuit is as shown in Figure 8.9.

$$Z_{eq} = \frac{X_{tH(s)}X_{C(s)}}{X_{tH(s)} + X_{C(s)}} = \frac{j5.44(-j26.80)}{j5.44 - j26.80} = \frac{145.79}{-j21.36} = j6.82 \ \Omega$$

$$V_C = I_L Z_{eq} = 30.32\angle-45.57° \cdot (j6.82) = 206.78\angle44.43° \ V$$

$$I_{C(5)} = \frac{V_{LN}}{X_{C(5)}} = \frac{206.78\angle44.43°}{-j26.80} = 7.71\angle134.43° \ A$$

$$I_{S(5)} = I_{C(5)} - I_{L(5)} = 7.71\angle134.43° - 30.32\angle-45.57° = 38.03\angle134.43° \ A$$

For the 7th harmonic (420 Hz):

$$Z_{eq} = \frac{X_{tH(7)}X_{C(7)}}{X_{tH(7)} + X_{C(7)}} = \frac{j7.62(-j19.15)}{j7.62 - j19.15} = \frac{145.92}{-j11.53} = j12.66 \ \Omega$$

$$V_C = I_{L(7)}Z_{eq} = 21.67\angle-45.57°(j12.66) = 274.25\angle44.43° \ V$$

$$I_{C(7)} = \frac{V_{LN}}{X_{C(7)}} = \frac{274.25\angle44.43°}{-j19.15} = 14.32\angle134.43° \ A$$

$$I_{S(7)} = I_{C(7)} - I_{L(7)} = 14.32\angle134.43° - 21.67\angle-45.57° = 36\angle134.43° \ A$$

The 11th harmonic is very critical since it is close to the resonance frequency. This is illustrated with the following equations:

$$f_r = f \sqrt{\frac{P_a}{P_{cap}}}$$

$$P_{cc} = \sqrt{3} V_{nom} I_{cc} = \sqrt{3}(13.2 \text{ kV})(7 \text{ kA}) = 160.04 \text{ MVA}$$

$$P_{cap} = 1.3 \text{ MVA}$$

$$f_r = 60 \sqrt{\frac{160.04 \text{ MVA}}{1.3 \text{ MVA}}} = 665.73 \text{ Hz}$$

This value is very close to resonance frequency which is at the 11th harmonic (660 Hz).

$$\frac{X_{TH(11)}X_{C(11)}}{X_{TH(11)} + X_{C(11)}} = \frac{j11.97(-j12.18)}{j11.97 - j12.18} = \frac{145.79}{-j0.21} = j694.26 \ \Omega$$

$$V_C = I_{L(11)}Z_{eq} = 13.79\angle-45.57°(j694.26) = 9573.84\angle44.43° \text{ V}$$

$$I_{C(11)} = \frac{V_{LN}}{X_{C(11)}} = \frac{9573.84\angle44.43°}{-j12.18} = 786.03\angle134.43° \text{ A}$$

$$I_{S(11)} = I_{C(11)} - I_{L(11)} = 786.03\angle134.43° - 13.79\angle-45.57° = 799.82\angle134.43° \text{ A}$$

For the 13th harmonic (780 Hz):

$$Z_{eq} = \frac{X_{TH(13)}X_{C(13)}}{X_{TH(13)} + X_{C(13)}} = \frac{j14.15(-j10.31)}{j14.15 - j10.31} = \frac{145.88}{j3.84} = -j38.0 \ \Omega$$

$$V_C = I_{L(13)}Z_{eq} = 11.67\angle-45.57°(-j38.0) = 443.46\angle-135.57° \text{ V}$$

$$I_{C(13)} = \frac{V_{LN}}{X_{C(13)}} = \frac{443.46\angle-135.57°}{-j10.31} = 43.01\angle-45.57° \text{ A}$$

$$I_{S(13)} = I_{C(13)} - I_{L(13)} = 43.01\angle-45.57° - 11.67\angle-45.57° = 31.34\angle-45.57° \text{ A}$$

Table 8.5 presents the values of currents, voltages, and impedances calculated for the fundamental and harmonics 5th, 7th, 11th, and 13th.

Total harmonic distortion values:

$$THD = \frac{\sqrt{A_2^2 + A_3^2 + \cdots}}{A_1} = \frac{\sqrt{A_5^2 + A_7^2 + A_{11}^2 + A_{13}^2}}{A_1}$$

Table 8.5 Summary of results for example 8.1

h	1	5	7	11	13
$I_L(A)$	151.6	30.32	21.67	13.79	11.67
$I_S(A)$	117.92 $\angle-25.84°$	$38.03\angle134.43°$	$36\angle134.43°$	$799.82\angle134.43°$	$31.34\angle45.57°$
$X_C(\Omega)$	$-j134.03$	$-j\,26.80$	$-j\,19.15$	$-j\,12.18$	$-j\,10.31$
$X_{TH}(\Omega)$	$j\,1.088$	$j\,5.44$	$j\,7.62$	$j\,11.97$	$j\,14.15$
$I_C(A)$	$56.86\angle90°$	$7.71\angle134.43°$	$14.32\angle134.43°$	$786.03\angle134.43°$	$43.01\angle-45.57°$
$V_C(V)$	$7621.02\angle0°$	$206.78\angle44.43°$	$274.25\angle44.43°$	$9573.84\angle44.43°$	$443.46\angle-135.57°$

For current:

$$\text{THD} = \frac{\sqrt{30.32^2 + 21.67^2 + 13.79^2 + 11.67^2}}{151.6} = 0.2732 \cong 27.32\%$$

$$\frac{I_{sc}}{I_L} = \frac{7000\text{ A}}{151.6\text{ A}} = 46.17$$

From the IEEE Standard 519-1992:

For $20 < \frac{I_{sc}}{I_L} < 50 \rightarrow \text{THD} = 8\%$

The obtained value of 27.32% → It does not fulfill the standard.

For voltages

$$\text{THD} = \frac{\sqrt{0.206^2 + 0.274^2 + 9.573^2 + 0.443^2}}{7.62} = 1.258 \cong 126\%$$

The standard allows 5% → It does not fulfill the standard.

$$V_{rms_{TOTAL}} = \sqrt{V_1^2 + V_5^2 + V_7^2 + V_{11}^2 + V_{13}^2}$$

$$= \sqrt{7.62^2 + 0.206^2 + 0.274^2 + 9.573^2 + 0.443^2} = 12.25\text{ kV}$$

This exceeds 10% regarding the nominal $\cong 7.62(1.1) = 8.382$

$$I_{rms_{TOTAL}} = \sqrt{I_1^2 + I_5^2 + I_7^2 + I_{11}^2 + I_{13}^2}$$

$$= \sqrt{56.86^2 + 7.71^2 + 14.32^2 + 786.03^2 + 43.01^2} = 789.42\text{ A}$$

$$\frac{789.42\text{ A}}{56.86\text{ A}} \cong 1388\%$$

$I_{L(rms)} = 157.15\text{ A}$
$V_{C(rms)} = 12.25\text{ kV}$
$I_{C(rms)} = 789.42\text{ A}$
$I_{S(rms)} = 810.77\text{ A}$

$$\text{THD}_{I_L} = 27.32\%$$
$$\text{THD}_{V_C} = 126\%$$
$$\text{THD}_{I_C} = 1385\%$$
$$\text{THD}_{I_S} = 680\%$$

The former results show that the capacitor bank has to be re-sized, relocated, or both, since there is a clear violation of the standards referred before. In fact all the THD values were exceeded and overvoltages and overcurrents registered at the capacitors.

The best approach in this case would be an iterative process until the standards are satisfied.

8.10   Derating transformers

Transformers serving nonlinear loads register an important rise in eddy current losses associated to the harmonic currents generated by those loads. This increases the temperature of the transformers which obliges to reduce or derate their nominal rating. In order to determine the derating caused by this phenomenon, the harmonic loss factor for winding eddy currents factor, known as F_{HL}, has been introduced.

In pu, the F_{HL} factor is defined in IEEE Standard C57.110-2008 as follows:

$$F_{\text{HL}} = \frac{\displaystyle\sum_{h=1}^{h=h_{\max}} I_h^2 h^2}{\displaystyle\sum_{h=1}^{h=h_{\max}} I_h^2} \tag{8.15}$$

where I_h is the rms current at harmonic h, in pu or rated rms load current.

UL has introduced other term called the K-factor which defines a rating optionally applied to a transformer indicating its suitability for use with loads that have nonsinusoidal currents. The K-factor is defined with the following expression:

$$K = \sum_{h=1}^{h=h_{\max}} I_h^2 h^2 \tag{8.16}$$

From (8.15) and (8.16) it is clear that F_{HL} in pu is equal to the K-factor divided by the summation of the squared of the fundamental and harmonic currents in pu.

Manufactures build special K-factor transformers. Standard K-factor ratings are 4, 9, 13, 20, 30, 40, and 50. For linear loads, the K-factor is always 1.

For nonlinear loads, if harmonic currents are known, the K-factor is calculated and compared against the transformer's nameplate K-factor. As long as the load K-factor is equal to or less than 1, the transformer K-factor, the transformer does not need to be derated.

For transformers, ANSI/IEEE Standard C57.110-2008 provides a method to derate the capacity when supplying nonlinear loads.

$$\text{Transformer Derating} = \sqrt{\frac{1 + P_{\text{ec}-\text{r}}}{1 + \dfrac{\displaystyle\sum_{h=1}^{h=h_{\text{max}}} I_h^2 h^2}{\displaystyle\sum_{h=1}^{h=h_{\text{max}}} I_h^2} P_{\text{ec}-\text{r}}}} \tag{8.17}$$

or

$$\text{Transformer Derating} = \sqrt{\frac{1 + P_{\text{ec}-\text{r}}}{1 + F_{\text{HL}} P_{\text{ec}-\text{r}}}} \tag{8.18}$$

where:

$P_{\text{ec}-\text{r}}$ Maximum transformer pu eddy current loss factor (typically, between 0.05 and 0.10 pu for dry-type transformers)

F_{HL} Harmonic loss factor for winding eddy currents

I_h Harmonic current, normalized by dividing it by the fundamental current

h Harmonic order

Example 8.2 Assume that the pu harmonic currents are 1.000, 0.016, 0.261, 0.050, 0.003, 0.089, 0.031, 0.002, 0.048, 0.026, 0.001, 0.033, and 0.021 pu for harmonic order of 1, 3, 5, 7, 9, 11, 13, 15, 17, 19, 21, 23, and 25, respectively. Also assume that the eddy current loss factor is 8%.
Determine the following:
(a) The F_{HL} and K factors of the transformer
(b) The transformer derating

Table 8.6 Currents for different harmonic orders

h	i (pu)	h^2	i^2	$i^2 \times h^2$
1	1	1	1	1
3	0.016	9	0.000256	0.002304
5	0.261	25	0.068121	1.703025
7	0.05	49	0.0025	0.1225
9	0.003	81	0.000009	0.000729
11	0.089	121	0.007921	0.958441
13	0.031	169	0.000961	0.162409
15	0.002	225	0.000004	0.0009
17	0.048	289	0.002304	0.665856
19	0.026	361	0.000676	0.244036
21	0.001	441	0.000001	0.000441
23	0.033	529	0.001089	0.576081
25	0.021	625	0.000441	0.275625
Total			1.084283	5.712347

Solution:

(a) The K and F_{HL} factors are obtained from Table 8.6.
Table 8.6 Results corresponding to example 8.2

$$K = \sum_{h=1}^{h=25} I_h^2 h^2 = 5.712$$

$$F_{HL} = \frac{\sum_{h=1}^{h=25} I_h^2 h^2}{\sum_{h=1}^{h=25} I_h^2} = \frac{5.712}{1.084} \cong 5.3$$

(b) The transformer derating is obtained as follows:

$$\text{Transformer Derating} = \sqrt{\frac{1 + P_{ec-r}}{1 + F_{HL}P_{ec-r}}} = \sqrt{\frac{1 + 0.08}{1 + 5.3 \times 0.08}}$$

$$\cong 0.87 \text{ pu} \quad \text{or} \quad 87\%$$

Chapter 9
Modern protection of distribution systems

Overcurrent relays are the most common form of protection used to operate only under fault conditions. They should not be installed purely as a means of protecting systems against overloads. The relay settings that are selected are often a compromise in order to cope with both overload and overcurrent conditions.

Overcurrent relays can be classified as definite current, definite time, and inverse time as shown in Figure 9.1(a–c). The time delay units can work in conjunction with the instantaneous units as shown in Figure 9.1(d).

9.1 Fundamentals of overcurrent protection

9.1.1 Protection coordination principles

Relay coordination is the process of selecting settings that will assure that the relays will operate in a reliable and selective way. In overcurrent relays the coordination is based on the relay time-current characteristics of instantaneous and/or time-delay units.

1. Instantaneous units should be set so they do not trip for fault levels equal or lower to those at busbars or elements protected by downstream instantaneous relays.
2. Time-delay units should be set to clear faults in a selective and reliable way, assuring the proper coverage of the thermal limits of the elements protected.

Figure 9.2 illustrates for a typical distribution substation, the process to carry out the coordination, starting from the relay associated to breakers 1, 2, and 3 all the way up to breaker 9.

9.1.2 Criteria for setting instantaneous units

Instantaneous units are set by adjusting the pickup level current at which the relays operate. Most numerical relays now have the possibility of setting an operating time, allowing the relay to behave as a definite time unit.

1. Distribution lines
   Between 6 and 10 times the maximum circuit rating
   50% of the maximum short circuit at the point of connection of the relay
2. Lines between substations
   125% to 150% of the short circuit current existing on the next substation

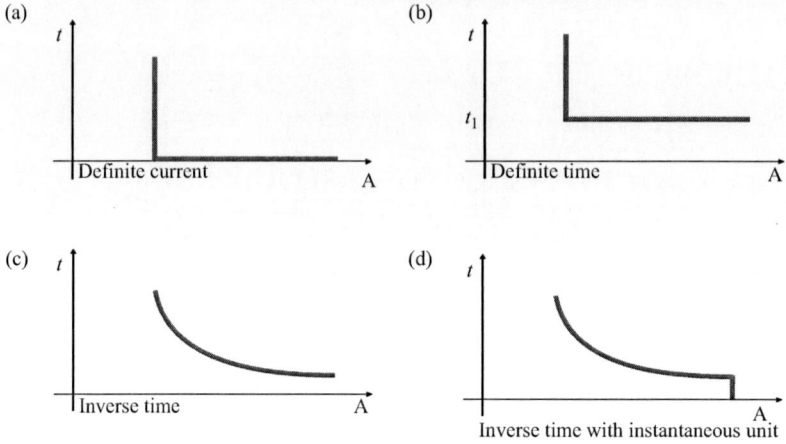

Figure 9.1   Time-current operating characteristics of overcurrent relays

3.   Transformer units
       125% to 150% of the short circuit current existing on the low voltage side
       The units at the low voltage side are overridden unless there is commu-
       nication with the relays protecting the feeders.

9.1.3   Setting time-delay relays

Time delay units are set by selecting the time-curve characteristic that is defined by
two parameters:

1.   Tap or pickup value: A value that defines the pickup current of the relay.
       Current values are expressed as multiples of this value in the time-current
       characteristic curves.
2.   Dial: Dial defines the time curve at which the relay operates for any tap value.
       Higher dial values represent higher operating times.

       Figure 9.3 shows overcurrent inverse time relay curves associated with two
breakers on the same feeder.
       A margin between two successive devices in the order of 0.2–0.4 seconds should
be used to avoid losing selectivity due to one or more of the following reasons:

1.   the breaker opening time,
2.   the overrun time after the fault has been cleared, and
3.   variations in fault levels, deviations from the characteristic curves of the relays,
       and errors in the current transformers (CTs).

       For phase relays, the tap or pickup value is determined by:

$$\text{Tap} = (\text{OLF} \times I_{\text{nom}}) \div \text{CTR}$$

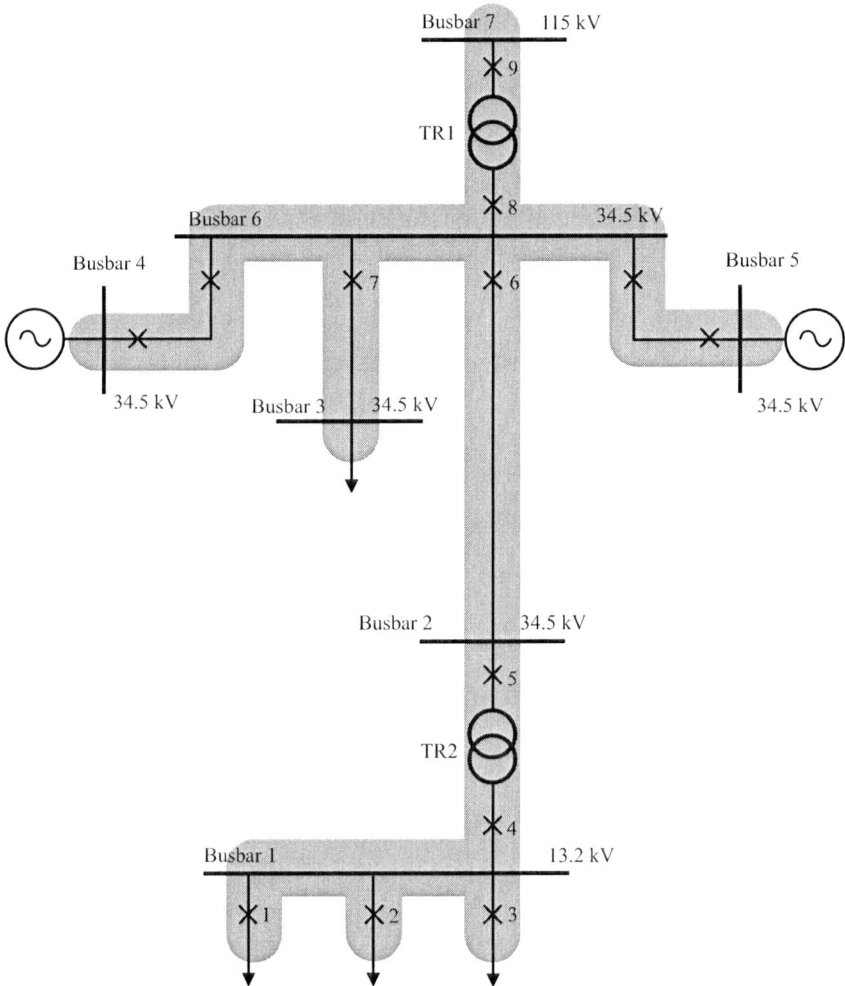

Figure 9.2   Overcurrent relay coordination procedure in a distribution system

For ground fault relays, the tap value is determined, with the maximum unbalance typically around 20%:

$$\text{Tap} = \left((0.2) \times I_{\text{nom}}\right) \div \text{CTR}$$

The overload factor recommended is as follows:

- Motors = 1.05
- High voltage lines, transformers, and generators = 1.25–1.5
- Distribution feeders = 2.0

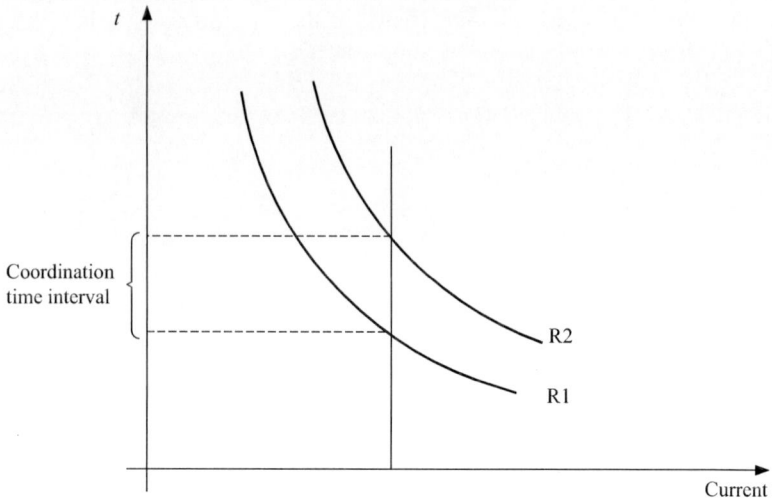

Figure 9.3   Overcurrent inverse time relay curves associated with two breakers on the same feeder

For phase relays, three-phase faults and maximum short time overload should be considered. For ground relays, line-to-ground faults and max $3I_o$ should be considered.

The procedure to determine the time dial settings is based on operating time targets corresponding to the multiples of pickup or tap values at the instantaneous values.

The process starts at the farthest downstream relay and finishes with the farthest up relay.

For the farthest downstream relays, the lowest time dial is chosen or that considering cold load pickup conditions.

Normally the settings are first carried out for phase relays and then for ground (neutral) relays. For the latter, the lowest Time Dial is selected whenever an open ground circuit is established, like that through Dy transformers.

The process to determine the time dial setting is rather elaborate and is summarized in the following steps:

1. Calculate the multiple of pickup value for the secondary short circuit current corresponding to the instantaneous setting of the relay where the process starts. If the instantaneous unit is overriden, the calculation is carried out with the total secondary short circuit current at the relay location.
2. With the value calculated above, determine the operating time t_1 of the relay for the given Time Dial.
3. Determine the operating time t_{2a} of the upstream relay with the expression $t_{2a} = t_1 + t_{margin}$.

4. Calculate the multiple of pickup value of the upstream relay using the same short circuit current used in the first relay (step 1).
5. Knowing t_{2a}, and having calculated the multiple of pickup value of the upstream relay, select the above nearest time dial for that relay.

The process follows the same steps for the next upstream relay and is repeated until the settings of the farthest up relay are calculated.

Operating time defined by IEC 60255 and IEEE C37.112 are:

$$t = \frac{k\beta}{\left(\frac{I}{I_s}\right)^{\alpha} - 1} + L$$

where:

   t Relay operating time in seconds
   k Time dial, or time multiplier setting
   I Fault current level in secondary amperes
   I_S Tap or pickup current selected
   L Constant
   α Slope constant
   β Slope constant

Table 9.1 shows the constant values of the parameters for curves defined by IEEE C37.112 and IEC 60255, and Figure 9.4 illustrates these curves.

9.1.4   Setting overcurrent relays using software techniques

1. Locate the fault and obtain the current for setting the relays.
2. Identify the pairs of relays to be set, first determining which one is farther away from the source.
3. Verify that the requirement's thermal capabilities are protected and devices operate for minimum short circuit levels.

Table 9.1   IEEE and IEC constants for standard, overcurrent relays

IDMT curve description	Standard	α	β	L
Moderately inverse	IEEE	0.02	0.0515	0.114
Very inverse	IEEE	2	19.61	0.491
Extremely inverse	IEEE	2	28.2	0.1217
Inverse	US-CO8	2	5.95	0.18
Short-time inverse	US-CO2	0.02	0.02394	0.01694
Standard inverse	IEC	0.02	0.14	
Very inverse	IEC	1	13.5	
Extremely inverse	IEC	2	80.0	
Long-time inverse	IEC	1	120	

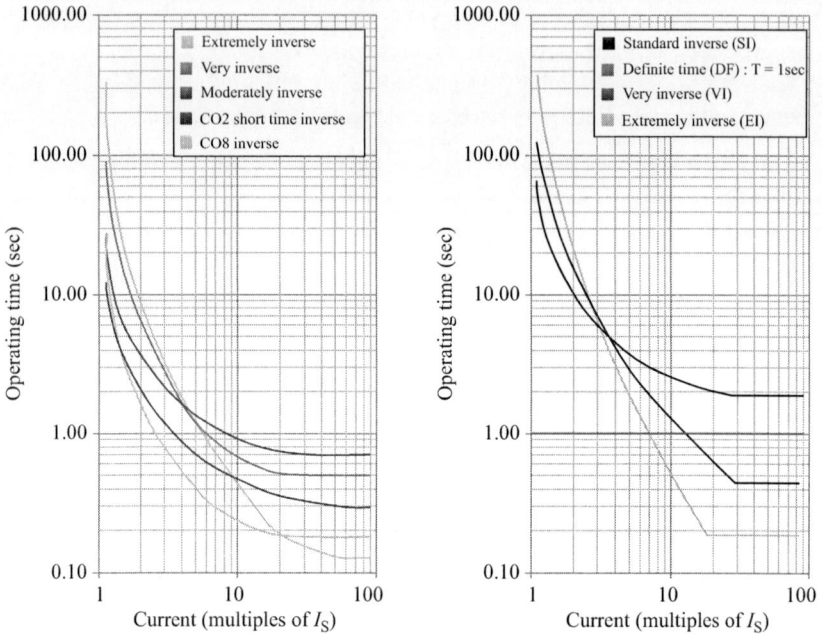

Figure 9.4   Typical ANSI/IEEE and IEC overcurrent relay curves

9.2   Coordination across Dy transformers

In the case of overcurrent relay coordination for Dy transformers, the distribution of currents in these transformers should be checked for three-phase, phase-to-phase, and single-phase faults on the secondary winding.

The results of the three cases are summarized in Table 9.2. Analyzing the results, it can be seen that the critical case for the coordination of overcurrent relays is the phase-to-phase fault. In this case, the relays installed in the secondary carry a current less than the equivalent current flowing through the primary relays that could lead to a situation where the selectivity between the two relays is at risk. For this reason, the discrimination margin between the relays is based on the operating time of the secondary relays at a current equal to $\sqrt{3}I_f/2$, and on the operating time for the primary relays for the full fault current value I_f, as shown in Figure 9.5.

Table 9.2   Summary of fault conditions

Fault	$I_{primary}$	$I_{secondary}$
Three phase	I	I
Phase-to-phase	I	$\sqrt{3}I/2$
Phase-to-earth	I	$\sqrt{3}I$

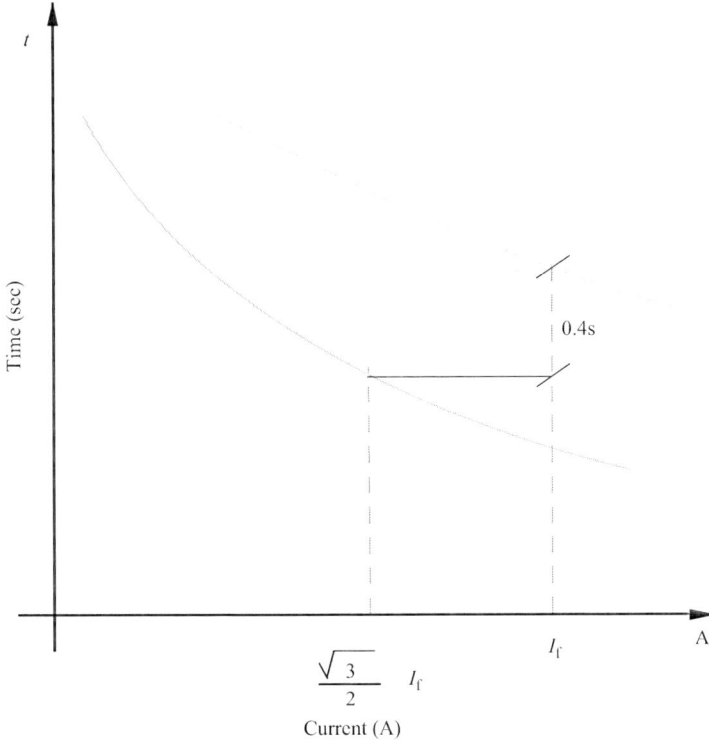

Figure 9.5 Coordination of overcurrent relays for a Dy transformer

Example 9.1 For the power system in Figure 9.6, calculate the following:

1. The three-phase short circuit levels on busbars 1 and 2.
2. The transformation ratios of the CTs associated with breakers 1–8, given that the number of primary turns is a multiple of 100. The CT for breaker number 9 is 250/5.
3. The settings of the instantaneous elements, and the tap and dial settings of the relays to guarantee a coordinated protection arrangement, allowing a discrimination margin of 0.4 sec.
4. The percentage of the 34.5 kV line protected by the instantaneous element of the overcurrent relay associated with breaker 6.

The pu impedances are calculated on the following bases:

$V = 34.5$ kV, $P = 100$ MVA

Consider all the relays to be set are numerical type Beckwith M-7651 with the characteristics as is shown in Figure 9.7. The relays have an extremely inverse time-current characteristic with the following constants:

$\alpha = 2.0$

$\beta = 80$

$L = 0$

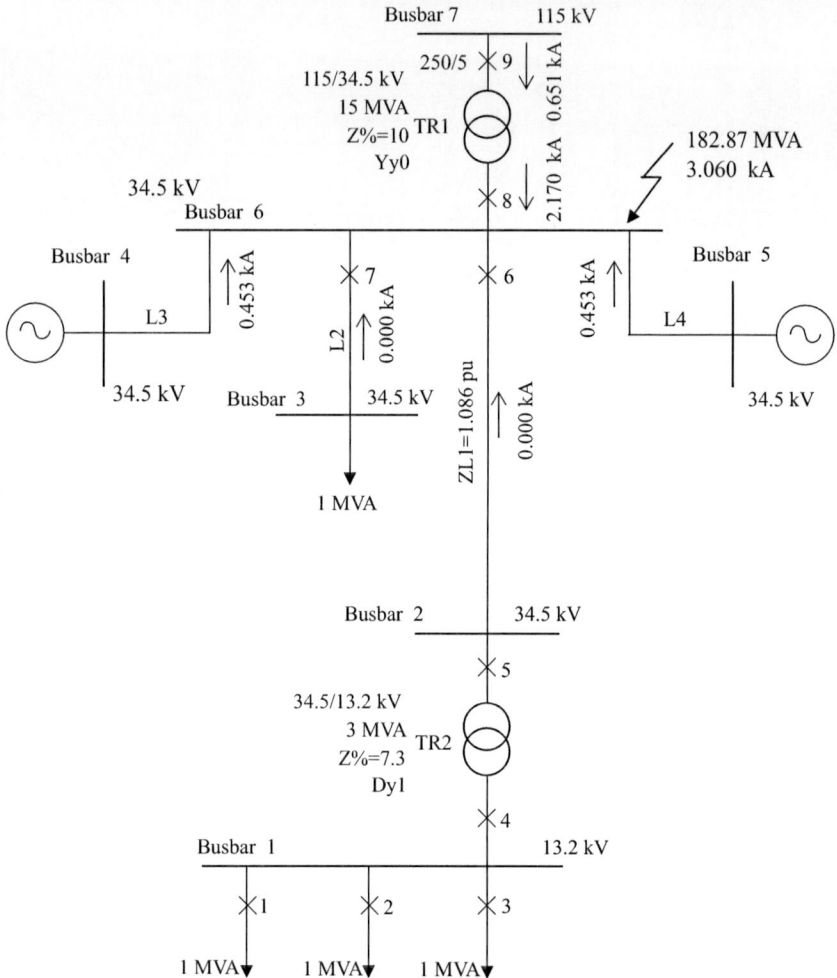

Figure 9.6   Power system of example 9.1

The Time current characteristic (TCC) is defined by $t = \dfrac{\text{Time Dial} \times 80}{(\text{MULT})^2 - 1}$

where MULT is fault current (in secondary amperes)/tap. The following considerations have to be taken into account:

- For setting of the instantaneous element a value of 10 times the maximum load current is used.
- The margin time for this relay can be 0.2 sec since it is of numerical type.
- Relay 7 is the same W CO 11 with tap $= 4$ A, dial $= 5$, and Inst $= 1100$ A.

The margin time for this relay can be 0.2 sec since it is of numerical type. All the parameters of the power system elements are shown in Table 9.3.

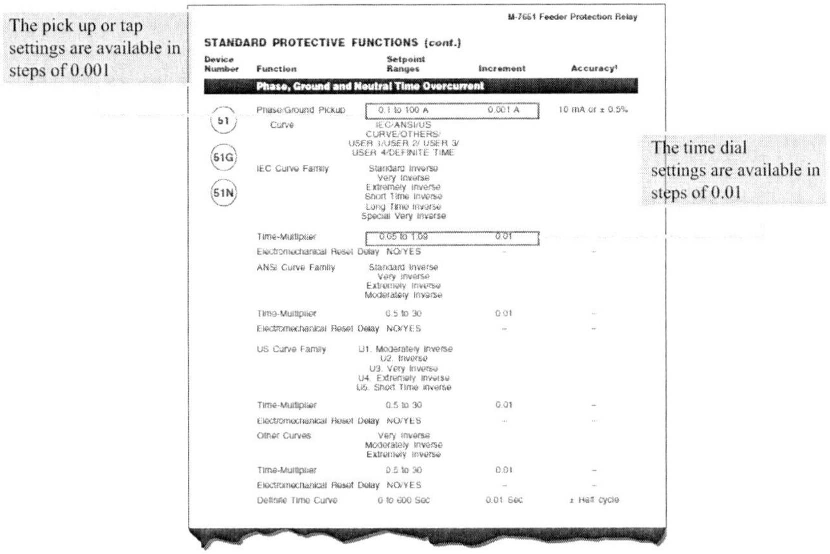

Modern protection of distribution systems   209

Figure 9.7   Specifications of relay Beckwith M-7651 (reproduced by permission of Beckwith Electric)

Defining the pickup for relays:

$I_{load1,2,3} = 43.74$ A; pickup$_{1,2,3} = (1.5)(43.74)(5/100) = 3.28$ A
$I_{load4} = 131,22$ A; pickup$_4 = (1.5)(131.22)(5/200) = 4.92$ A
$I_{load5} = 50.20$ A; pickup$_5 = (1.5)(50.20)(5/100) = 3.76$ A
$I_{load6} = 50.20$ A; pickup$_6 = (1.5)(50.20)(5/200) = 1.88$ A
$I_{load8} = 251.02$ A; pickup$_8 = (1.5)(251.02)(5/300) = 6.28$ A
$I_{load9} = 75.31$ A; pickup$_9 = (1.5)(75.31)(5/250) = 2.26$ A

For relays 1, 2, and 3:

$$I_{inst.\ trip} = 10 \times I_{nom} \times (1/CTR)$$
$$= 10 \times 43.74 \times (5/100) = 21.87 \text{ A}$$
$$I_{prim.\ trip} = 21.87(100/5) = 437.4 \text{ A}$$
$$MULT = 21.87/3.28 = 6.668 \text{ times}$$

with dial $= 0.05$:

$$t = \frac{0.05 \cdot 80}{(6.668)^2 - 1} = 0.092 \text{ sec}$$

To coordinate with relays 1, 2, and 3 at 43.74 A:

$$t_{4a} = 0.092 + 0.2 = 0.292 \text{ sec}$$

For relay 4:

$$MULT_{4a} = (437.4)(5/200)(1/4.92) = 2.223 \text{ times}$$

Table 9.3 *Short circuit calculation for power system of example 9.1*

Fault location		Distance from fault	Element name	Type	U_n (kV)	UL-E (RST) (kV)	AU L-E (RST) (°)	Ik″ (RST) (kA)
From node	To node							
Busbar6	Faulted	0			34.5	19.919	180	3.06
Busbar6	Busbar2		L1	Line				0
Busbar6	Busbar3		L2	Line				0
Busbar6	Busbar5		L4	Line				0.453
Busbar6	Busbar4		L3	Line				0.453
Busbar6	Busbar7		TR1	2W transformer				2.17
Busbar2		1			34.5	19.919	180	
Busbar2	Busbar6		L1	Line				0
Busbar2	Busbar1		TR2	2W transformer				0
Busbar3		1			34.5	19.919	180	
Busbar3	Busbar6		L2	Line				0
Busbar5		1			34.5	8.281	179.94	
Busbar5	Busbar6		L4	Line				0.453
Busbar5	Busbar5		G8	Synchronous machine				0.453
Busbar4		1			34.5	8.281	179.94	
Busbar4	Busbar6		L3	Line				0.453
Busbar4	Busbar4		G15	Synchronous machine				0.453
Busbar7		1			115	8.988	180	
Busbar7	Busbar6		TR1	2W transformer				0.651
Busbar7	Busbar7		Equivalent	Network feeder				0.651
Busbar1		2			13.2	7.621	180	
Busbar1	Busbar2		TR2	2W transformer				0

Table 9.4 Summary of relay settings for example 9.1

Relay no.	CT ratio	Pickup	Dial	Instantaneous
1,2,3	100/5	3.28	0.05	21.87 A
4	200/5	4.92	0.05	–
5	100/5	3.76	0.14	25.73 A
6	200/5	1.88	0.25	32.05 A
7	200/5	4	5	27.5 A
8	300/5	6.28	0.05	–
9	250/5	2.26	0.13	16.27 A

At 2.223 times, and $t_{4a} = 0.292$ sec:

$$\text{Dial} = \left(\frac{0.292}{80}\right)\left((2.223)^2 - 1\right) = 0.01$$

However, the dial 0.05 is the minimum that the relay has.

The operating time for a line-to-line fault is determined by taking 86% of the three-phase fault current.

$$\text{MULT}_{4b} = (0.86)(1076.06)(5/200)(1/4.92) = 4.702 \text{ times}$$

$$t_{4b} = \frac{0.05 \cdot 80}{(4.627)^2 - 1} = 0.190s$$

This relay has no setting for the instantaneous.

The procedure is the same for all the relays of the power. Table 9.4 shows the corresponding settings. The curves obtained are shown in Figure 9.8.

9.3 Protection equipment installed along the feeders

The devices most used for distribution system protection are overcurrent relays, reclosers, sectionalizers, and fuses.

The coordination of overcurrent relays was dealt with in detail in the previous section, and this section will cover the other three devices referred to above. The last three devices can be mounted as stand-alone or as part of other switchgear like those known as VFI. This type of switchgear can be arranged in different config-urations where the type of equipment can be different (breakers or fuses) and also the number of inputs and outputs can vary. Figure 9.9 illustrates a VFI with seven different configurations from S&C. Figure 9.10 is a picture of a typical Pad-Mounted Gear also from S&C.

9.3.1 Reclosers

9.3.1.1 General

A recloser is a device with the ability to detect phase and phase-to-earth overcurrent conditions, to interrupt the circuit if the overcurrent persists after a predetermined time, and then to automatically reclose to re-energize the line. If the fault which originated the operation still exists, then the recloser will stay open after a preset number of operations, thus isolating the faulted section from the rest of the system.

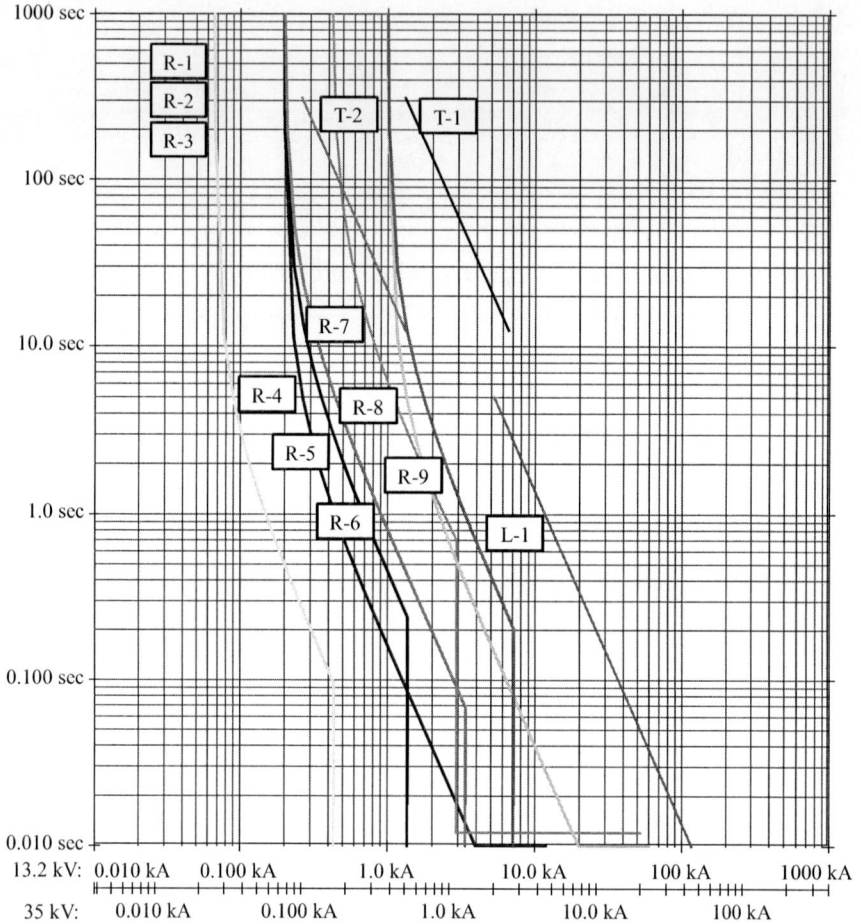

Figure 9.8   Relay coordination curves for example 9.1

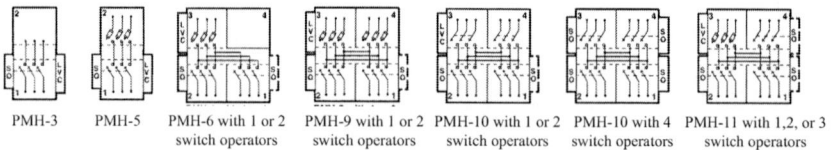

Figure 9.9   Remote supervisory PMH models (reproduced by permission of S&C)

In an overhead distribution system between 80% and 95% of the faults are of a temporary nature and last, at the most, for a few cycles or seconds. Thus, the recloser, with its opening/closing characteristic, prevents a distribution circuit being left out of service for temporary faults. Typically, reclosers are designed to have up to three open-close operations and, after these, a final open operation to

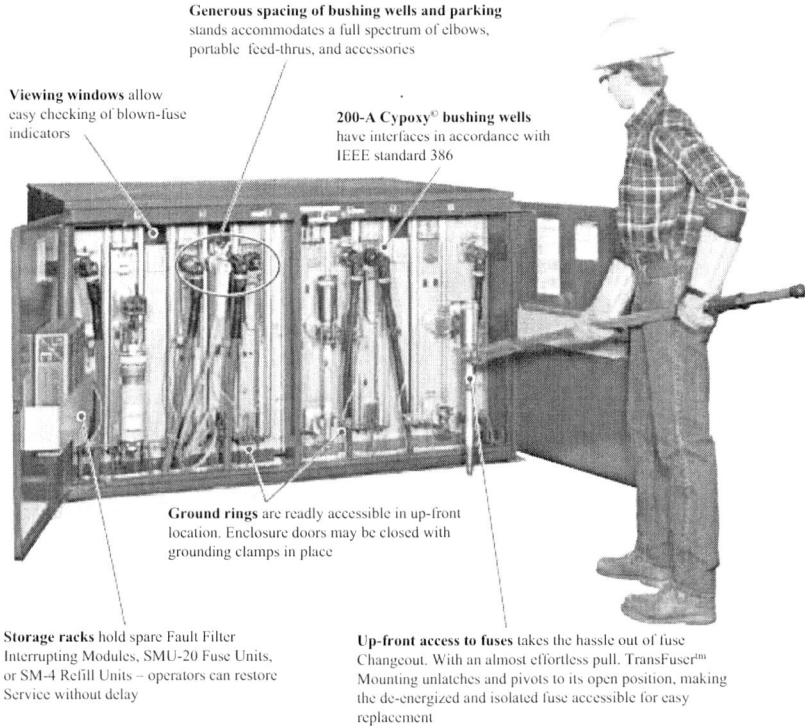

Generous spacing of bushing wells and parking
stands accommodates a full spectrum of elbows,
portable feed-thrus, and accessories

Viewing windows allow
easy checking of blown-fuse
indicators

200-A Cypoxy© bushing wells
have interfaces in accordance with
IEEE standard 386

Ground rings are readily accessible in up-front
location. Enclosure doors may be closed with
grounding clamps in place

Storage racks hold spare Fault Filter
Interrupting Modules, SMU-20 Fuse Units,
or SM-4 Refill Units – operators can restore
Service without delay

Up-front access to fuses takes the hassle out of fuse
Changeout. With an almost effortless pull. TransFuser™
Mounting unlatches and pivots to its open position, making
the de-energized and isolated fuse accessible for easy
replacement

Figure 9.10   PME Pad-Mounted Gear (reproduced by permission of S&C)

lock out the sequence. One further closing operation by manual means is usually allowed. The counting mechanisms register operations of the phase or earth-fault units which can also be initiated by externally controlled devices when appropriate communication means are available.

The operating time-current characteristic curves of reclosers normally incorporate three curves, one fast and two delayed, designated as A, B, and C, respectively. Figure 9.11 shows a typical set of time-current curves for reclosers. However, new reclosers with microprocessor-based controls may have keyboard-selectable time-current curves which enable an engineer to produce any curve to suit the coordination requirements for both phase and earth faults. This allows reprogramming of the characteristics to tailor an arrangement to a customer's specific needs without the need to change components.

Coordination with other protection devices is important in order to ensure that, when a fault occurs, the smallest section of the circuit is disconnected to minimize disruption of supplies to customers. Generally, the time characteristic and the sequence of operation of the recloser are selected to coordinate with mechanisms upstream toward the source. After selecting the size and sequence of operation of the recloser, the devices downstream are adjusted in order to achieve correct coordination. A typical sequence of a recloser operation for a permanent fault is

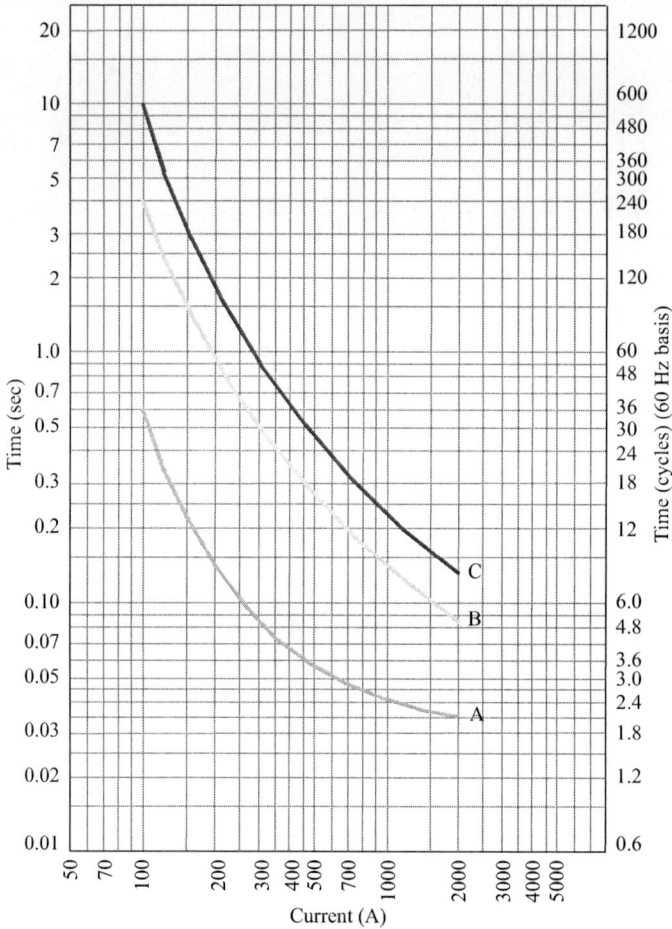

Figure 9.11   Time-current curves for reclosers

shown in Figure 9.12. The first shot is carried out in instantaneous mode to clear temporary faults before they cause damage to the lines. The three later ones operate in a timed manner with predetermined time settings. If the fault is permanent, the time delay operation allows other protection devices nearer to the fault to open, limiting the amount of the network being disconnected.

Earth faults are less severe than phase faults and, therefore, it is important that the recloser has an appropriate sensitivity to detect them. One method is to use CTs connected residually so that the resultant residual current under normal conditions is approximately zero. The recloser should operate when the residual current exceeds the setting value, as would occur during earth faults.

Typically, reclosers are designed to have up to three open-close operations and, after these, a final open operation to lock out the sequence. One further closing operation by manual means is usually allowed. The counting mechanisms register

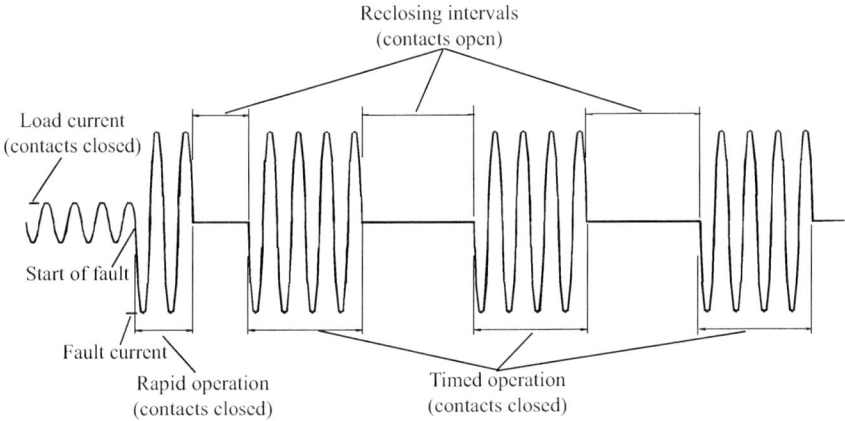

Figure 9.12   Typical sequence for recloser operation

operations of the phase or ground-fault units which can also be initiated by externally controlled devices when appropriate communication means are available.

9.3.1.2   Classification

Reclosers can be classified as follows:

- number of phases: single phase and three phase;
- arc-interrupting medium: oil or vacuum;
- type of insulation: oil, epoxy, or SF_6; and
- type of control: hydraulic, electronic, or magnetic actuator.

Single phase reclosers are used when the load is predominantly single phase. In such a case, when a single phase fault occurs the recloser should permanently disconnect the faulted phase so that supplies are maintained on the other phases. Three-phase reclosers are used when it is necessary to disconnect all three phases in order to prevent unbalanced loading on the system.

Reclosers with hydraulic operating mechanisms have a disconnecting coil in series with the line. When the current exceeds the setting value, the coil attracts a piston which opens the recloser main contacts and interrupts the circuit. The time characteristic and operating sequence of the recloser are dependent on the flow of oil in different chambers. The electronic type of control mechanism is normally located outside the recloser and receives current signals from a CT-type bushing. When the current exceeds the predetermined setting, a delayed shot is initiated which finally results in a tripping signal being transmitted to the recloser control mechanism. The control circuit determines the subsequent opening and closing of the mechanism, depending on its setting. Reclosers with electronic operating mechanisms use a coil or motor mechanism to close the contacts. Oil reclosers use the oil to extinguish the arc and also to act as the basic insulation. The same oil can be used in the control mechanism. Vacuum and SF_6 reclosers have the advantage of requiring less maintenance.

The following figures illustrate several features of reclosers. Figure 9.13 shows different types of single phase reclosers: (a) NOJA OSM – vacuum interrupting, epoxy insulated, and electronically controlled; (b) COOPER NOVA – oil interrupting, epoxy insulated, and electronically controlled; (c) COOPER D – oil interrupting, oil insulated, and hydraulically controlled.

Figure 9.14 shows different types of three-phase reclosers: (a) G&W Viper-LT – vacuum interrupting, epoxy insulated, and electronically controlled; (b) SCHNEI DER U – vacuum interrupting, epoxy insulated, and electronically controlled; (c) ABB OVR – vacuum interrupting, epoxy insulated, and electronically controlled; (d) Hawker Siddeley Switchgear Ltd's GVR – vacuum interrupting, SF_6 insulated, and magnetic actuator.

Figure 9.15 presents the internal components of a three-phase recloser – air-insulated, vacuum interrupting, electronically controlled recloser – manufactured by Cooper Systems.

The coordination margins with hydraulic reclosers depend upon the type of equipment used. In small reclosers, where the current coil and its piston produce the opening of the contacts, a separation greater than 12 cycles ensures non-simultaneous operation.

With large capacity reclosers, the piston associated with the current coil only actuates the opening mechanism. In such cases, a separation of more than eight cycles guarantees non-simultaneous operation.

The principle of coordination between two large units in series is based on the time of separation between the operating characteristics in the same way as for small units.

Electronically controlled reclosers can be coordinated more closely since there are no inherent errors such as those which exist with electromechanical mechanisms (due to overspeed, inertia, etc.). The downstream recloser must be faster than the upstream recloser, and the clearance time of the downstream recloser plus its tolerance should be lower than the upstream recloser clearance time less its tolerance.

(a)   (b)   (c)

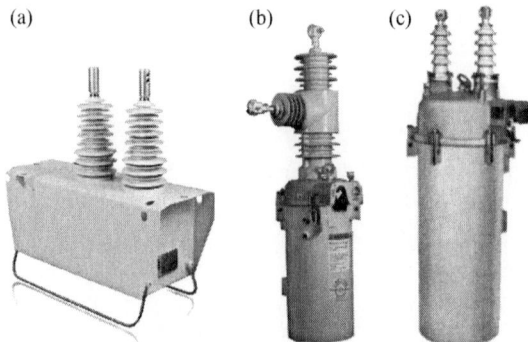

Figure 9.13   Single phase reclosers

(a)　(b)

(c)　(d)

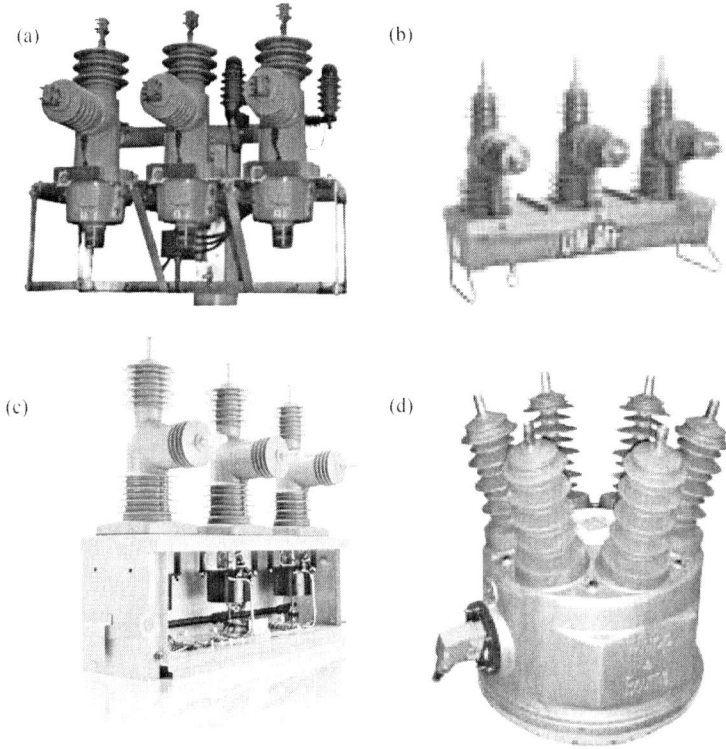

Figure 9.14　Three-phase reclosers

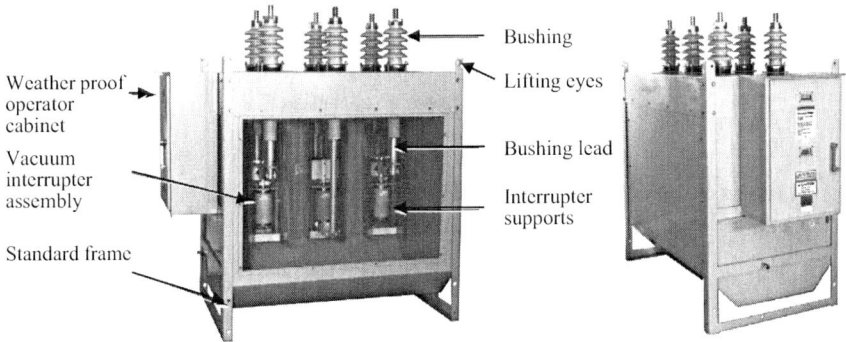

Weather proof operator cabinet

Vacuum interrupter assembly

Standard frame

Bushing

Lifting eyes

Bushing lead

Interrupter supports

Figure 9.15　COOPER Kyle Type VSA20A

Normally, the setting of the recloser at the substation is used to achieve at least one fast reclosure in order to clear temporary faults on the line between the substation and the load recloser. The latter should be set with the same, or a larger, number of rapid operations as the recloser at the substation.

It should be noted that the criteria of spacing between the time-current characteristics of electronically controlled reclosers are different to those used for hydraulically controlled reclosers.

9.3.1.3   Applications

Reclosers are used at the following points on a distribution network:

*   in substations, to provide primary protection for a circuit;
*   in main feeder circuits, in order to permit the sectioning of long lines and thus prevent the loss of a complete circuit due to a fault toward the end of the circuit; and
*   in branches or spurs, to prevent the tripping of the main circuit due to faults on the spurs.

Figure 9.16 illustrates several options to locate reclosers in a distribution system.

9.3.1.4   Specifications

The voltage rating and the short circuit capacity of the recloser should be equal to, or greater than, the values which exist at the point of installation. The same criteria should be applied to the current capability of the recloser in respect of the maximum load current to be carried by the circuit. It is also necessary to ensure that the

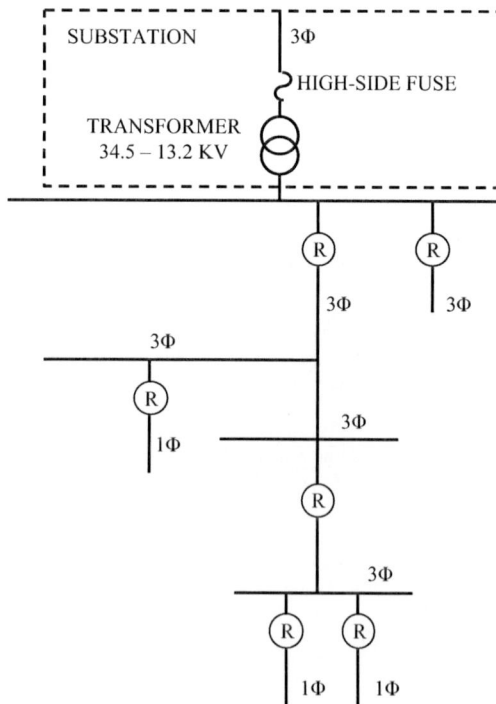

Figure 9.16   Option to locate reclosers

fault current at the end of the line being protected is high enough to cause operation of the recloser.

To properly apply automatic circuit reclosers, five major factors must be considered:

1. System voltage: System voltage will be known. The recloser must have a voltage rating equal to (or greater than) system voltage.
2. Maximum fault current available at the recloser location: Maximum fault current will be known or can be calculated. The recloser interrupting must be equal to (or greater than) the maximum available fault current at the recloser location.
3. Maximum load current: The recloser continuous current rating must be equal to (or greater than) anticipated circuit load. For series-coil-type reclosers, the coil size can be selected to match the present load current, the anticipated future load current, or the substation transformer capacity. Minimum-trip current is nominally twice the coil continuous-current rating. For electronically controlled reclosers, minimum-trip current must be greater than any anticipated peak load. Generally, a trip-current value of at least twice the expected load current is used.
4. Minimum-fault current within the zone to be protected: Minimum fault current that might occur at the end of the line section must be checked to confirm that the recloser will sense and interrupt this current.
5. Coordination with other protective devices on both the source and the load sides of the recloser.

After the first four application factors have been satisfied, coordination of the recloser with both the source and the load-side devices must be determined. Proper selection of time delays and sequences is vital to assure that any momentary interruption or longer term outage due to faults is restricted to the smallest possible section of the system.

Generally, recloser timing and sequences are selected to coordinate with the source-side devices. After the size and sequence of the required recloser has been determined, the protective equipment farther down the line is selected to coordinate with it.

9.3.2   Sectionalizers

9.3.2.1   General

A sectionalizer is a device which automatically isolates faulted sections of a distribution circuit once an upstream breaker or recloser has interrupted the fault current and is usually installed downstream of a recloser. Since sectionalizers have no capacity to break fault current, they must be used with a backup device which has fault current breaking capacity.

Sectionalizers will open to isolate a faulted area after a predefined number of "counts" with one count defined as seeing passage of fault current followed by a loss of voltage. They can be closed to pick up load, either remotely or via a local control panel or an operating handle actuated with a switching stick.

Sectionalizers count the number of operations of the recloser during fault conditions. After a preselected number of recloser openings, and while the recloser is open, the sectionalizer opens and isolates the faulty section of line. This permits the recloser to close and re-establish supplies to those areas free of faults. If the fault is temporary, the operating mechanism of the sectionalizer is reset. Sectionalizers are constructed in single- or three-phase arrangements with hydraulic or electronic operating mechanisms. A sectionalizer does not have a current-time operating characteristic, and can be used between two protective devices whose operating curves are very close and where an additional step in coordination is not practicable.

Sectionalizers with hydraulic operating mechanisms have an operating coil in series with the line. Each time an overcurrent occurs the coil drives a piston which activates a counting mechanism when the circuit is opened and the current is zero by the displacement of oil across the chambers of the sectionalizer. After a pre-arranged number of circuit openings, the sectionalizer contacts are opened by means of pretensioned springs.

This type of sectionalizer can be closed manually. Sectionalizers with electronic operating mechanisms are more flexible in operation and easier to set. The load current is measured by means of CTs and the secondary current is fed to a control circuit which counts the number of operations of the recloser or the associated interrupter and then sends a tripping signal to the opening mechanism. This type of sectionalizer is constructed with manual or motor closing.

9.3.2.2   Classification

Reclosers can be classified as follows:

- number of phases: single phase and three phase;
- type of insulation: oil, epoxy, or SF_6; and
- type of control: hydraulic, electronic, or magnetic actuator.

Figure 9.17 shows different types of sectionalizers: (a) COOPER GH – single phase oil insulated and hydraulic controlled; (b) Hawker Siddeley Switchgear Ltd's GVS 38 – three-phase SF_6 insulated and magnetic actuator; (c) COOPER GN3VE – three-phase oil insulated and hydraulic controlled. Figure 9.18 shows the components of a three-phase sectionalizer.

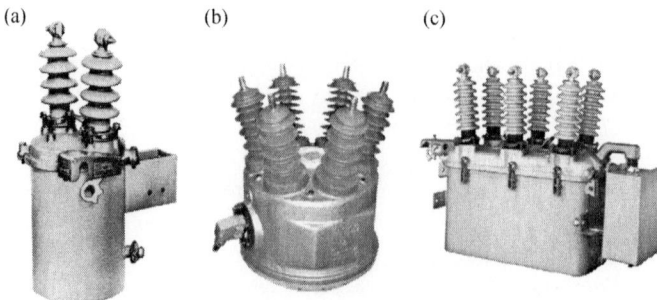

(a)   (b)   (c)

Figure 9.17   Illustration of sectionalizers

Universal Clamp – Type Terminals
Accept #6 solid through 350 MCM
stranded copper or aluminum conductors
In horizontal or vertical position.

Cover-Clamped 27 kv Bushing
Wet process porcelain, can be
replaced in the field.

Manual Operating Handle
Permits manual opening and closing,
indicates contact position.

Head casting
Cast aluminum, supports bushing
and operating mechanism.
o-Ring Gasket
Confined under controlled
compression to provide an
effective seal between head
casting and tank.

Sleet Hood
Protects manual operating handle,
allows easy access with switch stick.

Mechanism Trip Rod
Easily adjusted to open sectionalizer
after 1, 2, or 3 overcurrent counts.

Support Plate
Provides mounting for contact
mechanism.

By-pass Gap
Protects series-operating coil from
lighting surges.

Stationary Contacts
With arcing tips, easily
accessible for inspection,
cleaning, or replacement.

Moving Contacts
Wedge-shaped for positive engagement, easily
accessible for inspection and maintenance.

Actuating Coil and Counting
Mechanism Cover-Clamped
27 kv Bushing

On each phase, counts
overcurrent interruptions and
opens sectionalizer after
preselected number of counts.

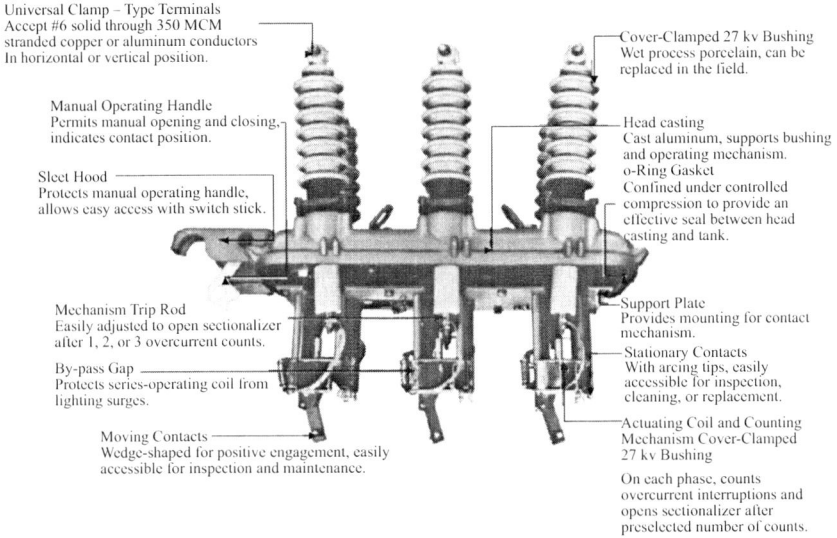

Figure 9.18   Untanked view of a sectionalizer

9.3.2.3   Specifications

The nominal voltage and current of a sectionalizer should be equal or greater than the maximum values of voltage or load at the point of installation. The short circuit capacity (momentary rating) of a sectionalizer should be equal to or greater than the fault level at the point of installation. The maximum clearance time of the associated interrupter should not be permitted to exceed the short circuit rating of the sectionalizer. Coordination factors that need to be taken into account include the starting current setting and the number of operations of the associated interrupter before opening.

The following basic coordinating principles should be observed in the application of the electronically controlled sectionalizer.

1. The minimum actuating current setting of the sectionalizer should not be greater than 80% of the minimum trip current setting of the backup recloser or reclosing breaker for both phase and ground currents.

2. The counts-to-open setting of the sectionalizer must be at least one less than the number of operations to lock out of the backup device.

3. The count reset time of the sectionalizer control must be greater than the reset time of the backup protective device.

4. The minimum clearing time of the backup device must be greater than the charging time of the trip-energy storage capacitors to assure that the capacitors are fully charged before the backup trips.

5. Three-phase sectionalizers must be used with backup breakers or reclosers in which all three phases open simultaneously. The counting functions of the sectionalizer do not recognize a signal as originating in a particular phase, but

total the overcurrent interruptions in all three phases. Non-simultaneous three-phase tripping of the backup could result in the sectionalizer interrupting fault current in one or more phases.

6.   Application on multigrounded wye systems generally requires ground fault sensing and inrush current restraint. Setting the phase actuating level to the ground setting of the backup device may result in erroneous counts due to inrush currents and incorrect opening of the sectionalizer for source-side faults.

7.   Ground fault actuating settings should be set at or above peak loading at the sectionalizer location. This will prevent sectionalizer sensing system unbalance during upline phase-to-ground faults which generate system imbalances.

9.3.2.4   Applications

Sectionalizers can be applied where the main feeder is divided into two feeders close to the substation as shown in Figure 9.19.

Here, loads are likely to be reasonably equally divided. Sectionalizers may also be applied to protect an important branch which may be carrying only a small portion of the total load as shown in Figure 9.20.

The inrush current restraint has been designed to provide optimum protection from inrush currents. No settings are required as pickup values will be blocked during inrush conditions. This allows flexibility in the application without concern for increased connected loading on the system.

The sectionalizer is blocked for three seconds to allow for decay of magnetizing inrush, system imbalance, and initial inductive loading after long reclose intervals.

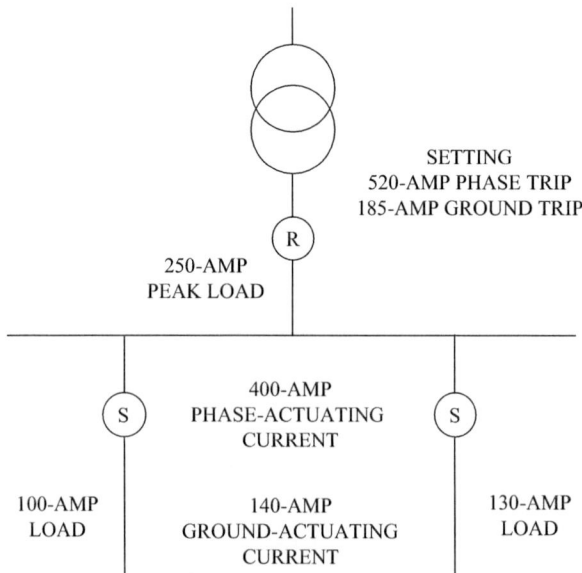

Figure 9.19   Sectionalizer application where two feeders of system are protected

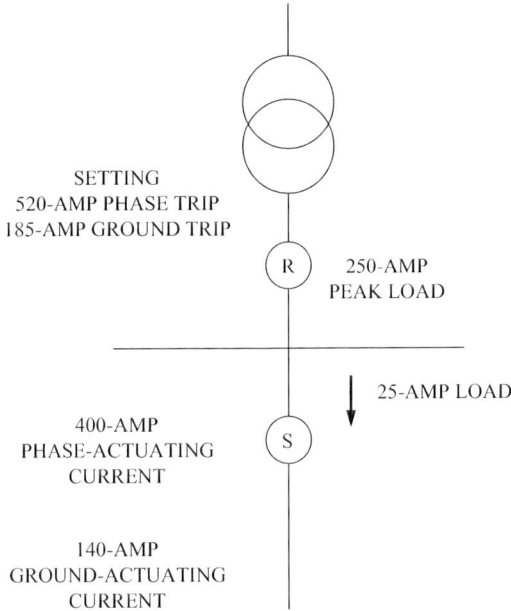

Figure 9.20   Sectionalizer application where one branch system is protected

9.3.3   Fuses

9.3.3.1   General

A fuse is an overcurrent protection device; it possesses an element which is directly heated by the passage of current and which is destroyed when the current exceeds a predetermined value. A suitably selected fuse should open the circuit by the destruction of the fuse element, eliminate the arc established during the destruction of the element, and then maintain circuit conditions open with nominal voltage applied to its terminals (i.e., no arcing across the fuse element).

The majority of fuses used in distribution systems operate on the expulsion principle, i.e. they have a tube to confine the arc, with the interior covered with de-ionizing fiber, and a fusible element. In the presence of a fault, the interior fiber is heated up when the fusible element melts and produces de-ionizing gases which accumulate in the tube. The arc is compressed and expelled out of the tube; in addition, the escape of gas from the ends of the tube causes the particles which sustain the arc to be expelled. In this way, the arc is extinguished when current zero is reached. The presence of de-ionizing gases, and the turbulence within the tube, ensures that the fault current is not re-established after the current passes through zero point. The zone of operation is limited by two factors: the lower limit based on the minimum time required for the fusing of the element (minimum melting time) with the upper limit determined by the maximum total time which the fuse takes to clear the fault.

9.3.3.2   Applications

Fuses are the most popular protective device used in electrical systems. In distribution systems in particular they are applied in most elements. Specific applications have been developed for the following elements:

- distribution transformers,
- capacitors, and
- feeders.

9.3.3.3   Type

Power fuses: Power fuses provide reliable and economical protection for transformers and capacitor banks. They are normally in outdoor substations served, and they incorporate silver or nickel–chrome fusible elements.

Fuse cutouts: A fuse cutout is a combination of a fuse and a switch, used in primary overhead feeder lines and distribution transformers to protect from current surges and overloads. The main components of fuse cutouts are the cutout body, the fuse holder, and the fuse element.

The cutout body is an open "C"-shaped frame that supports the "fuse holder" and a porcelain insulator that electrically isolates the conductive portions of the assembly from the support to which the insulator is fastened.

The fuse holder, often called the "fuse tube", contains the interchangeable fuse element and also acts as a simple knife switch. When the contained fuse operates or blows, the fuse holder drops open, disengages the knife switch, and hangs from a hinge assembly.

The fuse element, or "fuse link," is the replaceable portion of the assembly that operates due to high electrical currents.

Current-limiting fuse: A current-limiting fuse is a fuse that abruptly introduces a high resistance to reduce current magnitude and duration, resulting in subsequent current interruption. This type of fuse significantly reduces the current amplitude and the energy released in the event of a short circuit.

These fuses develop a positive internal gap of high dielectric strength after circuit interruption, thus precluding destructive re-ignitions when exposed to full system voltage – such as are experienced with current-limiting fuses after clearing under low recovery-voltage conditions.

Type SM Power Fuses have helically coiled silver fusible elements that are of solderless construction and are surrounded by air. Because of this construction, the fusible element is free from mechanical and thermal stress and confining support, and therefore is not subject to damage – even by inrush currents that approach but do not exceed the fuse's minimum melting time-current characteristic curve. Current-limiting fuses, in contrast, have fusible elements which consist of a number of very fine diameter wires, or one or more perforated or notched ribbons, surrounded by, and in contact with, a filler material such as silica sand. Because of this construction, current-limiting fuses are susceptible to element damage caused by current surges that approach the fuse's minimum melting time-current characteristic curve. This damage may occur in one or more of the following ways: the fusible element may melt, but not completely separate because the molten metal is constrained by the filler material – resulting, possibly, in resolidification of the element with a different cross-sectional area.

One or more, but not all, of the parallel wires or ribbons of the fusible element may melt and separate.

The fusible element may break as a result of fatigue brought about by current cycling that can cause localized buckling from thermal expansion and contraction.

Damage to fusible elements of current-limiting fuses, as described above, may shift or alter their time-current characteristics, resulting in a loss of complete coordination between the fuse and other downstream overcurrent protective devices. Moreover, a damage current-limiting fuse element may melt due to an otherwise harmless inrush current, but the fuse may fail to clear the circuit due to insufficient power flow – with the fuse continuing to arc and burn internally due to load-current flow. Because of the potential for damage to the fusible element from inrush currents, and because of the effects of loading and manufacturing tolerances, current-limiting fuse manufacturers typically require that when applying such fuses, adjustments be made to the minimum melting time-current characteristic curves. These adjustments are referred to as "safety zones" or "set back allowances," and range from 25% in terms of time to 25% in terms of current. The latter can result in an adjustment of 250% or more in terms of time, depending on the slope of the time-current characteristic curve at the point where the safety zone or setback allowance is measured. Furthermore, most current-limiting fuses inherently have steep, relatively straight time-current characteristic curves which, together with the required large safety-zone or setback-allowance adjustments, force the selection of a current-limiting fuse ampere rating substantially greater than the transformer full-load current in order to withstand combined transformer-magnetizing and load inrush currents, and also to coordinate with secondary-side protective devices. The selection of such large fuse ampere ratings results in reduced protection for the transformer and possible impairment of coordination with upstream protective devices. Also, since high-ampere-rated current-limiting fuses typically require the use of two or three lower ampere-rated fuses connected in parallel, increased cost and space requirements may be countered.

The value of the limited cutoff current is determined as a function of the prospective current for current values in first half cycle where the short circuit current is limited as is shown in Figure 9.21.

The fuse elements or fuse links used in most distribution cutouts are tin or silver alloy fuse links that melt (or operate) when exposed to high current conditions. Ampere ratings of fuse elements vary from 1 A to 200 A.

In distribution systems, it is common to designate fuse links as K and T for fast and slow types, respectively, depending on the speed ratio.

The speed ratio is the ratio of minimum melt current which causes fuse operation at 0.1 sec to the minimum-melt current for 300 sec operation.

For the K link, a speed ratio (SR) of 6–8 is defined, and for a T link, 10–13.

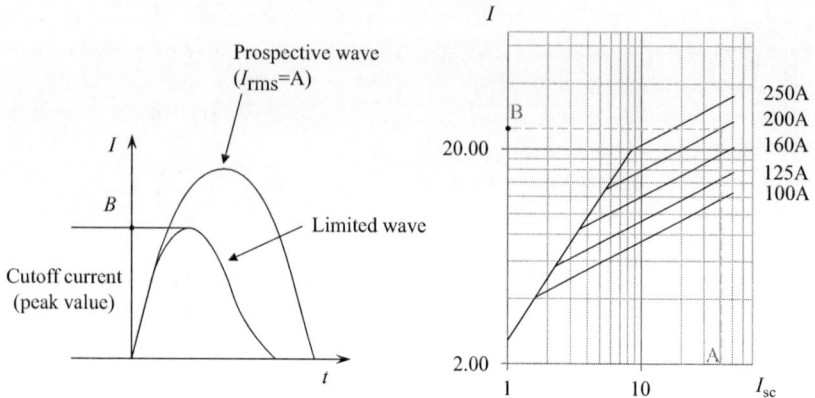

Figure 9.21   Sectionalizer application where one branch system is protected

9.3.3.4 Classification

There are a number of standards to classify fuses according to the rated voltages, rated currents, time-current characteristics, manufacturing features, and other considerations. For example, there are several sections of ANSI/UL 198-1982 Standards which cover low-voltage fuses of 600 V or less. For medium- and high-voltage fuses within the range 2.3–138 kV, standards such as ANSI/IEEE C37.40, 41, 42, 46, 47, and 48 apply. Other organizations and countries have their own standards; in addition, fuse manufacturers have their own classifications and designations.

In distribution systems, the use of fuse links designated K and T for fast and slow types, respectively, depending on the speed ratio, is very popular. The speed ratio is the ratio of minimum melt current which causes fuse operation at 0.1 sec to the minimum-melt current for 300 sec operation. For the K link, a speed ratio (SR) of 6–8 is defined and for a T link, 10–13. Figure 9.22 shows the comparative operating characteristics of type 200K and 200T fuse links. For the 200K fuse a 4400 A current is required for 0.1 sec clearance time and 560 A for 300 sec, giving an SR of 7.86. For the 200T fuse, 6500 A is required for 0.1 sec clearance, and 520 A for 300 sec; for this case, the SR is 12.5.

9.3.3.5 Specifications

The following information is required in order to select a suitable fuse for use on the distribution system:

1. voltage and insulation level,
2. type of system,
3. maximum short circuit level, and
4. load current.

The above four factors determine the fuse nominal current, voltage, and short circuit capability characteristics.

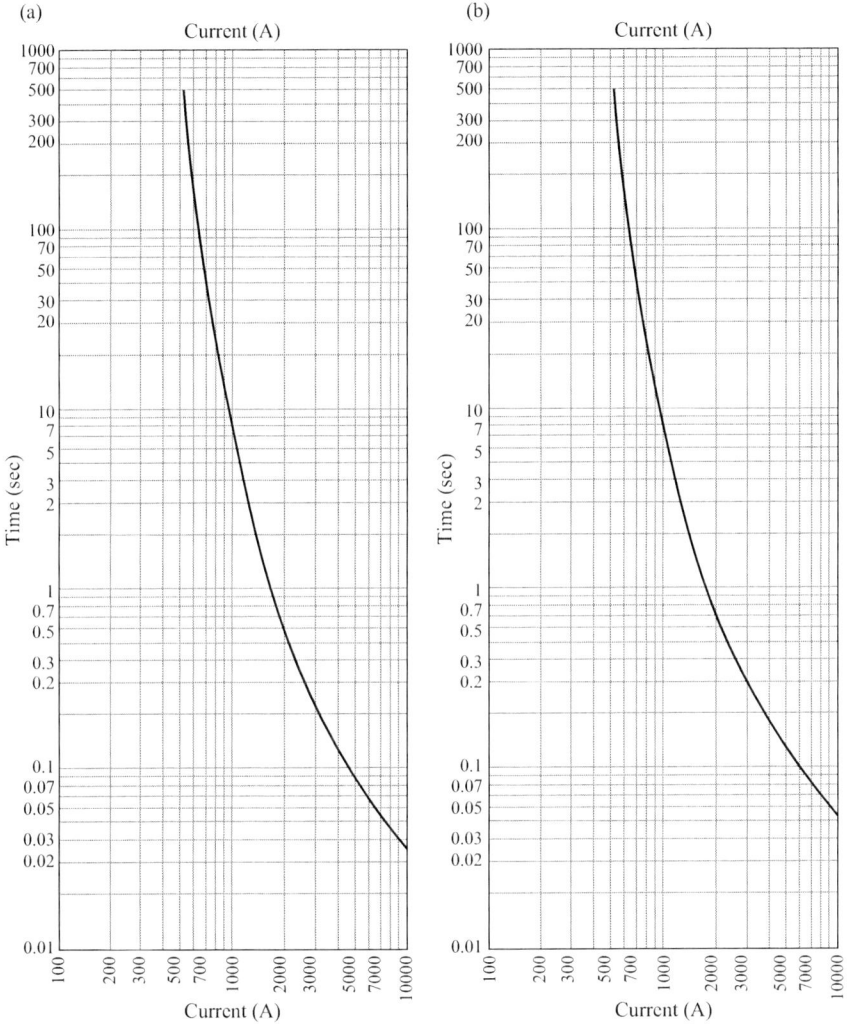

Figure 9.22　Characteristics of typical fuse links: (a) 200K fuse link; (b) 200T fuse link

Selection of nominal current

The nominal current of the fuse should be greater than the maximum continuous load current at which the fuse will operate. An overload percentage should be allowed according to the protected-equipment conditions. In the case of power transformers, fuses should be selected such that the time-current characteristic is above the inrush curve of the transformer and below its thermal limit. Some manufacturers have produced tables to assist in the proper fuse selection for different ratings and connection arrangements.

Selection of nominal voltage

The system characteristics determine the voltage seen by the fuse at the moment when the fault current is interrupted. Such a voltage should be equal to, or less than, the nominal voltage of the fuse. Therefore, the following criteria should be used:

- In unearthed systems, the nominal voltage should be equal to, or greater than, the maximum phase-to-phase voltage.
- In three-phase earthed systems, for single phase loads the nominal voltage should be equal to, or greater than, the maximum line-to-earth voltage and for three-phase loads the nominal voltage is selected on the basis of the line-to-line voltage.

Selection of short circuit capacity

The symmetrical short circuit capacity of the fuse should be equal to, or greater than, the symmetrical fault current calculated for the point of installation of the fuse.

Fuse notation

When two or more fuses are used on a system, the device nearest to the load is called the main protection, and that upstream, toward the source, is called the backup. The criteria for coordinating them will be discussed later.

9.4   Setting criteria

The following basic criteria should be employed when coordinating time-current devices in distribution systems:

1. The main protection should clear a permanent or temporary fault before the backup protection operates, or continue to operate until the circuit is disconnected. However, if the main protection is a fuse and the back-up protection is a recloser, it is normally acceptable to coordinate the fast operating curve or curves of the recloser to operate first, followed by the fuse, if the fault is not cleared (see section 9.4.2).
2. Loss of supply caused by permanent faults should be restricted to the smallest part of the system for the shortest time possible.

    Further in the section criteria and recommendations are given for the coordination of different devices used on distribution systems.

9.4.1   Fuse-fuse coordination

The essential criterion when using fuses is that the maximum clearance time for a main fuse should not exceed 75% of the minimum melting time of the backup fuse, as indicated in Figure 9.23. This ensures that the main fuse interrupts and clears the fault before the back-up fuse is affected in any way. The factor of 75% compensates for effects such as load current and ambient temperature, or fatigue in the fuse element caused by the heating effect of fault currents which have passed through the fuse to a fault downstream but which were not sufficiently large enough to melt the fuse. Keeping the coordination between fuses is the idea as is shown in Figure 9.24.

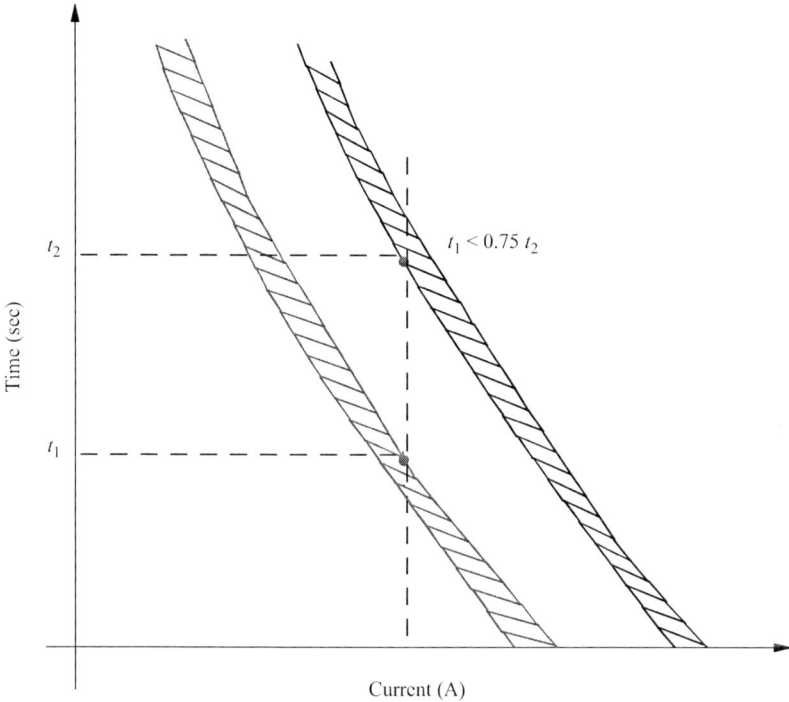

Figure 9.23   Criteria for fuse-fuse coordination

The series fuse-fuse coordination is given by manufacturers based on information which is presented normally in tables that list maximum fault current values that represent the intersection of the total clearing time-current characteristic curve of the load-sided fuse link with the minimum melting time-current characteristic curve of the source-side fuse link.

Typically the tables present the maximum fault current with source-side fuse links arranged horizontally and the load-side fuse links arranged vertically. Coordination between two fuses is guaranteed when the short circuit values are equal or lower to that given in the table for the intersection of the values corresponding to the two fuses.

Some manufacturers give the tables considering both preload and no preload. In the first case the curve is steeper as shown in the figure. Table 9.5 presents the coordination of S&C Standard Speed Fuse Links, based on preloading of source-side fuse links.

9.4.2   Recloser-fuse coordination

The criteria for determining recloser-fuse coordination depend on the relative locations of these devices, i.e. whether the fuse is at the source side and then backs up the operation of the recloser which is at the load side or vice versa. These possibilities are treated in the following paragraphs.

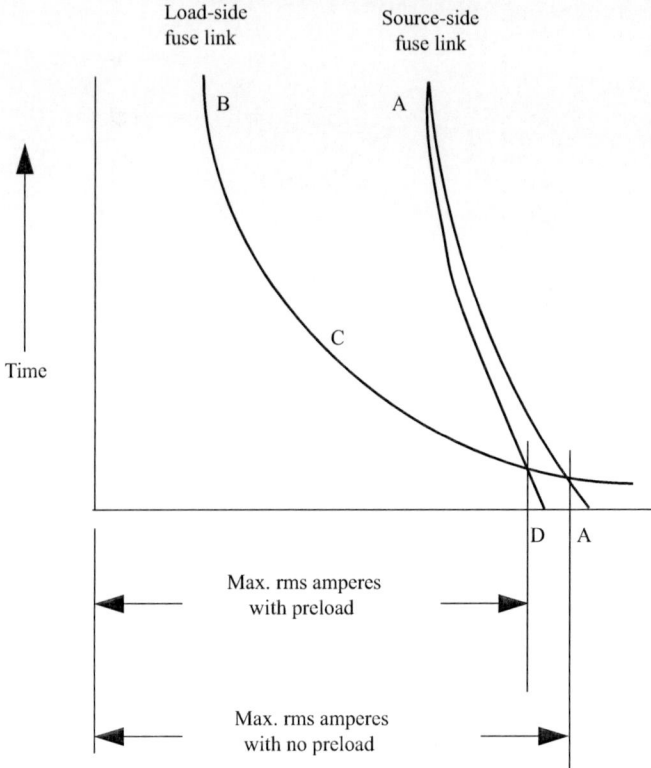

Figure 9.24   Time-current characteristics for fuse-fuse coordination

9.4.2.1   Fuse at the source side

When the fuse is at the source side, all the recloser operations should be faster than the minimum melting time of the fuse. This can be achieved through the use of multiplying factors on the recloser time-current curve to compensate for the fatigue of the fuse link produced by the cumulative heating effect generated by successive recloser operations. The recloser opening curve modified by the appropriate factor then becomes slower but, even so, should be faster than the fuse curve. This is illustrated in Figure 9.25.

The multiplying factors referred to above depend on the reclosing time in cycles and on the number of the reclosing attempts. Some values proposed by Cooper Power Systems are reproduced in Table 9.6.

It is convenient to mention that if the fuse is at the high-voltage side of a power transformer and the recloser at the low-voltage side, either the fuse or the recloser curve should be shifted horizontally on the current axis to allow for the transformer turns ratio. Normally it is easier to shift the fuse curve based on the transformer tap which produces the highest current on the high-voltage side. On the other hand, if the transformer connection group is delta-star, the considerations given in section 9.2 should be taken into account.

Table 9.5 Maximum fault current in amperes, rms in S&C Standard Speed Fuse Links

Source – Side fuse link ampere rating

5	7	10	15	20	25	30	40	50	65	80	100	101*	102*	103*	125	150	200
120	220	370	590	750	890	1100	1500	1850	2250	2800	3700	5200	8900	15000	4300	5500	7100
95	205	360	580	750	890	1100	1500	1850	2250	2800	3700	5200	8900	15000	4300	5500	7100
	175	335	570	740	880	1100	1500	1850	2250	2800	3700	5200	8900	15000	4300	5500	7100
	60	280	530	700	850	1050	1450	1800	2250	2800	3700	5200	8900	15000	4300	5500	7100
		170	490	680	830	1050	1450	1800	2250	2750	3700	5200	8900	15000	4300	5500	7100
			330	560	740	970	1400	1750	2200	2750	3700	5200	8900	15000	4250	5500	7100
				120	480	780	1250	1650	2100	2700	3650	5200	8900	15000	4250	5500	7100
						520	1100	1600	2050	2600	3600	5100	8900	15000	4200	5500	7100
							920	1450	1950	2550	3500	5100	8900	15000	4150	5400	7000
							560	1200	1750	2450	3450	5100	8900	14500	4050	5400	7000
								285	1300	2100	3200	5000	8900	14500	3850	5200	7000
									290	1550	2850	4800	8800	14500	3650	5100	6800
										365	2400	4650	8800	14000	3250	4800	6700
											1350	4300	8600	14000	2550	4350	6500
												3750	8500	13000		3100	6200
												2300	8200	6200			5200
													7600	13000			1650
														11500			3150

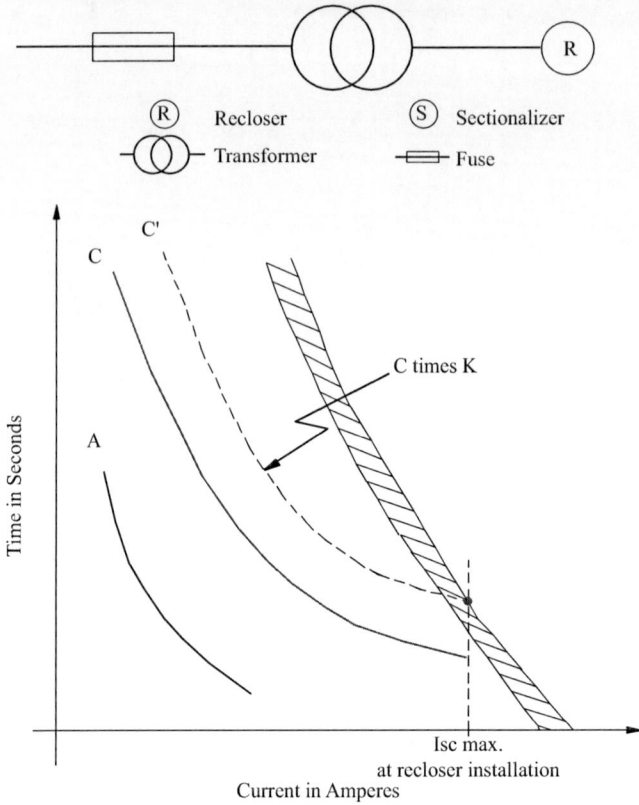

Figure 9.25   Criteria for source-side fuse and recloser coordination

Table 9.6   K factor for the source-side fuse link

Reclosing time in cycles	Multipliers for		
	Two fast, two delayed sequence	One fast, three delayed sequence	Four delayed sequence
25	2.7	3.2	3.7
30	2.6	3.1	3.5
50	2.1	2.5	2.7
90	1.85	2.1	2.2
120	1.7	1.8	1.9
240	1.4	1.4	1.45
600	1.35	1.35	1.35

Table 9.7 K factor for the load-side fuse link

Reclosing time in cycles	Multipliers for	
	One fast operations	Two fast operations
25–30	1.25	1.80
60	1.25	1.35
90	1.25	1.35
120	1.25	1.35

9.4.2.2 Fuses at the load side

The procedure to coordinate a recloser and a fuse, when the latter is at the load side, is carried out with the following rules:

- The minimum melting time of the fuse must be greater than the fast curve of the recloser times the multiplying factor, given in Table 9.7.
- The maximum clearing time of the fuse must be smaller than the delayed curve of the recloser without any multiplying factor; the recloser should have at least two or more delayed operations to prevent loss of service in case the recloser trips when the fuse operates.

The application of the two rules is illustrated in Figure 9.26.

Better coordination between a recloser and fuses is obtained by setting the recloser to give two instantaneous operations followed by two timed operations. In general, the first opening of a recloser will clear 80% of the temporary faults, while the second will clear a further 10%. The load fuses are set to operate before the third opening, clearing permanent faults. A less effective coordination is obtained using one instantaneous operation followed by three timed operations.

9.4.3 Recloser-sectionalizer coordination

Since the sectionalizers have no time-current operating characteristic, their coordination does not require an analysis of these curves.

The coordination criteria in this case are based upon the number of operations of the backup recloser. These operations can be any combination of rapid or timed shots as mentioned previously, e.g., two fast and two delayed. The sectionalizer should be set for one shot less than those of the recloser, e.g., three disconnections in this case. If a permanent fault occurs beyond the sectionalizer, the sectionalizer will open and isolate the fault after the third opening of the recloser. The recloser will then re-energize the section to restore the circuit. If additional sectionalizers are installed in series, the furthest recloser should be adjusted for a smaller number of counts. A fault beyond the last sectionalizer results in the operation of the recloser and the start of the counters in all the sectionalizers. Figure 9.27 shows an example of coordination between three sectionalizers and their setting.

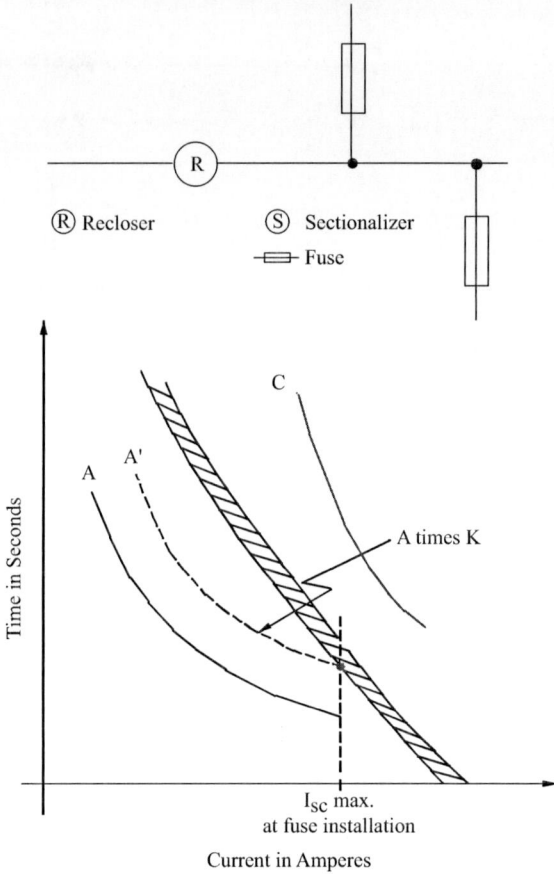

Figure 9.26   Criteria for load-side fuse and recloser coordination

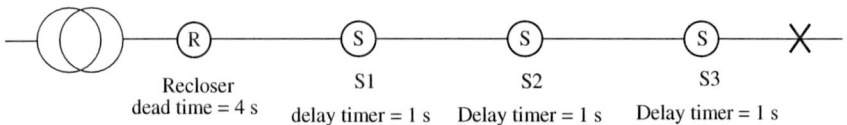

Figure 9.27   Coordination of one recloser with three sectionalizers

9.4.4   Recloser-sectionalizer-fuse coordination

Each one of the devices should be adjusted in order to coordinate with the recloser. In turn, the sequence of operation of the recloser should be adjusted in order to obtain the appropriate coordination for faults beyond the fuse by following the criteria already mentioned.

Figure 9.28 shows a portion of a 13.2 kV distribution feeder which is protected by a set of overcurrent relays at the substation location. A recloser and a sectionalizer have been installed downstream to improve the reliability of supply to customers.

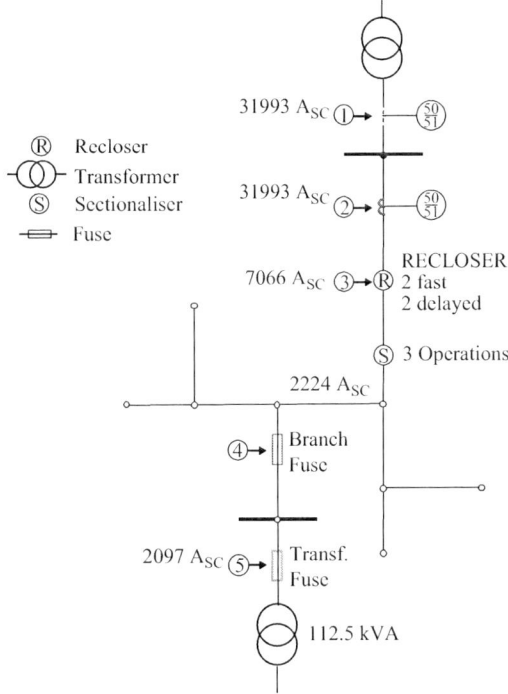

Figure 9.28   Portion of a distribution feeder

The recloser chosen has two fast and two delayed operations with 90 cycles intervals.

The time-current curves for the transformer and branch fuses, the recloser, and the relays are shown in Figure 9.28. For a fault at the distribution transformer, its fuse should operate first, being backed up by the recloser fast operating shots. If the fault is still not cleared, then the branch fuse should operate next followed by the delayed opening shots of the recloser and finally by the operation of the feeder relay. The sectionalizer will isolate the faulted section of the network after the full number of counts has elapsed, leaving that part of the feeder upstream still in service.

As the nominal current of the 112.5 kVA distribution transformer at 13.2 kV is 4.9 A, a 6T fuse was selected on the basis of allowing a 20% overload. The fast curve of the recloser was chosen with the help of the following expression based on the criteria already given, which guarantees that it lies in between the curves of both fuses:

$$t_{\text{recloser}} \times k \le t_{\text{MMT of branch fuse}} \times 0.75$$

where $t_{\text{MMT of branch fuse}}$ is the minimum melting time. The 0.75 factor is used in order to guarantee the coordination of the branch and transformer fuses, as indicated in section 9.4.1.

At the branch fuse location the short circuit current is 2224 A, which results in operation of the branch fuse in 0.02 sec. From Table 9.7, the K factor for two fast operations and a reclosing time of 90 cycles is 1.35. With these values, the maximum time for the recloser operation is $(0.02 \times 0.75/1.35) = 0.011$ sec.

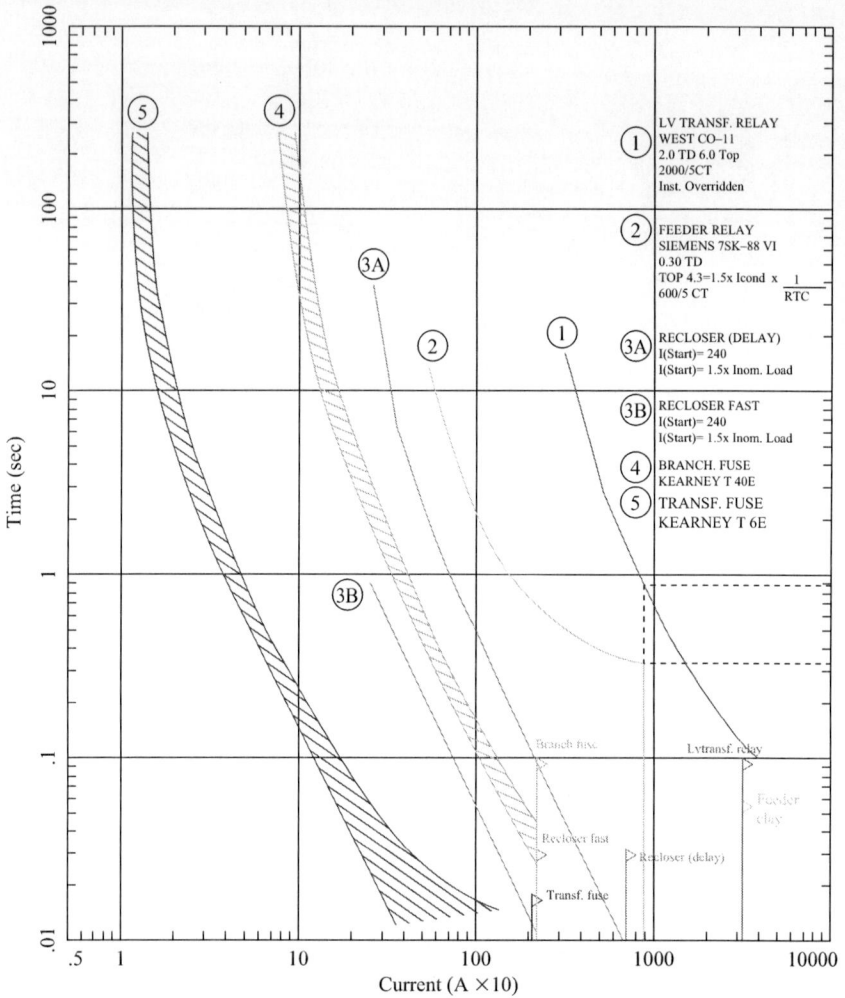

Figure 9.29   Phase-current curves

This time, and the pickup current of the recloser, determines the fast curve of the recloser.

The feeder relay curve is selected so that it is above that of the delayed curve of the recloser, and so that the relay reset time is considered. The curves of Figure 9.29 show that adequate coordination has been achieved.

9.4.5   Recloser-recloser coordination

The coordination between reclosers is obtained by appropriately selecting the amperes setting of the trip coil in the hydraulic reclosers or of the pickups in electronic reclosers.

9.4.6   *Recloser-relay coordination*

Two factors should be taken into account for the coordination of these devices; the interrupter opens the circuit some cycles after the associated relay trips, and the relay has to integrate the clearance time of the recloser. The reset time of the relay is normally long and, if the fault current is reapplied before the relay has completely reset, the relay will move toward its operating point from this partially reset position.

For example, consider a recloser with two fast and two delayed sequence with reclosing intervals of 2 sec, which is required to coordinate with an inverse time delay overcurrent relay which takes 0.6 sec to close its contacts at the fault level under question, and 16 sec to completely reset. The impulse margin time of the relay is neglected for the sake of this illustration. The rapid operating time of the recloser is 0.030 sec, and the delayed operating time is 0.30 sec. The percentage of the relay operation during which each of the two rapid recloser openings takes place is (0.03 sec/0.6 sec) × 100% = 5%. The percentage of relay reset which takes place during the recloser interval is (2 sec/16 sec) × 100% = 12.5%. Therefore, the relay completely resets after both of the two rapid openings of the recloser.

The percentage of the relay operation during the first time delay opening of the recloser is (0.3 sec/0.6 sec) × 100% = 50%. The relay reset for the third opening of the recloser = 12.5%, as previously, so that the net percentage of relay operation after the third opening of the recloser = 50%−12.5 % = 37.5%. The percentage of the relay operation during the second time delay opening of the recloser takes place = (0.3 sec/0.6 sec) × 100% = 50%, and the total percentage of the relay operation after the fourth opening of the recloser = 37.5%+50% = 87.5%.

From the above analysis it can be concluded that the relay does not reach 100% operation by the time the final opening shot starts, and therefore coordination is guaranteed.

9.5   Protection considerations when distributed generation is available

As previously indicated, distributed generation (DG) brings about important benefits to the operation of distribution systems. The main benefit, of course, is the possibility of having generation at the user level that increases service reliability. If the source comes from green power, not only the price is lowered but also the pollution emission. However, important considerations have to be done to make sure that the generation is properly handled as follows.

9.5.1   *Short circuit levels*

Short circuit levels increase along the feeders with the DG, which increase the withstand capabilities of breakers, sectionalizers, reclosers, capacitors, etc. In particular extra care has to be exercised with breakers and reclosers to make sure that their duties are above the maximum total short circuit currents.

9.5.2   Synchronization

The reclosing functionality at the substation that feeds distribution lines should be disabled if there is not a proper way to open the generators connected along the feeders upon the occurrence and clarification of a fault. This prevents the possibility of energizing a line with generation at the other end without following an appropriate synchronization procedure. This of course reduces the flexibility of the operation but is required to avoid accidents that could be fatal.

9.5.3   Overcurrent protection

The use of overcurrent relays should be examined as the short circuit currents can flow in and out of the substation if distributed generators are present. In this case it could be convenient to replace them with directional overcurrent protection to achieve a better coordination.

9.5.4   Adaptive protection

Due to the inherent nature of distributed generation, machines can be on and off during normal operation. This is even more possible if the generation is from solar or wind power. In these cases, the relays should accommodate to different topologies that make their operation risky or too slow unless adaptive protection is used. Most numerical relays have four or more setting groups that could be selected according to equal number of operating scenarios.

Microgrids or grids that contain secondary local generation are a great case where adaptive protection should be applied. From the diagram below you can see that two sources are present. During normal operation the utility grid that operates the relays require one set of protection settings. When the utility grid is interrupted the diesel generator starts. In order to protect the system the relays must change settings in order to be coordinated correctly. Having adaptive protection will allow for continuous protection of the distribution system regardless of the source (Figure 9.30).

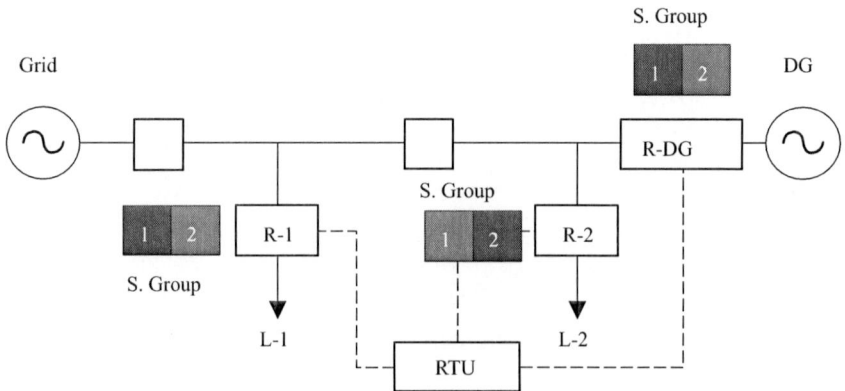

Figure 9.30   Example of adaptive protection setting with an RTU device

Proposed exercises

1. Calculate the pickup setting, time dial setting, and the instantaneous setting of the phase relays installed in the high-voltage and low-voltage sides of the 115/13.2 kV transformers T1 and T3 in the substation illustrated in Figure 9.31.

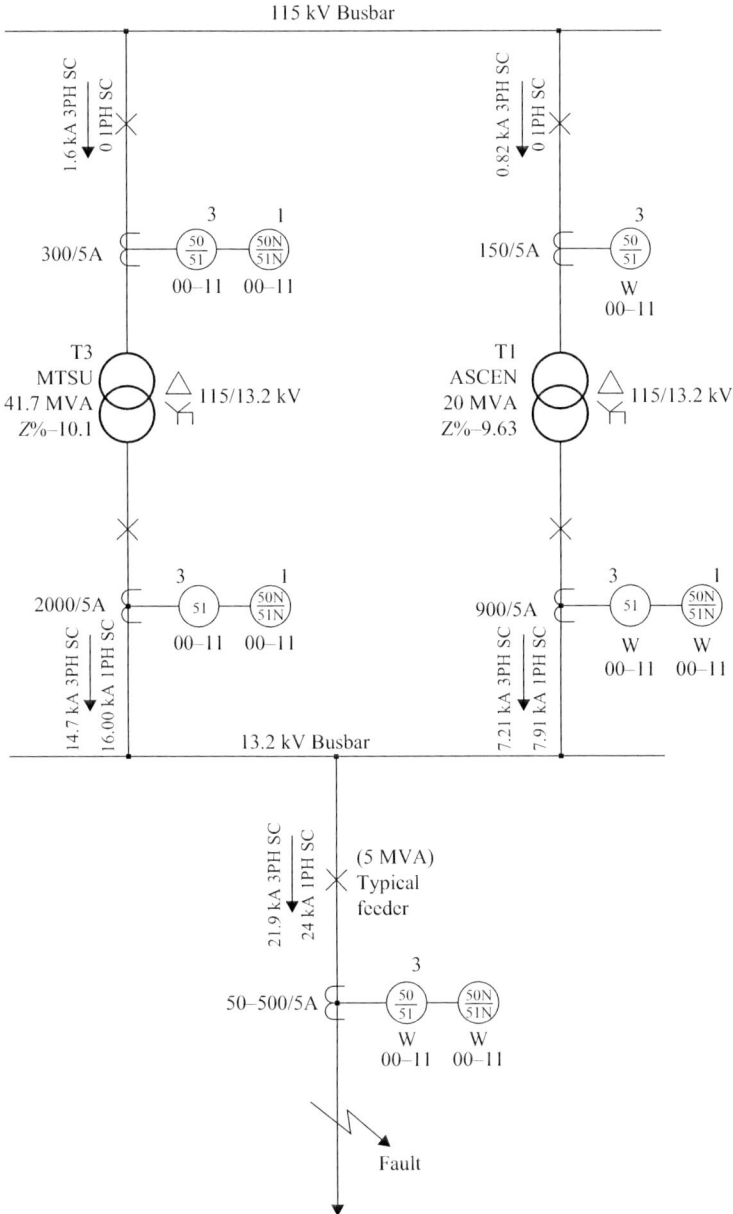

Figure 9.31 Single line diagram for exercise 9.1

The short circuit levels, CT ratios, and other data are shown in the same diagram.

2. For the system shown in Figure 9.32, carry out the following calculations:
    (a) The maximum values of short circuit current for three-phase faults at busbars A, B, and C, taking into account that busbar D has a fault level of 12906.89 A rms symmetrical (2570.87 MVA).
        • The maximum peak values to which breakers 1, 5, and 8 can be subjected.
        • The rms asymmetrical values which breakers 1, 5, and 8 can withstand for five cycles for guarantee purposes.
          For these calculations assume that the *L/R* ratio is 0.2.
    (b) The turns ratios of the CTs associated with breakers 1–8. The CT in breaker 6 is 100/5. Take into account that the secondaries are rated at 5 A

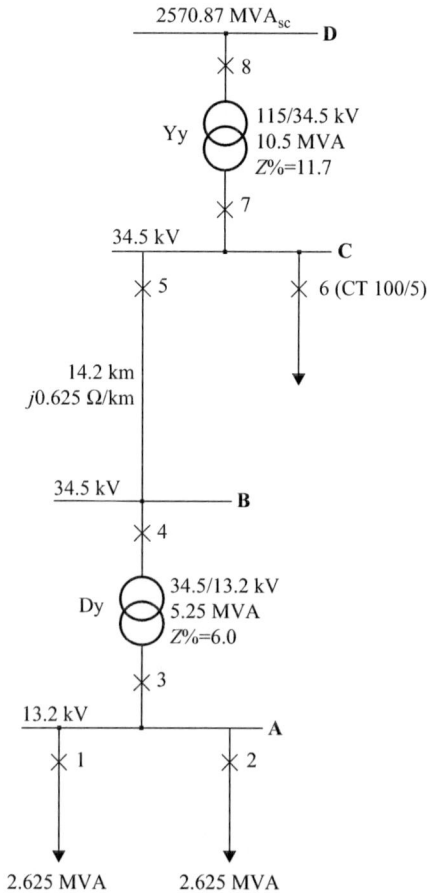

2570.87 MVA$_{sc}$ D

8

Yy   115/34.5 kV
      10.5 MVA
      Z%=11.7

7

34.5 kV

5   C

6 (CT 100/5)

14.2 km
j0.625 Ω/km

34.5 kV   B

4

Dy   34.5/13.2 kV
      5.25 MVA
      Z%=6.0

3

13.2 kV   A

1   2

2.625 MVA   2.625 MVA

Figure 9.32   Single line diagram for exercise 2

and that the ratios available in the primaries are multiples of 50 up to 400, and from then on are in multiples of 100.

(c)   The instantaneous, pickup and time dial settings for the phase relays in order to guarantee a coordinated protection system, allowing a time discrimination margin of 0.4 sec.

(d)   The percentage of the 34.5 kV line which is protected by the instantaneous element of the overcurrent relay associated with breaker 5.

Bear in mind the following additional information:

- The settings of relay 6 are as follows: pickup 7 A, time dial 5, instantaneous setting 1000 A primary current.
- All the relays are inverse time type, with the following characteristics:
  Pickup: 1–12 in steps of 2 A
  Time dial: as in Figure 9.4
  Instantaneous element: 6–144 in steps of 1 A

Calculate the setting of the instantaneous elements of the relays associated with the feeders assuming 0.5 I_{sc} on busbar A.

Chapter 10

Communications in Smart Grids

For years, analog communication networks have provided channels that allowed long-distance information exchange. However, with the arrival of digital communications it was possible to employ physical mediums that have their own capacity to transfer large data packets in a reliable and efficient manner.

Digital communication networks and data services have provided solutions that optimize thousands of processes worldwide. In the smart network environments such solutions have proved successful. A good example of this is the application on protective relaying. It is known that protection functionalities have maintained the theoretical foundations since they were first developed.

However, the operation of protective relaying has been greatly improved in time response due to the availability of much faster relays based on the numerical technology as well as the wonderful development of communication capabilities that have been attained in recent years.

The standardization of protocols brought a convergent platform to implement solutions based on open and flexible solutions. However, since high levels of reliability, availability, and security are required by electrical system networks, the protocols needed to be adapted to offer lower levels of latency and larger broadband.

In the following sections reference will be made to important topics pertaining to the effect of communications on distribution automation and in general on Smart Grid, including the OSI model, the most popular protocols, and the transmission mediums. Special attention is given at the end of the chapter to the IEC 61850 Standard, considering its huge impact in all aspects of power system automation handling.

10.1 ISO OSI model

The open system interconnection reference model (the OSI model) was developed in the late 1970s by the International Organization for Standardization (ISO) for open systems interface. It has served as a conceptual model to represent interactions and events necessary for establishing a communication channel between two data terminals.

Figure 10.1 shows the various layers. Each layer constitutes a stage where an activity takes place to store the information in segments, packets, branches, and bits until reaching the physical layer where the analog wave is transmitted with the

Data type   OSI layer

Data	Application
Data	Presentation
Data	Session
Segments	Transport
Packets	Network
Frames	Data Link
Bits	Physical

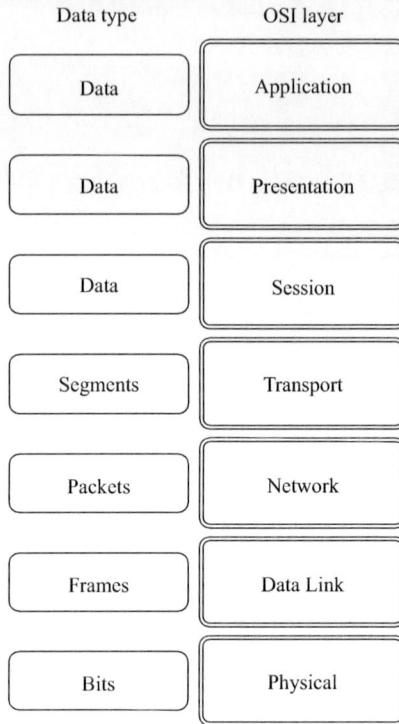

Figure 10.1   OSI model

corresponding digital information to its destination. The opposite process is used to interpret the message.

Figure 10.2 illustrates how the OSI model is applied to an Internet network. It is based on the TCP/IP protocols that are closely related to the OSI model. The intermediate communication elements such as switches and routers are illustrated in the center row. They enable addressing and using efficiently the network elements like switches and routers.

10.2   Communication solutions for the power system world

The spectrum of Smart Grid solutions increases and each new solution involves additional communication solutions for each particular case. Traditionally, communications like wired communication between field equipment, auxiliary services, and controllers, serial communication between intelligent electronic devices (IEDs), communication between equipment in different places, and communication through wide area network (WAN) for the integration of substations in a central control station are considered in the power system environment. Communications have been converted to the convergence of multiple designs appropriate to each

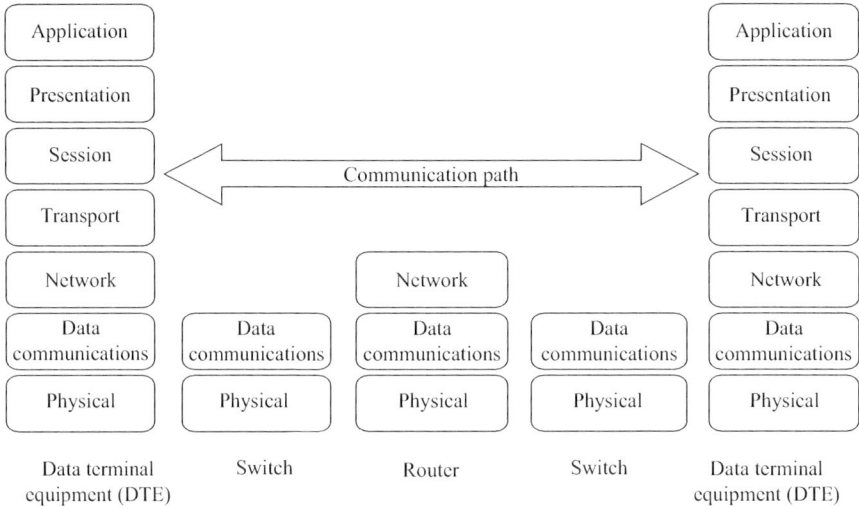

Figure 10.2   TCP/IP link applying OSI model

domain in an information model offering interoperability between systems and a distributed computing system to reach optimum levels of usage. Consequently, standards such as Common Information Model (CIM), specified in IEC 61968 and IEC 61970, were developed.

Figure 10.3 illustrates sample communication solutions applied to Smart Grid domains. This figure shows a communication infrastructure for Smart Grid enables utilities to interact with their devices and grid systems as well as with customers and facilities for distributed generation and energy storage. To fully achieve the vision of a Smart Grid, companies need to support multiple communication networks: home area network (HAN) for energy efficiency on the client side, neighborhood area network (NAN) for advanced measurement applications, and wide area network (WAN) for distributed automation and Smart Grid backbone.

Communication applications have evolved rapidly in Smart Grid, particularly in two areas: intelligent metering infrastructure and distribution network, and substation communications.

10.2.1   Communication solutions in AMI

Advanced metering infrastructure (AMI) provides multiple solutions like phone lines, GPRS/3G/4G, PLC, fiber optic, Ethernet which could employ traditional proprietary protocols, and due to the Smart Grid demands, turns them into DLMS/COSEM open standards for data interchange with metering equipment.

DLMS (Device Language Message Specification) is an application level protocol integrated in IEC 62056 for device message specification. Similarly, COSEM is defined as companion specification for energy metering.

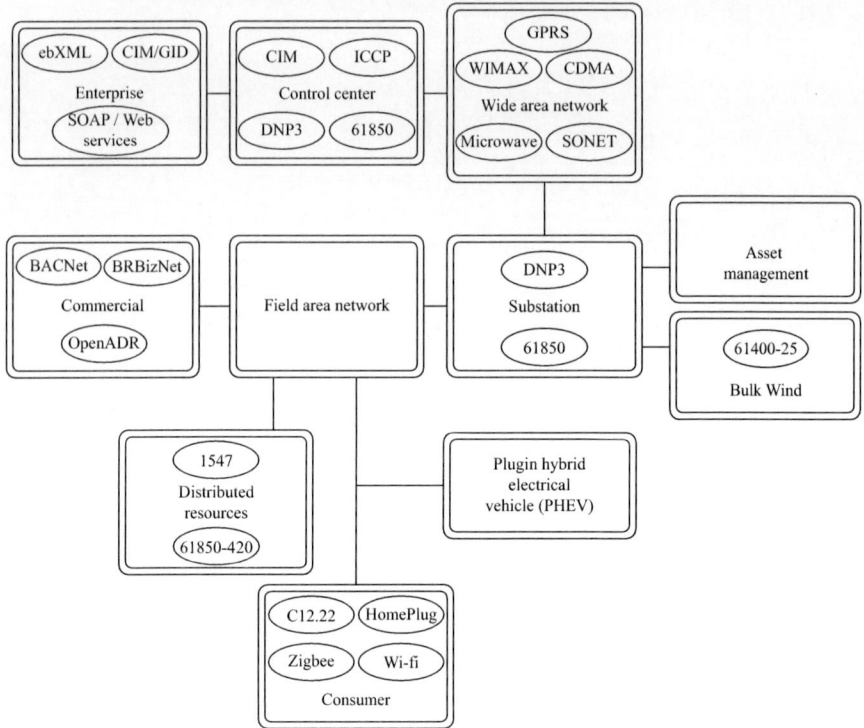

Figure 10.3   Communication options in the Smart Grids

10.2.2   Distribution network communications

10.2.2.1   IEC 61850

One of the smart network objectives is the interoperability among systems, as is defined by the National Institute of Standards and Technology (NIST) Framework and Roadmap for Smart Grid Interoperability Standards, Release 1.0. Substation automation has been restricted to the solution offered by the various protection and control equipment manufacturers that needed to interface and integrate their units without restrictions. IEC started to develop a common standard for substation communications in 1994. Similarly, IEEE developed a common communication standard known as UCA (Utility Communication Architecture). In 1997, IEEE and IEC agreed to work in conjunction to create a common substation communication standard IEC 61850.

The IEC 61850 Standard in section 8-1 discusses the critical and non-critical data interchange method in time across local area networks (LANs) employing Abstract Communication Service Interface (ACSI) mapping in multimedia messaging (multimedia messaging service, MMS) over frames ISO/IEC 8802-3. Similarly, in section IEC 61850-9-2 the specific communication service mapping (SCSM) is defined for the transmission of sampled values according to the IEC 61850-7-2, also using frames ISO/IEC 8802-3.

Raw data samples	GOOSE	Time synchronization	Client server	Application

		UDP	TCP	Transport

		IP		Network

Ethernet				Link

Physical medium				Physical

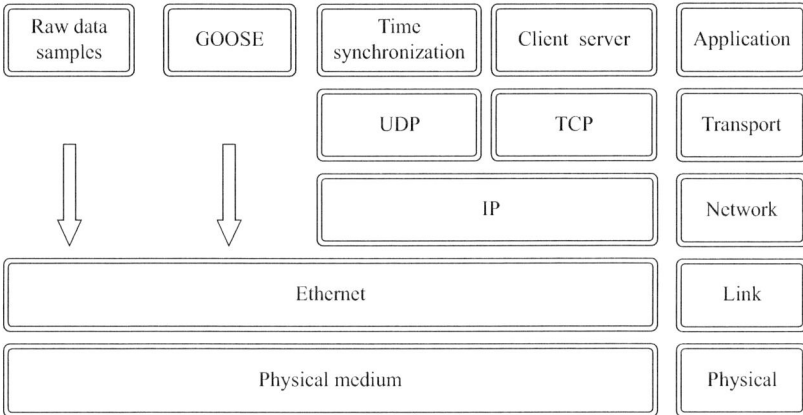

Figure 10.4   Message communication OSI-7 stack

Figure 10.4 illustrates how critical data like Generic Object Oriented Substation Events (GOOSE) messages and sampled values are directly mapped to lower Ethernet layers and how metered values or time synchronizations pass through the stack of TCP/IP protocols.

IEC 61850 provides the multicast-based generic substation events (GSE) as a way to quickly transfer event data over an entire substation network. Furthermore, part 9-1 and 9-2 specify a process bus for use with IEDs.

10.2.2.2   DNP3-IEEE Standard 1815

The IEEE 1815-2012 Standard defines the DNP3 protocol used for communication media employed in utility automation systems where it describes the structures, functions, or applications. The revision of the IEEE 1815-2010 Standard includes improved protocols that aid in preventing cyber security concerns particular to the communication media employed in utility automation systems.

10.2.2.3   IEC 60870-5 as the standard for remote control

The trend is to adopt IEC 61850 architecture for substation communications and remote control. IEC 60870 has experienced a massive expansion and wide application.

IEC 60870-5 is the standard covering telecontrol equipment and systems. Part 5 covers transmission protocols. The standard is well established worldwide.

The IEC 60870-5-104 or IEC 104 protocol is a standard based on IEC 60870-5-101 or IEC 101. It employs the TCP/IP network interface to provide LAN and WAN connectivity. The original IEC 101 itself is also retained for some particular data and services utilized.

The IEC 60870-5-101 and 104 parts are mainly used for exchanging information between substations and control centers. Application areas range from primary substations (high and medium levels of voltage) to the secondary substations (medium or low voltage).

10.3   Transmission mediums

The transmission methods supporting data transfer continue to be similar to what has been used for decades: wired, electric, wireless, and fiber optic. More recently, modulation technologies have come to be used as data highways although they were not designed for this purpose.

10.3.1   Wired and electric mediums

These are the traditional mediums used for transferring digital data like electricity across a conductor. Wired serial interfaces such as RS-232, RS-485, RS-449, V.35, G.703, RS-530, RS-422, although proved to be robust and reliable, are being replaced by IEC 61850-8-2 where the binary signals are sent across Ethernet cables in GOOSE messages utilizing IEC 61850-8-2 Standards. Similarly, metering data of electrical parameters collected and integrated in serial interfaces are being replaced in applications.

This classification includes the power system networks utilized for data transmission employing BPL (broadband over power lines), PLC (power line carrier), or the power line communication technology patented by Aclara Technologies where, instead of using wave carrier technology, the data are transmitted around the power wave zero crossing enabling data to flow across the entire distribution network without requiring bridges or jumpers in transformers or capacitor banks.

10.3.2   Wireless mediums

Frequencies from 2 to 16 GHz, usually called as microwaves, constitute a reliable route for data transmission from multiple integrated substations to control centers employing modern digital modulation and multiplication technologies.

Technologies like WiFi, Wimax, and other wireless applications are used in communication systems where data transmission is not time-critical. Even though there have been developments in modern digital modulation techniques at fast data transmission speeds. Their application is still limited for reliability considerations, which is influenced by propagation medium.

WiFi has been able to improve computer security in wireless networks using the IEEE 802.11i Standard (also known as WPA-2) with the implementation of Advanced Encryption Standards (AES).

Technologies like Zigbee and Bluetooth are considered viable for short-distance tasks like domotics, control applications in small industries or homes, and non-sensible data collection.

Zigbee works on top of the IEEE 802.15.4 Standard, in the unlicensed 2.4 GHz or 915/868 MHz bands. An important feature of Zigbee is the possibility of handling mesh-networking, thereby extending the range and making a Zigbee network self-healing.

Bluetooth is a communication protocol especially designed for low consumption devices that require a low emitting range and based on low-cost transceivers.

Bluetooth devices are divided into class 1, 2, and 3 in regards to their transmission power, where class 1 devices cover a range of 100 meters.

Solutions operating with cellular networks have been accepted by the power system sector to reach remote locations where using wired methods or even wireless methods have proved challenging. Due to their low reliability and cost, they are mainly used for telemetering or non-critical equipment control.

Satellite communications could eventually offer new remote supervision and control solutions in the transmission and distribution level.

10.3.3   Optical mediums

Traditionally, fiber optics has been employed in the electrical sector for fast data transfer. However, as a result of new developments, fiber optics could replace serial or point-to-point communications within substations implementing IEC 61850.

The use of optical fiber media has dramatically increased in the last 15 years. This is due to multi-vendor availability, much lower purchase and installation costs, dielectric characteristics, increased bandwidth potentials, and communication speed. Optical fibers are available as part of the overhead ground wire, in a self-supported dielectric conductor, as a messenger conductor, or designed for an underground environment. Utilities are using a variety of installation options to build a very reliable communication backbone with minimum common mode failures.

10.4   Information security as the crucial element in smart networks

It is understandable that in an interconnected world, reliability risks increase and security varies according to protocols and Smart Grid domains where they are implemented. Some security standards are applicable to certain protocols while others are applied to particular profiles.

When extremely sensible and critical data is transferred, security levels should be elevated to meet basic information security like reliability, confidentiality, and integrity.

In the electric field, associations like North American Electric Reliability Corporation (NERC) have developed the NERC 1300 Standard for power system security. These standards are divided into eight specific areas from CIP-002-1 to CIP-009-2 focused on establishing requirements for power system owners, operators, and users to guard critical assets with the best security practices. Also, IEC 62351 or FIPS180-4, 186-3 must be mentioned. For example, IEC 62351 is developed for handling the security of protocols including IEC 60870-5 series, IEC 60870-6 series, IEC 61850 series, IEC 61970 series and IEC 61968 series.

It must be noted that cyber security standards in the application levels are part of applicable models in the computer sector, such as ISO 27002 or when employing the common critical model known as Common Criteria for Information Technology Security Evaluation, ISO 15408.

10.5   IEC 61850 overview

Relay applications but also to all aspects of power system automation handling.

Substations designed in the past made use of protection and control schemes implemented with single-function, electromechanical or static devices, and hard-wired relay logic. SCADA functions were centralized and limited to monitoring of circuit loadings, bus voltages, aggregated alarms, control of circuit breakers and tap changers, etc. Disturbance recording and sequence-of-event data if available were centralized and local to the substation.

With the advent of microprocessor-based multi-function IEDs, more func-tionality into fewer devices was possible, resulting in simpler designs with reduced wiring. In addition, owing to communication capabilities of the IEDs more infor-mation could be accessed remotely, translating into fewer visits to the substation.

Microprocessor-based protection solutions have been successful because they offered substantial cost savings while fitting very well into pre-existing frameworks of relay application. A modern microprocessor-based IED replaces an entire panel of electromechanical relays with external wiring intact, and internal DC wiring replaced by integrated relay logic. Users retained total control over the degree of integration of various functions, while interoperability with the existing environment (instrument transformers, other relays, control switches, etc.) has been maintained using traditional hard-wired connections.

In terms of SCADA integration, the first generation of such systems achieved moderate success especially in cases where the end user could lock into a solution from a single vendor. Integrating systems made up of IEDs from multiple vendors invariably led to interoperability issues on the SCADA side. Integration solutions tended to be customized.

Owners of such systems were faced with long-term support and maintenance issues. During this period two leading protocols emerged: DNP 3.0 and IEC 60870.

Beginning in the early 1990s, initiatives were undertaken to develop a com-munication architecture that would facilitate the design of systems for protection, control, monitoring, and diagnostics in the substation. The primary goals were to simplify development of these multi-vendor substation automation systems and to achieve higher levels of integration reducing even further the amount of engi-neering and wiring required.

In 1994 EPRI/IEEE started a work UCA2 with focus on the station bus. In 1996, IEC TC57 (technical committee 57) began work on IEC 61850 to define a station bus. This led to a combined effort in 1997 to define an international standard that would merge the work of both groups that were focused on the development of a standard in which devices from all vendors could be connected together to share data, services, and functions. The result was the international IEC 61850 Standard Edition 1 "Communication Networks and Systems in Substation Automation."

The International IEC 61850 Standard was issued in 2005 and it was developed to control and protect power systems by standardizing the exchange of information between all IEDs within an automated substation and a remote control link. IEC 61850 provides a standardized framework for substation integration that specifies the communication requirements, the functional characteristics, the structure of data in devices, the naming conventions for the data, how applications interact and control the devices, and how conformity to the standard should be tested.

The development of the IEC 61850 Standard is continuing. This work was resulting in Edition 2 of the standard, which was published during 2010. IEC 61850 was originally defined exclusively for substation automation systems, but has since been extended to other application areas – as is reflected in its changed title "IEC 61850 Ed. 2. Communication Networks and Systems for Power Utility Automation."

The IEC 61850 is increasingly being used for the integration of electrical equipment into distributed control systems in process industries. The fact that new application areas such as hydro and wind power are being added is yet another indication of its success.

Some of the benefits of the IEC 61850 Standard are:

- reduced dependence on multiple protocols,
- higher degree of integration,
- reduced construction cost by eliminating most copper wiring,
- flexible programmable protection schemes,
- communication networks replacing hard-wired connections,
- advanced management capability,
- high-speed peer-to-peer communications,
- improved security/integrity, and
- reduced construction and commissioning time.

Eliminating copper wiring with Ethernet/fiber cables mediums no more binary inputs and outputs for control and protection functions. The traditional method of tripping a breaker via a contact could be replaced by GOOSE messages sent via Ethernet or fiber optic cables.

10.5.1   Standard documents and features of IEC 61850

The standard is composed of 10 standard documents that cover all the requirements that have to be fulfilled by a substation automation system (SAS). It is important to keep in mind that the evolution of this technology is very rapid and therefore more changes and additions are feasible. The following are the parts of the standard at the moment of writing this book:

- IEC 61850-1: Introduction and overview,
- IEC 61850-2: Glossary,
- IEC 61850-3: General requirements,
- IEC 61850-4: System and project management,
- IEC 61850-5: Communication requirements for functions and device models, Ed. 2,
- IEC 61850-6: Configuration language for communication in electrical substations related to IEDs – Ed. 2,
- IEC 61850-7: Basic communication structure for substation and feeder equipment,

- IEC 61850-7-1: Principles and models – Ed. 2,
- IEC 61850-7-2: Abstract communication service interface (ACSI) – Ed. 2,
- IEC 61850-7-3: Common data classes – Ed. 2,
- IEC 61850-7-4: Compatible logical node classes and data classes – Ed. 2,
- IEC 61850-8: Specific communication service mapping (SCSM),
- IEC 61850-8-1: Mappings to MMS (ISO/IEC9506-1 and ISO/IEC 9506-2) – Ed. 2,
- IEC 61850-9: Specific communication service mapping (SCSM),
- IEC 61850-9-2: Sampled values over ISO/IEC 8802-3 – Ed. 2,
- IEC 61850-10: Conformance testing.

In IEC 61850 Ed. 2, Part 9-1 is deprecated as per conformance table. Ethernet-based sampled value transmission (IEC 61850 9-2) will take over process bus communication. Now all IEC 61850 Specific communication service mappings are using Ethernet technology. Some of the features of IEC 61850 include:

1. data modeling,
2. reporting schemes,
3. fast transfer of events, GSE – GOOSE and GSSE,
4. commands,
5. sampled data transfer,
6. setting groups, and
7. data storage – SCL (substation configuration language).

1. *Data modeling*: In IEC 61850, series functionalities that the real devices comprise are decomposed into the smallest entities, which are used to exchange information between different devices. These entities are called logical nodes. Complete functionality of the substation is modeled into different standard logical nodes. The logical nodes are the virtual representation of the real functionalities. The intent is that all data that could originate in the substation can be assigned to one of these logical nodes. Several logical nodes from different real devices build up a logical device.

Logical devices, logical nodes, and data objects are all virtual terms. They represent real data, which are used for communication. One device (e.g., a control unit) only communicates with the logical nodes or its data objects of another device (e.g., an IED). The real data, which the logical nodes represent, are hidden and they are not accessed directly. This approach has the advantage that communication and information modeling is not dependent on operating systems, storage systems, and programming languages. The concept of virtualization is shown in Figure 10.5, where the real device on the right-hand side is modeled as a virtual model in the middle of the figure. The logical nodes (e.g., XCBR and circuit breaker) defined in the logical device (Bay) correspond to well-known functions in the real devices. In this example the logical node XCBR represents a specific circuit breaker of the bay to the right.

Based on their functionality, a logical node contains a list of data (e.g., position) with dedicated data attributes. The data have a structure and a pre-defined semantic. The information represented by the data and their attributes are exchanged by the communication services according to the well-defined rules and the requested performances.

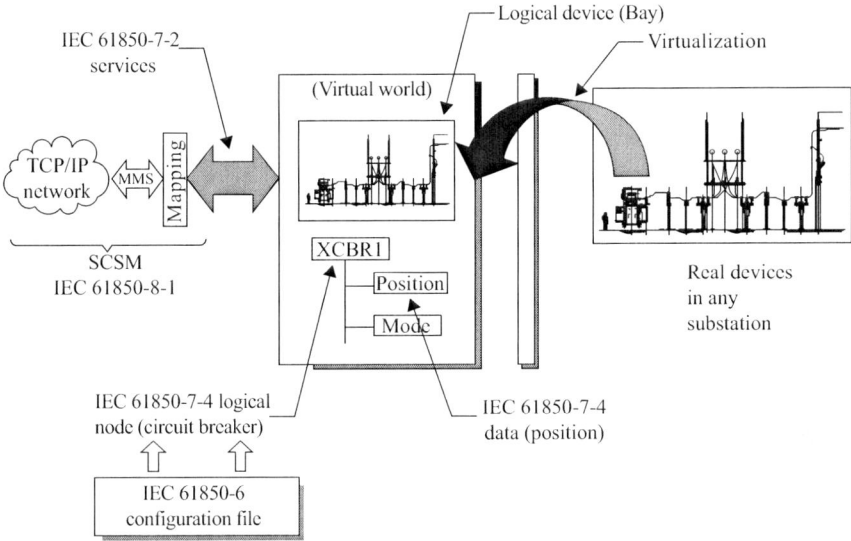

Figure 10.5   Virtual and real world

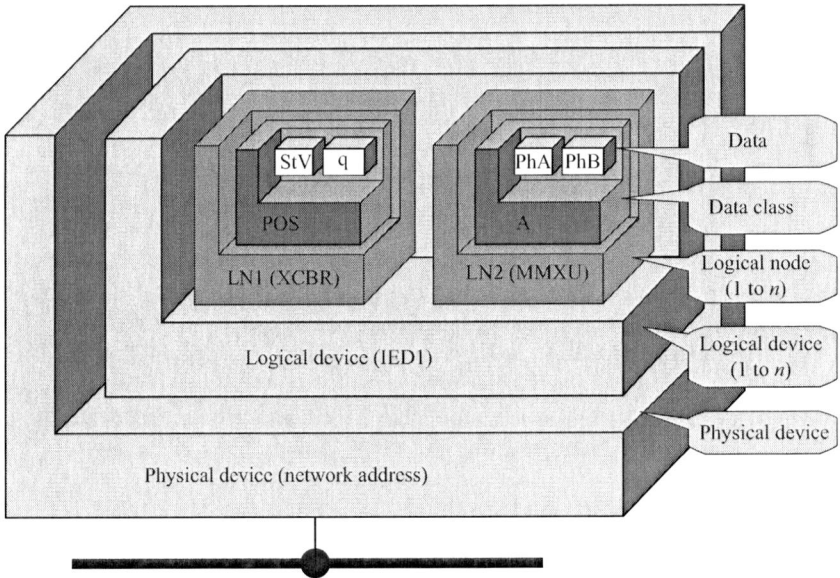

Figure 10.6   Physical and logical device

To illustrate more clearly how the logical devices, logical nodes, classes, and data concepts map to the real world, imagine an IED that is a container as shown in Figure 10.6.

The container is the physical device, which is containing one or more logical devices. Each logical device contains one or more logical nodes, each of which contains a pre-defined set of data classes. Every data class contains many data attributes (status value, quality, etc.).

IEC 61850-7-4 Ed. 2 defines a list of logical node groups as shown in Table 10.1.

In Ed. 2 LN list has been extended from 92 to 208 based on new requirements. New system logical nodes are introduced to indicate the status of GOOSE and sampled value subscription. Grouping of logical node is also extended to include relevant areas.

IEC 61850-5 Ed. 2 defines the protection logical nodes which are shown in Table 10.2.

2. *Reporting schemes*: There are various reporting schemes (BRCB and URCB) for reporting data from server through a server–client relationship which can be triggered based on pre-defined trigger conditions.

3. *Fast transfer of events*: Generic substation events (GSE) are defined for fast transfer of event data for a peer-to-peer communication mode. This is subdivided

Table 10.1   List of Logical node groups defined by IEC 61850-7-4 Edition 2

Group indicator	Logical node groups
A	Automatic Control
B	Reserved
C	Supervisory Control
D	Distributed Energy Resources
E	Reserved
F	Functional Blocks
G	Generic Function References
H	Hydro Power
	Interfacing and Archiving
J	Reserved
K	Mechanical and Non-electrical Primary Equipment
L	System Logical Nodes
M	Metering and Measurement
N	Reserved
O	Reserved
P	Protection Functions
Q	Power Quality Events Detection Related
R	Protection Related Functions
Sa	Supervision and Monitoring
Ta	Instrument Transformer and Sensors
U	Reserved
V	Reserved
W	Wind Power
Xa	Switchgear
Ya	Power Transformer and Related Functions
Za	Further (Power System) Equipment

[a] LNs of this group exist in dedicated IEDs if a process bus is used. Without a process bus, LNs of this group are the I/Os in the hardwired IED one level higher (e.g., in a bay unit), representing the external device by its inputs and outputs (process image).

Table 10.2   Protection logical nodes defined by IEC 61850-5 Edition 2

Functionality allocated to LN	IEC	IEEE	LN Function	LN Class	LN Class naming
Transient earth fault protection			**PTEF**	**PTEF**	Transient earth fault
Sensitive directional earth fault		(37) (67N)	**PSDE**	**PSDE**	Sensitive directional earth fault
Thyristor protection			**PTHF**	**PTHF**	Thyristor protection
Protection trip conditioning			**PTRC**	**PTRC**	Protection trip conditioning
Checking or interlocking relay		3	**CILO**	**CILO**	Interlocking
Over speed protection	$\omega>$	12	**POVS**		
Zero speed and under speed protection	$\omega<$	14	**PZSU**	**PZSU**	Zero speed or underspeed
Distance protection	$Z<$	21	**PDIS**	**PDIS**	Distance protection
				PSCH	Protection Scheme
Volt per Hz protection		24	**PVPH**	**PVPH**	Volts per Hz
Synchronism check		25	**RSYN**	**RSYN**	Synchronism-check
Over temperature protection	$>$	26	**PTTR**	**PTTR**	Thermal overload
(Time) Undervoltage protection	$U<$	27	**PTUV**	**PTUV**	Undervoltage
Directional power /reverse power protection	$\overrightarrow{P}>$	32	**PDPR**	**PDOP**	Directional over power
				PDUP	Directional under power
Undercurrent/underpower protection	$P<$	37	**PUCP**	**PTUC**	Undercurrent
				PDUP	Directional under power
Loss of field/Under-excitation protection		40	**PUEX**	**PDUP**	Directional under power
				PDIS	(Distance) Impedance
Reverse phase or phase balance current protection,Negative sequence current relay	$I_2>$	46	**PPBR**	**PTOC**	Time overcurrent
Phase sequence or phase-balance voltage protection,Negative sequence voltage relay	$U_2>$	47	**PPBV**	**PTOV**	Overvoltage protection
Motor start-up protection		48, 49, 51LR66	**PMSU**	**PMRI**	Motor restart inhibition
				PMSS	Motor starting time supervision
Thermal overload protection	$\Theta>$	49	**PTTR**	**PTTR**	Thermal overload
		49R	**PROL**	**PTTR**	Thermal overload

(Continues)

Table 10.2   (Continued)

Functionality allocated to LN	IEC	IEEE	LN Function	LN Class	LN Class naming
Rotor thermal overload protection					
Rotor protection		49R 64R (40) 50 51	**PROT**	**PTTR**	Thermal overload
				PTOC	Time overcurrent
				PHIZ	Ground detector
				PDUP	Directional under power
				PDIS	Distance (impedance)
Stator thermal overload protection		49S	**PSOL**	**PTTR**	Thermal overload
Instantaneous overcurrent or rate of rise protection	I≫	50	**PIOC**	**PIOC**	Instantaneous overcurrent
AC time overcurrent protection	I>, t	50TD 51	**PTOC**	**PTOC**	Time overcurrent
Voltage controlled/dependent time overcurrent protection		51V	**PVOC**	**PVOC**	Voltage controlled time overcurrent
Power factor protection	cos φ> cos φ<	55	**PPFR**	**POPF**	Over power factor
				PUPF	Under power factor
(Time) Overvoltage protection	U>	59	**PTOV**	**PTOV**	Overvoltage
DC-overvoltage protection		59DC	**PDOV**	**PTOV**	Overvoltage
Voltage or current balance protection		60	**PVCB**	**PTOV**	Overvoltage
				PTOC	Time overcurrent
Earth fault protection, Ground detection	I$_E$>	64	**PHIZ**	**PTOC**	Time overcurrent
				PHIZ	Ground detector
Rotor earth fault protection		64R	**PREF**	**PTOC**	Time overcurrent
				PHIZ	Ground detector
Stator earth fault protection		64S	**PSEF**	**PTOC**	Time overcurrent
				PHIZ	Ground detector
Interturn fault protection		64W	**PITF**	**PTOC**	Time overcurrent
AC directional overcurrent protection	$\vec{I}$>	67	**PDOC**	**PTOC**	Time overcurrent
Directional protection		87B	**PDIR**	**PDIR**	Direction comparison
Directional earth fault protection	$\vec{I}_E$>	67N	**PDEF**	**PTOC**	Time overcurrent
DC time overcurrent protection		76	**PDCO**	**PTOC**	Time overcurrent

Table 10.2 (*Continued*)

Functionality allocated to LN	IEC	IEEE	LN Function	LN Class	LN Class naming
Phase angle or out-of-step protection	$\phi>$	78	PPAM	PPAM	Phase angle measuring
Frequency protection	81 7		PFRQ	PTOF	Overfrequency
				PTUF	Under-frequency
				PFRC	Rate of change of frequency
Differential protection		87	PDIF	PDIF	Differential (Impedance)
Busbar protection[a]		87B	PBDF	PDIF	Differential
				PDIR	Direction comparison
Generator differential protection[b]		87G	PGDF	PDIF	Differential
Differential line protection		87L	PLDF	PDIF	Differential
Motor differential protection[b]		87M	PMDF	PDIF	Differential
Restricted earth fault protection		87N	PNDF	PDIF	Differential
Phase comparison protection		87P	PPDF	PDIF	Differential
Differential transformer protection		87T	PTDF	PDIF	Differential
Harmonic restraint			PHAR	PHAR	Harmonic restraint

into GOOSE (Generic Object Oriented Substation Events) and GSSE (Generic Substation Status Event) which provide backward compatibility with the UCA GOOSE.

In IEC 61850 Ed. 2 GSSE is deprecated and moved to Annex of 7 2. This will cause older less flexible GSSE-based systems move out from IEC 61850 scope. Hence, GOOSE will have added importance in Inter Bay Communication.

4. *Setting groups*: The setting group control blocks (SGCB) are defined to handle the setting groups so that user can switch to any active group according to the requirement.

5. *Sampled data transfer*: Schemes are also defined to handle transfer of sampled values using sampled value control blocks (SVCB).

6. *Commands*: Various command types are also supported by IEC 61850 which include direct and select before operate (SBO) commands with normal and enhanced securities.

7. *Data storage*: SCL (substation configuration language) is defined for complete storage of configured data of the substation in a specific format. SCL originally was the abbreviation for substation configuration language. Now that the

use of IEC 61850 beyond substations has become reality, the name is changed to system configuration language.

10.5.2 System configuration language

The fact that IEC 61850 is a worldwide standard makes it also possible that relays from different manufacturers are able to exchange information as long as the relays conform to the IEC 61850 Standard.

To guarantee interoperability and enhance the configuration phase the IEC 61850-6 introduced a common language which can be used to exchange information between different manufactures using system configuration language (SCL).

Each proprietary tool must have a function which allows the export of the IED's description into this common, XML-based language. The ICD (IED Capability Description) file contains all information about the IED, which allows the user now to configure a GOOSE message.

The development process of a project based on IEC 61850 depends on the availability of software tools that make use of the SCL language. The SCL specifies a common file format for describing IED capabilities, a system-specific caption that can be viewed in terms of a single line diagram, and a substation automation system description. IEC 61850-6 introduces four types of common files. These files are the IED Capability Description (ICD), Configured IED Description (CID), Substation Configuration Description (SCD), and System Specification Description (SSD). IEC 61850-6 Ed. 2 introduced the new "Instantiated IED Description" (IID) file. Figure 10.7 describes the complete engineering process using SCL language.

The configuration can be performed by a manufacturer independent tool, the so-called IEC 61850 System Configurator. Some manufacturers developed their proprietary tools in a way that they can be used as IEC 61850 System Configurator; however, there are also some third-party tools available. All ICD files get imported into the IEC 61850 System Configurator and the GOOSE messages can be programmed by specifying the sender (publisher) and the receiver (subscriber) of a message. On the end the whole description of the system, including the description of the GOOSE messages, get stored in the SCD (Substation Configuration

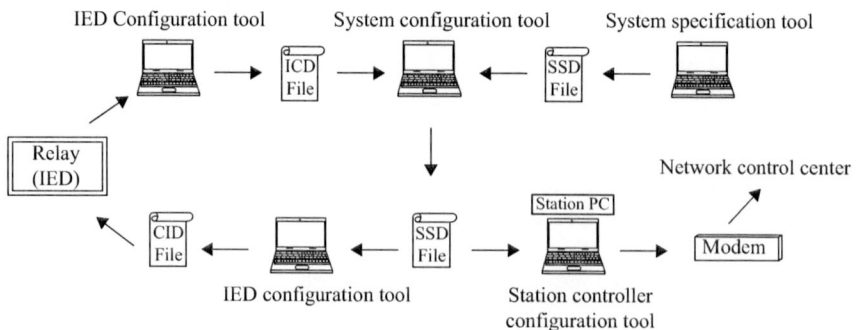

Figure 10.7 Substation engineering process using SCL language

Description) file. Each proprietary tool must be able to import this SCD file and extract the information needed for the IED. The information of the SCD file is typically the list of the GOOSE messages.

SCL language does not cover all features of today's IEDs. In fact it was not the intention of the editors of IEC 61850 Standard to standardize all aspects of IEDs. This is because of the wide variety of functionality provided by each manufacturer.

It is important to know that in addition to the configuration information specified by part 6 of IEC 61850, the complete configuration of an IED should be completed with a proprietary IED configuration tool provided by the manufacturer in order to configure the device parameters and gain access to all of the internal functions supported.

The following list shows examples of features that can only be set up with a vendor's tool:

- logic and trip equations,
- graphical display on an IED's HMI,
- internal mappings, and
- non-IEC 61850 and vendor-specific parameters.

Although the present practice of connecting a notebook computer directly to the front port of the IED will likely be eventually phased out in favor of remote access solutions via the LAN, it is assumed that some activities will continue to need to be performed at the relay, particularly during initial commissioning. It is also assumed that each individual vendor's IED configuration methodology will continue to be unique to each device or product family. Continuous innovation in IED design and competitive market forces would tend to preclude standardization of individual IED configuration software, though some activity is underway in this regard.

The final configuration done by the proprietary tool is individualized per IED and can have either a proprietary format or a standard CID format. Nowadays many vendors have decided to have a CID file using a proprietary format while a few vendors create the CID file in a similar way as the ICD file.

10.5.3 Configuration and verification of GOOSE messages

This section will describe how to configure and verify GOOSE messages for an automated IEC 61850 substation. The type of scheme being configured to illustrate the process is a communication-based breaker failure scheme. Figure 10.8 shows the single line diagram of the system.

The system has four IEDs from different vendors which are shown as relays A, B, C, and D. Relays B, C, and D are the main feeder protection and relay A is the backup protection.

Not shown are the IEDs in the breaker cabinet that take the GOOSE messages from each relay and convert them to a physical output to energize the breaker coil. In this example, a fault has been placed at location A, marked in the single line diagram. The instantaneous overcurrent element in relay B will transmit a trip GOOSE to breaker B. Simultaneously, relay A will receive the trip GOOSE from

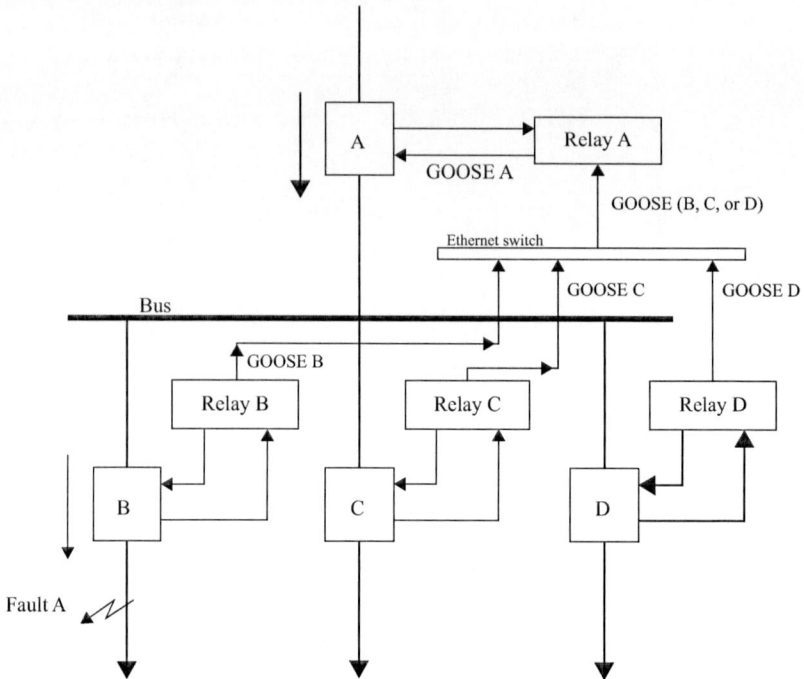

Figure 10.8 System single line diagram

relay B as the breaker failure initiate signal. Relay A will receive this GOOSE and it will initiate the breaker failure timer, which will expire in 10 cycles. If for any reason the breaker failure initiate signal is not received and breaker B does not open, relay A will trip on a definite time overcurrent in 15 cycles. This section will only show the configuration between relays A and B. This same process is repeated for the other IEDs on the system. Figure 10.9 shows the logic of the breaker failure scheme.

The programming of this scheme can be achieved in two parts. The first part is the programming of the individual IEDs in order to obtain the SCL file required for the configuration of the system. These files will contain the published GOOSE messages for each IED. With this information the full system configuration can be accomplished. Before any configuration can occur, it is good practice to create a virtual wiring map. This map will show what messages are going to be required in order to configure the scheme of interest. The map provides such information as to what GOOSE messages each IED will publish and subscribe.

The IED is configured via its own proprietary configuration tool. One of the biggest challenges in configuring an IEC 61850-based protection scheme is becoming familiar with the different IED configuration software packages. It is important to get familiar with how each configuration tool works. This will save time in the long run, especially when trying to configure the system scheme. The primary goal of configuring the IED is to obtain the SCL file types necessary for

67 OC time delay = 15 Cy

Max phase —|+

50BF pickup —|−

50BF time delay = 10 Cy

GOOSE Trip B

GOOSE Trip C

GOOSE Trip D

GOOSE trip A

Max phase —|+

10C pickup —|−

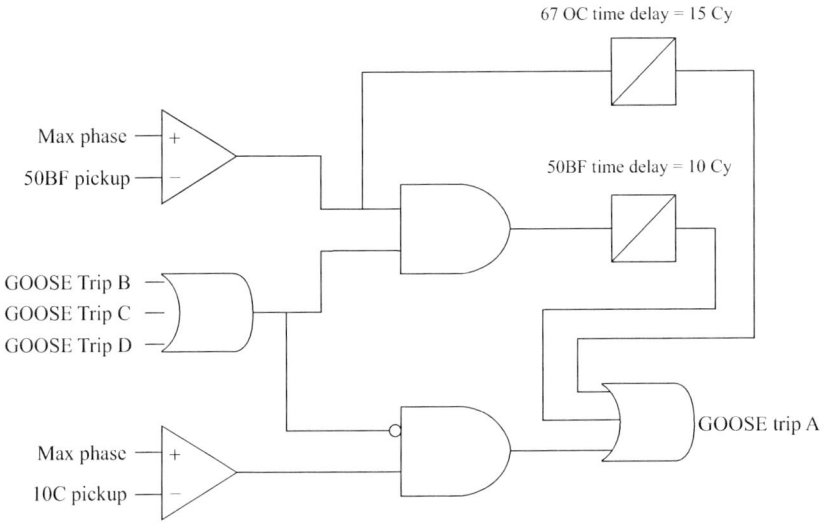

Figure 10.9   Logic of breaker failure scheme

the system configuration. Figure 10.11 shows examples of the configuration tools used to configure the individual IEDs.

The SCL files provide the overall information as to how the relay is configured, but most importantly, what GOOSE messages the particular IED will publish or subscribe. The configuration process would begin by configuring the IEDs that will only be publishing and subscribing to the least number of GOOSE messages.

When the IED is configured an SCL file is exported. This same process is repeated until all IEDs of interest have been configured and an SCL file is generated.

If desired, a simple test can be performed to verify that the published GOOSE messages are correct. This can be done by connecting some form of network analyzer and capturing the GOOSE messages on the network. An illustration of proprietary configuration tools used for the configuration of IEDs is shown in Figure 10.10.

Some of the details of the GOOSE configuration were left out due to differences in the configuration of the IEDs. For example, in some IEDs the GOOSE messages are going to be published through GGIO (generic GOOSE I/O), while others will be published through the protection node such as PDIS or PTOC. This difference can sometimes lead to the GOOSE message being delayed by some fixed time set in the IED. To avoid these problems, it is recommended to read the IED's manual and see where the IED will publish its high-speed GOOSE.

10.5.4   Configuration of the system

In order to configure the system, it is necessary to have all ICD files available. These files are going to be used by a system configuration tool or the IED's

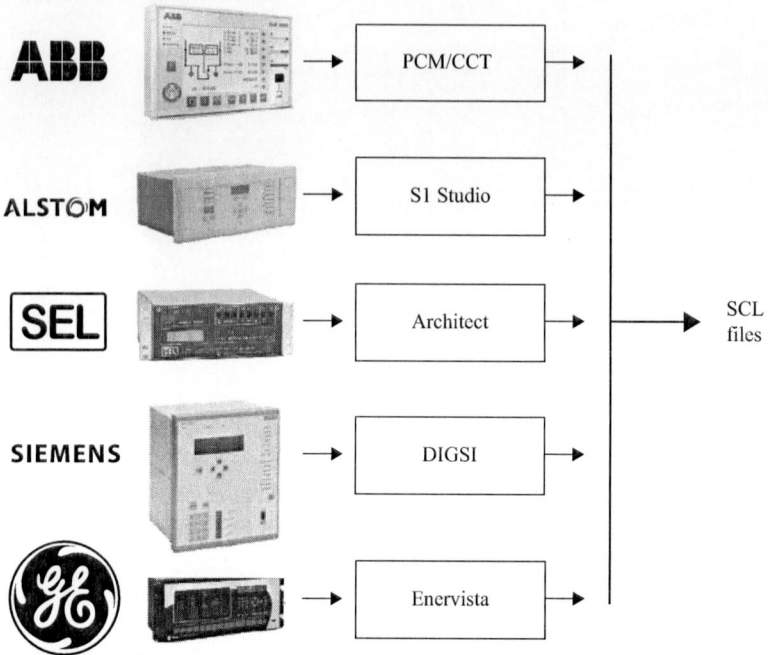

Figure 10.10   Proprietary configuration tools used for the configuration of IEDs

individual configuration tool. The system configuration can be done by a substation configuration tool. This type of tool imports all the ICD files required for the system configuration. The substation configuration tool will generate an SCD file. Another method is by opening up the configuration tool of the IEDs that will subscribe to a GOOSE message.

For example, the configuration of relay A is done by importing all the ICD files into its configuration tool. The ICD file is intended to be a template for an IED type, which could be instantiated multiple times – e.g. per feeder – to build a system.

When the system is configured, a system verification test should be performed. This test will help identify any problems with the system configuration. The first time a system is configured, some configuration problems may occur. Many of these problems are due to the differences in the configuration of the IEDs. Some of the IEDs require a more manual approach, while others are a bit more automatic.

10.5.5   System verification test

The system verification test will identify any configuration conflicts in the system. The tools required are going to be a network analyzer (software) and a modern test set that is IEC 61850 compliant. The test set must be able to receive and send GOOSE messages via the substation LAN. This would require the test system to interrogate the network, acquire the right GOOSE message, and stop a timer with minimal effect on time.

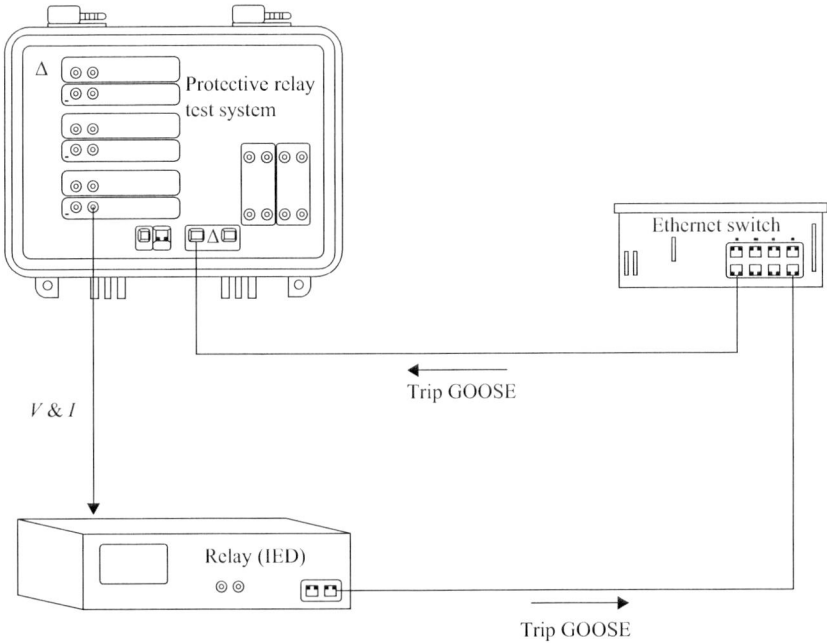

Figure 10.11   Test connections used for standalone IEC 61850–based IED

Figure 10.11 presents the test connections used for standalone IEC 61850 and illustrates how the trip-GOOSE messages are sent from a numerical relay to an Ethernet switch and from this to a protective relay test system.

Modern test systems must be able to receive and send GOOSE messages via the substation LAN. This would require the test system to be able to interrogate the network, acquire the right GOOSE message, and stop injections or timer in less than 2 milliseconds. Also the test system would have to be able to read SCL files and map inputs to the various GOOSE messages available in the SCL file. If an SCL file is not available then the test system would have to be able to interrogate the network, and display all available GOOSE messages on the network, to allow the user to be able to map these messages to binary inputs on the test system.

10.5.6   Substation IT network

As communications in the substation take on more critical roles in the protection and control task of the utility, it is important for the protection engineer to understand the basics of the IT network. The protection engineer must also understand the behavior and characteristics of components like Ethernet switches, Ethernet ports, and router, as well as being familiar with terminology such as LAN, VLAN, Mac address, network topology, latency, priority tag, firewalls, RSTP, HSR/PRP, etc.

Many experienced protection engineers find discussion of IT network issues to be dense and perhaps intimidating, because until now they have not faced the need to understand the behavior and performance characteristics of IT networks. In the modern substations, Ethernet switches are as important to understand as protective relays in order to achieve availability, dependability, security, and maintainability goals of the substation.

10.5.7 Process bus

Process bus is defined by IEC 61850 Ed. 2 Part 9-2. Process bus is the digitalization of all analog signals in the substation. This is achieved by connecting all current transformers, potential transformers, and control cables to merging units. These units convert the analog signals to binary signals and send the information via the process bus to all the devices that subscribe for that information. There are some pilot projects installed around the world with this technology reporting successful results. Figure 10.12 shows a full implementation of IEC 61850 – process bus and station bus.

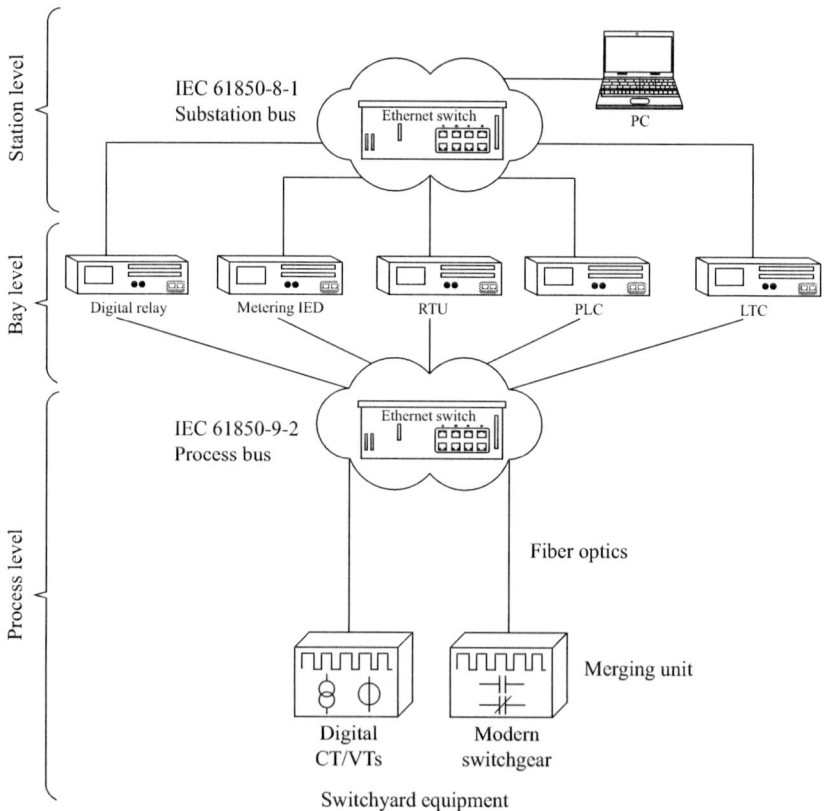

Figure 10.12   Substation network – process bus and station bus

Chapter 11

Interoperability concepts in power electric systems

To achieve the objectives of the Smart Grid efficiently, it is necessary to integrate all the components of the power system from generation to the end user. They should relate to each other in a transparent manner, i.e. independent of their technologies and protocols.

Currently there are many limitations on communication among the components of a power system, sometimes not even allowing any communication at all, because of incompatibilities between protocols and technologies.

Additionally, there can be difficulties with the information systems because its components do not interpret the information in the same way. Interfaces solve this problem but have the disadvantage that they limit the information system's growth. Therefore, it is required to standardize the representation of the power system's information in such a way that a component accesses the data independently of the organization and meaning of the data from the source. A proper solution is achieved when power systems fulfill the conditions of interoperability.

IEEE Standard 2030-2011 defines interoperability as the capability of two or more networks, systems, devices, applications, or components to externally exchange and readily use information securely and effectively. The same standard defines a methodology to guarantee the interoperability in power systems from the perspectives of three systems: the power system itself, the information system, and the communication system. The process has to be done in such a way as to guarantee a long-lived update.

This concept applied to Smart Grid networks ensures efficient communication whether the information systems are used on different types of infrastructure or even at a distance.

Also to be considered are the ideas of interoperability of the hardware and software of the system, and in general of all content that is exchanged between systems. There must be common definitions of data.

For this, the use of the concept of ontology from computer science helps greatly. Ontology is the philosophical study of the nature of the basic categories and their relationships. Ontology-based strategies applied to interoperability constitute a framework for organizing information and are used in the representation of components of power systems.

11.1   Elements required for interoperability

Entities such as the EPRI and NIST have defined the concept of interoperability by using architectural principles. Table 11.1 enumerates the elements that are required to define the concept of interoperability.

Table 11.1   Architecture of the principles of interoperability (taken from EPRI and NIST)

Principle	Description
Standardization	The elements of the infrastructure and the ways in which they interrelate are clearly defined, published, useful, open, and stable over time
Openness	The infrastructure is based on technology that is available to all qualified stakeholders on a nondiscriminatory basis. Providers of the technology have the ability to modify access to technology through time to ensure continuous openness and standardization
Interoperability	The standardization of interfaces within the infrastructure is organized such that (1) the system can be easily customized for particular geographical, application-specific, or business circumstances, but (2) customization does not prevent necessary communications between elements of the infrastructure
Security	The infrastructure is protected against unauthorized access and interference with normal operation. The utility continuously implements information privacy and other security policies as required
Extensibility	The infrastructure is not designed with built-in constraints against extending its capabilities as new applications are discovered and developed. Toward this goal, (1) its data are defined and structured according to a CIM, (2) it separates the definition of data from the methods used to deliver it, and (3) its components can announce and describe themselves to other components
Scalability	The infrastructure can be expanded with no inherent limitations on its size
Manageability	The components of the infrastructure can have their configuration assessed and managed. Faults can be identified and isolated. The components are remotely manageable
Upgradeability	The configuration, software, algorithms, and security credentials of the infrastructure can be upgraded safely and securely with minimal remote-site visits. This also touches on manageability
Shareability	The infrastructure uses shared resources that offer economies of scale, minimize duplicative efforts, and, if appropriately organized, encourage the introduction of competing innovative solutions
Ubiquity	Authorized users of the Smart Grid can readily take advantage of the infrastructure and what it provides regardless of geographic or other barriers
Integrity	The infrastructure operates at a high level of availability, performance, and reliability. It reroutes communications automatically
Ease of use	There are logical, consistent, and preferably intuitive rules and procedures for the use and management of the infrastructure. The system maximizes the information and choices available to users of the Smart Grid, while minimizing the actions they must take to participate if they choose to do so

The various functions in IT have the responsibility of distributing safely and reliably information to all points of the network where this is required for monitoring or decision making. Thanks to these information technologies, Smart Grid networks will become much more dynamic in their configuration and will approach operating conditions that give greater opportunities for real-time analysis and optimization technologies. The companies in the electricity sector must have large amounts of operational information, both real-time and stored, in either static or dynamic form. They also receive large amounts of information remotely, directly from the analysis done at fault sites and events, and AMI deployments along the system, thanks to the installation of intelligent electronic devices (IED) and other Smart Grid devices.

11.2 Information exchange processes

Interoperability in the exchange of information ought to allow two applications to exist whenever each one can use the interchange information to develop its own results. This exchange should be carried out in a data flow sequence along different data paths. The applications will use a link data series (or communication systems technology) that employs sufficient protocols to bear the data flow requests. In addition, the structure and the meaning of the information exchange should be understood by both systems. This implies that they share an understanding of the unit of measure, the context of use, indices of information, and validity. A scheme of information exchange can be seen in Figure 11.1.

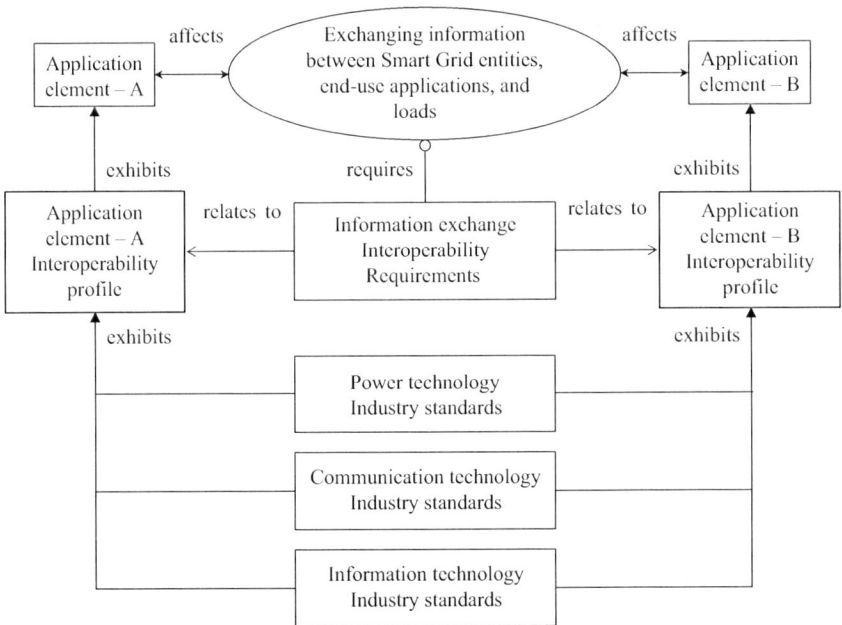

Figure 11.1   Information exchange between two application elements (taken from IEEE 2030-2011)

11.3   Data models and international standards

A model of well-defined data elements not only allows the simple exchange of information but also assists in fulfilling requests from development and security. The existence of a data model for the architecture is almost always a good indicator of a well-governed process of facilitating the use of the data by multiple applications because of the establishment of a common semantics.

Two global standards exist for the integration of businesses in the electricity sector. One of them is MultiSpeak®, sponsored by National Rural Electric Cooperative Association (NRECA), and the other is the Common Information Model (CIM), an international standard maintained by Technical Committee 57 (TC57) of the International Electrotechnical Commission (IEC).

Currently, MultiSpeak®-compatible connectivity data exchange is in operation at dozens of utilities.

Working Group 14 (WG14) of TC57, which deals with work processes germane to distribution utilities, has recently defined the NetworkDataSet message that specifies how the CIM should be used to exchange detailed models for distribution engineering analysis. CIM-compatible NetworkDataSet model exchanges are in operation at several utilities.

Figure 11.2(a) depicts a single line diagram of a system that has a connection to a grid via a 115 kV line. The system has an internal generator of 13.2 kV and a 13.2/115 kV step-up transformer. The load is fed at 34.5 kV through a 115/34.5 kV step-down transformer. Figure 11.2(b) represents all the elements of the power system shown in Figure 11.2(a) converted to CIM objects, which have different attributes in a common base for each element.

Figure 11.2(a) shows an arbitrary circuit of the power system. Figure 11.2(b) shows the same figure mapped in CIM.

Further specifications may provide service descriptions and protocol specifications for electricity metering. Good data models will enhance the return on investment of the Smart Grid by enabling more applications to use the data, and improving its value for newly developed analytics that may examine the information in novel ways.

Ontology-based strategies are commonly used with success in creating and manipulating data models since they provide easy export or translation to Unified Modeling Language (UML), which provides for a great deal of interoperability. Data are of little use if their meaning is not clear. Within the context of a single user interface application, developers strive to make the meaning clear, but when data are transferred to another system, the meaning may be lost.

The objects of the real world are represented by means of classes, which conform the basic unit that encompasses all their information. Thanks to these classes, the elements of power systems (transformers, sectionalizers, generators, etc.) can be modeled. UML language is used to represent the different relations among the classes that the model uses. These relations among classes are known as association.

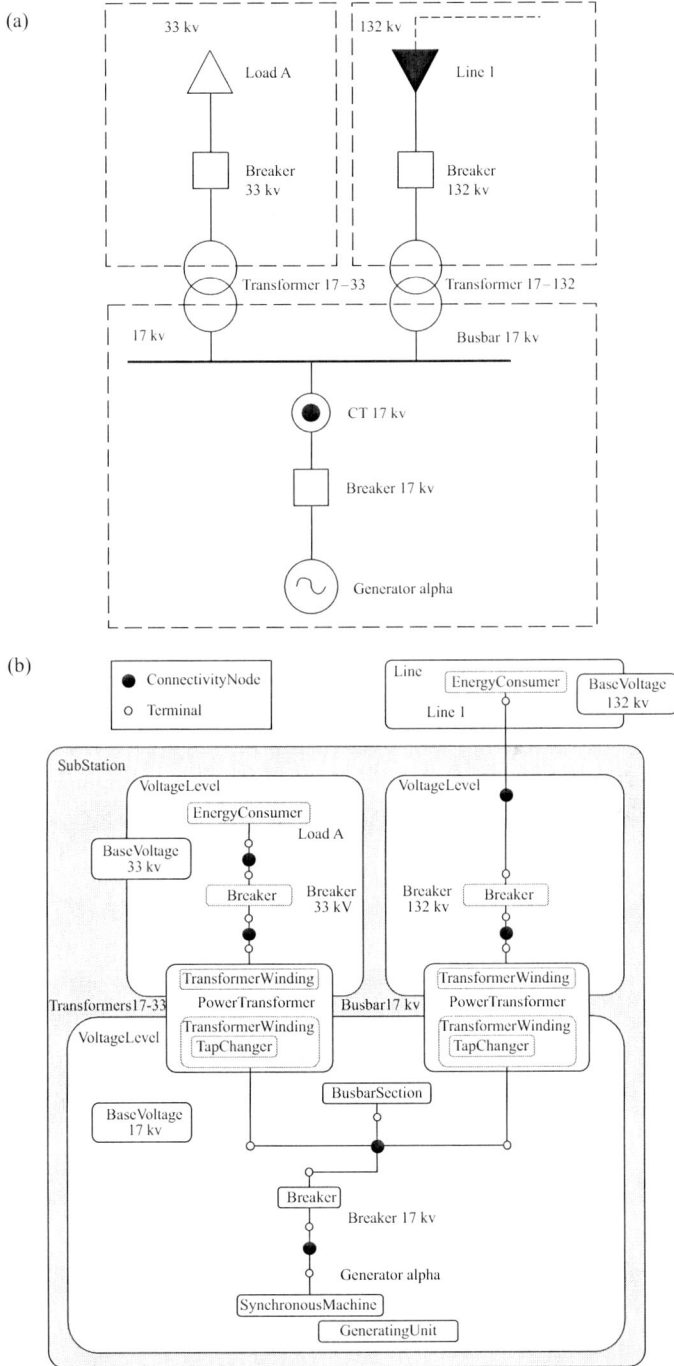

Figure 11.2   Example circuit (a) as a line diagram and (b) with full CIM mappings

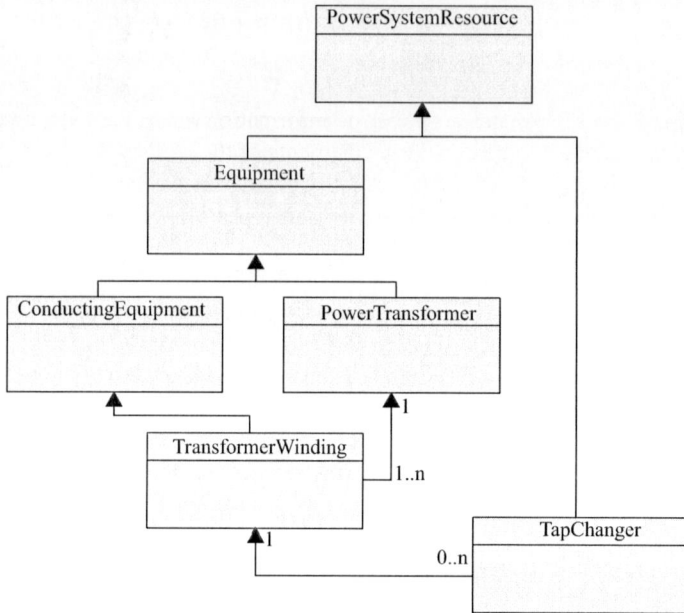

Figure 11.3   Transformer class diagram

For each class, the attributes can be observed, i.e., the information or data that represent the object. On the other hand, when the elements to be modeled have been already defined, the Extensible Markup Language (XML) language is utilized to create instances, which constitute the physical model of the elements of the real-world power network.

Figure 11.3 is an example of a diagram of the classes of a transformer represented in UML. The relationship of the different classes can be seen for some of the subclasses like the tap changer, transformer winding, and the conducting equipment. In order to be simpler, the attributes of each class were hidden but they could be visible if required.

Figure 11.4 shows the XML representation of some elements of the power system. In this case it is clear that the modeled elements have been already defined, and it is possible to create the instances in the physical model of the power system. The illustration here represents a 400 kV Node 2 in XML language. The rectangle in of Figure 11.4(a) indicates the node that will be represented in XML. The rectangle in red of Figure 11.4(b) is a portion of the code generated in XML to illustrate how the information is represented with this format.

11.4 Implementation of information models in the Power Electric System (PES)

To implement an interoperable model of information as in the case of the CIM model, it is necessary to identify all the elements of the physical infrastructure into which all the applications and functions will be integrated. Examples of this are the configuration of the network, the quantity and types of servers, and very importantly, their availability.

It should be verified that the utility has a data bus that permits it to exchange messages among the different functions and services and that facilitate the integration of the system. Thus, it is of vital importance to implement at least one business bus. Usually, these buses utilize the service-oriented architecture (SOA), which is an assembly of guidelines designed to support user-defined functional interfaces. These interfaces take care of the marketing activities, the connection of the service, and the connection of new clients. Normally, for the messages exchanged among services the XML language is utilized.

SOA only establishes the structure, architecture, media, and formats of messages. Nevertheless, a message sent between applications can contain any sequence of information in any order, structure, or hierarchy. Semantic service-oriented architecture (SSOA) is the application in the SOA service that provides a model of information and a common language to facilitate the integration of applications into the established infrastructure, so that the messages sent between applications follow a plan, or common semantic model.

In Figure 11.5 one can see that the implementation of an ESB together with the SSOA allows integration of all the functions within the CIM model, significantly reducing the number of interfaces to implement.

After having characterized all the existing resources, the required function implementations are designed. It is of vital importance that the purpose of the system be completely documented and defined through a case of use. For example, to guarantee integration within a system that generates work orders to a team within a GIS system, it is necessary to have well-defined reasons for this integration. In this case, it is necessary to generate automatic work orders to the local personnel. This helps in defining the data that will be interchanged, e.g., the name of the worker, the type of event, the place of the event, the status, and availability.

With these elements identified, the CIM mapping is performed. The relationships among the characteristics that will be exchanged among the applications in the model are expressed by a good approximation of their condition/value in the system. This procedure is followed for each case of use, that is to say, for each one of the elements that have to be integrated. The integration and the characteristics shared with the other systems should be well documented.

After all the aspects of the design are defined, the phase of development starts, where the classes that will be utilized in order to create the CIM profile are

(a)

(b)

```xml
<?xml version="1.0" encoding="UTF-8"?>
<rdf:RDF xmlns:profile="http://ucte.org/2009/profile1_v11#" xmlns:neplan="http://www.neplan.ch/Neplan_CIM-schema-cim14#"
xmlns:cim="http://iec.ch/TC57/2009/CIM-schema-cim14#" xmlns:rdf="http://www.w3.org/1999/02/22-rdf-syntax-ns#">
  <cim:IEC61970CIMVersion rdf:ID="_fa3b52ce132e11deb8cf000c295e15ce">
    <cim:IEC61970CIMVersion.version>IEC61970CIM14v02</cim:IEC61970CIMVersion.version>
    <cim:IEC61970CIMVersion.date>2009-02-04</cim:IEC61970CIMVersion.date>
  </cim:IEC61970CIMVersion>
  <cim:BaseVoltage rdf:ID="_fa3b52d8132e11deb8cf000c295e15ce">
    <cim:BaseVoltage.nominalVoltage>110.0</cim:BaseVoltage.nominalVoltage>
    <cim:BaseVoltage.isDC>false</cim:BaseVoltage.isDC>
  </cim:BaseVoltage>
  <cim:VoltageLevel rdf:ID="_fa3b52d2132e11deb8cf000c295e15ce">
    <cim:IdentifiedObject.name>110.0 kV</cim:IdentifiedObject.name>
    <cim:VoltageLevel.MemberOf_Substation rdf:resource="#_fa3b52d0132e11deb8cf000c295e15ce"/>
    <cim:VoltageLevel.BaseVoltage rdf:resource="#_fa3b52d8132e11deb8cf000c295e15ce"/>
  </cim:VoltageLevel>
  <cim:Substation rdf:ID="_fa3b52d0132e11deb8cf000c295e15ce">
    <cim:IdentifiedObject.name>NODE 1</cim:IdentifiedObject.name>
    <cim:Substation.Region rdf:resource="#_fa3b52d6132e11deb8cf000c295e15ce"/>
  </cim:Substation>
  <cim:BaseVoltage rdf:ID="_fa3b52e0132e11deb8cf000c295e15ce">
    <cim:BaseVoltage.nominalVoltage>400.0</cim:BaseVoltage.nominalVoltage>
    <cim:BaseVoltage.isDC>false</cim:BaseVoltage.isDC>
  </cim:BaseVoltage>
  <cim:VoltageLevel rdf:ID="_fa3b52df132e11deb8cf000c295e15ce">
    <cim:IdentifiedObject.name>400.0 kV</cim:IdentifiedObject.name>
    <cim:VoltageLevel.MemberOf_Substation rdf:resource="#_fa3b52e0132e11deb8cf000c295e15ce"/>
    <cim:VoltageLevel.BaseVoltage rdf:resource="#_fa3b52e0132e11deb8cf000c295e15ce"/>
  </cim:VoltageLevel>
  <cim:Substation rdf:ID="_fa3b52dd132e11deb8cf000c295e15ce">
    <cim:IdentifiedObject.name>NODE 2</cim:IdentifiedObject.name>
    <cim:Substation.Region rdf:resource="#_fa3b52d6132e11deb8cf000c295e15ce"/>
  </cim:Substation>
  <cim:BaseVoltage rdf:ID="_fa3b52e5132e11deb8cf000c295e15ce">
    <cim:BaseVoltage.nominalVoltage>220.0</cim:BaseVoltage.nominalVoltage>
    <cim:BaseVoltage.isDC>false</cim:BaseVoltage.isDC>
  </cim:BaseVoltage>
```

Properties of BaseVoltage element

Figure 11.4   Example of (a) power system representation and (b) CIM/XML representation

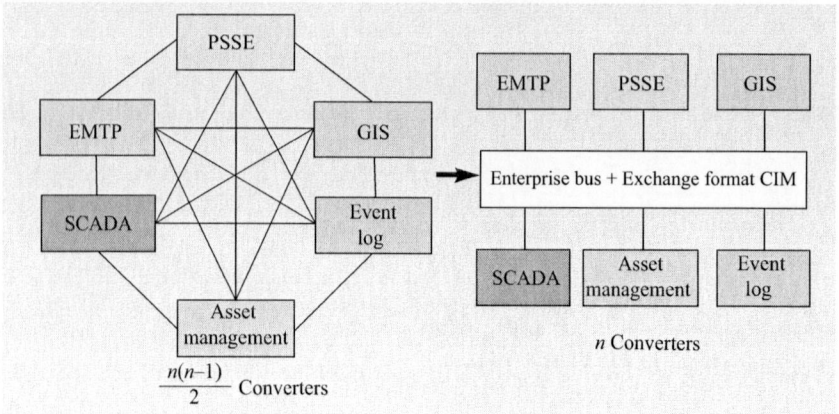

Figure 11.5   Elements for implementing interoperability in information systems

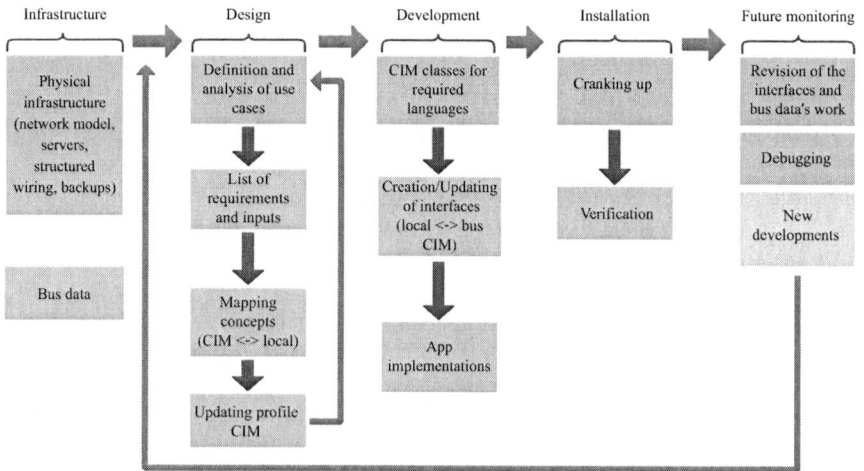

Figure 11.6   Stages in a CIM implementation

determined. This profile is nothing more than the representation of the function or system in terms of the model. In it, all the elements that are to be integrated fit perfectly in this reference that is common to them all. With this profile the applications are installed. As in all software development, it is necessary to verify that the interfaces function adequately and that they are revised constantly to correct errors that are found later.

The stages required in a CIM implementation are shown in Figure 11.6.

Chapter 12

Maturity models

Distribution organizations are aware of the need to implement distribution automation and Smart Grid programs. It is easy to appreciate the various individual efforts being made, given the number of existing applications. However, there is a lack of definite procedures and recommended practices to help utility organizations establish an order of priority in their implementation and development of these technologies.

Developing a maturity model is important for modern grid management and for selecting the best way of creating a sustainable path toward integration. This involves tools such as having a common strategy and vision, which not only helps in the development of an organized work plan but also allows for the implementation of profitable projects as is needed.

The maturity model helps utilities to implement Smart Grid application by prioritizing tasks and measuring the progress achieved. It also helps to identify the characteristics of the organization by designing a roadmap and by promoting the exchange of common terms among internal and external actors. All share experiences with the community and prepare the organization to undertake the required changes.

Maturity models, very common in IT organizations, help an organization assess its methods and processes according to management criteria. The key to achieving a maturity model is a good strategy and a good vision in the Smart Grid context. The maturity model covers three major elements: communications, IT, and electrical components.

This chapter discusses the joint efforts among different entities which have come together to produce interesting procedures and schemes to aid organizations in using the most appropriate solutions for their requirements.

12.1 Smart Grid maturity model definition

The Smart Grid maturity model (SGMM) is a tool that provides the basis to help an organization guide, evaluate, and improve its efforts to best select applications of Smart Grid in order to achieve a proper transformation and modernization. From a methodological standpoint, the model allows to create a map defining the task and technologies, to identify gaps in the strategy and execution, to support business opportunities that promote Smart Grid projects, to delineate the organization vision and strategy, and to evaluate alternative solutions and future goals that will help guide the future of the electric network.

12.2   Benefits of using a Smart Grid maturity model

In today's competitive world, the industry demands that organizations strive to achieve sustainable improvements and repeatable and scalable procedures and to promote improvements within an organization. Maturity models were initially intended to be applied in the software development industry, driven by the necessity of evaluating various organizations under identical parameters. These models gave the possibility of developing improved planning, engineering, and governing practices to guarantee higher quality levels in both processes and results.

Many organizations approved the use of the maturity models in electrical utilities, considering the successful experience in the software industry. It allowed then to determine the development level of networks and to visualize the gap between the current and the future situation. From this the best solutions can be proposed.

Maturity models must be combined with the entire work methodology to identify the standards and technical solutions that will be considered in developing the Smart Grid roadmap. Maturity models also support the implementation of applications.

The methodology has three goals: first, to identify the entity's current developmental state from a Smart Grid perspective and the desired state expressed as a maturity level. Gap analysis is then employed to obtain a simplified list of required steps; second, a cost-benefit analysis to determine which Smart Grid solutions are financially feasible; third, to arrange and describe the user requirements with use-cases based on the financial evaluations previously approved by the executives.

A robust model needs to recognize not only management activities being carried out at the individual project level but also those activities within an organization that build and maintain a framework of effective project approaches and management practices.

By undertaking a maturity assessment against an industry standard model, an organization will be able to verify what they have achieved, where their strengths and weaknesses are, and then to identify a prioritized action plan to take them to an improved level of capability.

12.3   Genesis and components of SGMM

The best known examples of maturity models are those developed by Carnegie Mellon University and IBM. In this chapter, the SGMM, which is one of the most accepted globally, will be explained.

The SGMM was originally proposed by four electrical utilities of the United States (Center Point Energy, Progress Energy, Pepco Holdings, and Sempra Energy); DONG Energy from Denmark; NDPL from India; Country Energy from

Australia; and IBM and APQC. In 2009, with the support of the Department of Energy (DOE), the development of the SGMM was given to Carnegie Mellon. That university still is responsible for its administration and update. The SGMM is maintained by the Carnegie Mellon Software Engineering Institute (SEI) as a resource for utility industry transformation. Carnegie Mellon encourages utilities to leverage SGMM to ensure that all aspects of transformation planning are considered, their options are prioritized, and their progress measured as the utilities implement a Smart Grid structure.

The main organizations using maturity models are depicted in Figure 12.1.

12.4 Development process of an SGMM

The maturity model development process can be carried out with the following steps:

1. *Information gathering in the eight domains of the related organization*: In this stage, the consulting engineer must explore all organization characteristics from a strategic point of view. Since the navigator assists the organization in comprehending the questions included in the evaluation, it is vital that he has a full understanding of the model.

2. *Smart Grid and SGMM concept awareness*: The navigator must be familiar with the organization; likewise the organization itself must understand the aspects to be evaluated. The navigator must prepare the organization to understand the concepts of the SGMM. Next, the criteria must be adjusted and the evaluation process must be established.

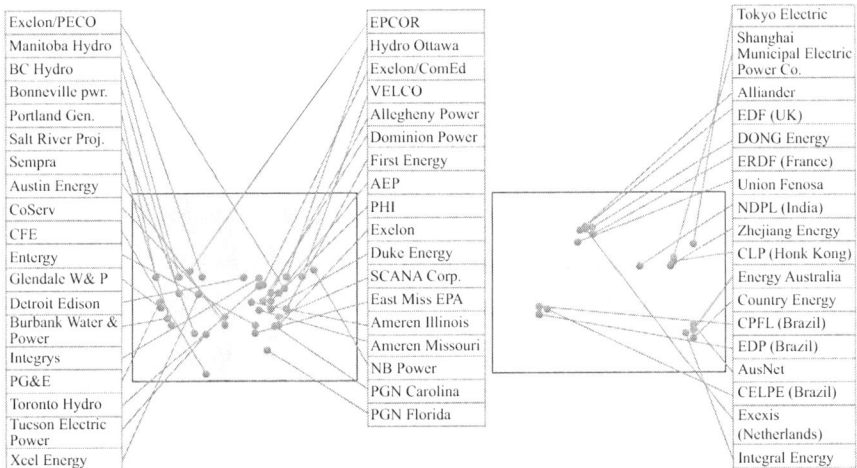

Figure 12.1 Some organizations that use the SGMM (taken from the Carnegie Mellon Software Engineering Institute)

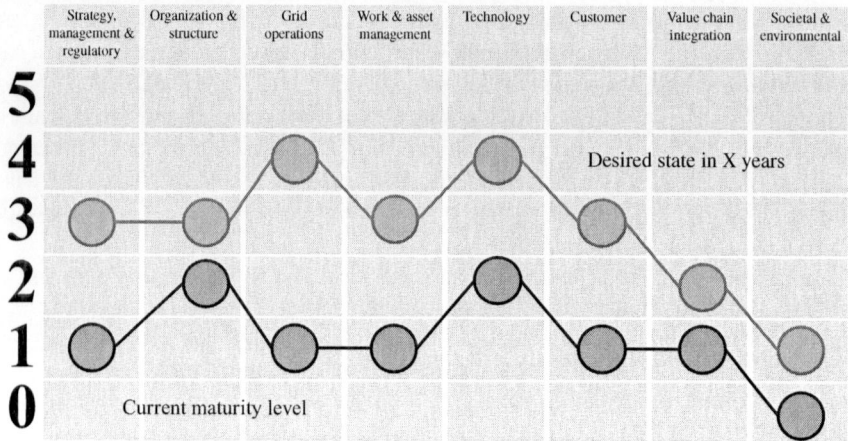

Figure 12.2   Example of results obtained by applying the SGMM

3. *SGMM application*: In this stage, the navigator researches the questions included in the maturity model. The idea is to evaluate the current condition of the organization as well as the desired future condition. It is vitally important that the organization responds to questions in a critical, objective, and honest manner. Also, it is important to involve individuals from the various areas of the organization, since the model covers all aspects of the organization and requires that answers are given with an exhaustive view of the problem. If this is not the case, key points will be missed and efforts may be focused on irrelevant aspects. It is recommended to re-evaluate after the first model attempt.
4. *Final results*: Results are obtained regarding the actual condition of the organization and the future condition as illustrated in Figure 12.2. The results help identify gaps that must be filled in order to reach the desired condition. The cost and time required to do so should be carefully analyzed to assure the viability of the project. If the gap is large, a viable economic solution may be to extend the time or reduce the scope of the future vision.

12.5   Levels and dominions of an SGMM

The characteristics and capabilities of an organization can be determined with the help of the SGMM Model Definition as proposed in a document entitled "SGMM Compass Assessment Survey", Version 1.2 of September 2011 of the Software Engineering Institute of the Carnegie Mellon University. SGMM Model Definition is composed of six levels in eight maturity domains.

12.5.1   Maturity levels of SGMM

The SGMM has six maturity levels that represent well-defined states. Each one describes the capabilities and the characteristics of the organization to achieve the

Smart Grid vision regarding efficiency, automation, reliability, energy savings, interaction with the user, integration of distribution energy resources, and access to new opportunities of the business.

The lowest level (level 0) represents the default position of the organization when the study starts. An organization operating in a traditional way without modernization will be at this level. It is important for an organization to evaluate its condition to establish the future vision in a predefined time interval. Since level 0 represents the starting condition, the model does not have precise characteristics for this level.

Level 1 shows that the organization is in the starting process and exploring the Smart Grid technologies. At this level, the organization has a vision but does not have a clear strategy. At this level, the organization is capable to communicate its vision to the community and industry.

Level 2 shows the organization has a defined strategy and that it is already investing to attain the modernization of the electrical network. At this level tests are performed with the business case already implemented to assess the changes in the organization.

At level 3, the organization integrates its Smart Grid program with the operating departments. The procedures should be made repeatable and the information should be shared within the whole organization.

At level 4 the functionality and benefits of Smart Grid can be assessed. The organization performs analysis and makes corrections on real time.

At level 5 the organization is on a permanent innovation state, develops standards, and improves procedures. The organization becomes a leader of the industry. The vision and strategies of the organization fulfill national, regional, and local interests.

The organization will be in a higher maturity level when the implementation of changes starts. Each organization has to establish its own target maturity levels based on its own operating system, strategy, and timeline. It is obvious that higher levels of maturity in the model indicate a successful adoption from its grid modernization efforts. The target level is not the same for each organization. So achieving any level could be appropriate to an organization but not to other.

12.5.2   *Domains of SGMM*

SGMM has eight domains whose description is presented in the following paragraphs.

The Strategy, Management, and Regulatory (SMR) domain establishes the internal procedures of performance and governance of the organization and encourages support relationships with the groups responsible to implement the vision and strategy.

The Organization and Structure (OS) domain represents the capabilities and characteristics that allow an organization to plan and operate in order to achieve a Smart Grid in place.

For the transformation efforts to be successful, the organizational structure must promote and reward the planning and operation in various functions. This domain focuses on changes in communications, culture, structure, training, and education within the organization. Maturity in this particular domain evidences an increased organizational proficiency to develop their decision making initiatives focus on fact-based decision to meet its Smart Grid goals. Additionally, it exhibits the general vision of individuals committed to reach the Smart Grid goals.

The Grid Operations (GO) domain represents the functional practices that support reliable, safe, and efficient electric network operations. The organization employs new automation and communication solutions to improve key network elements' view and decrease control action response time. The information gathered from implemented Smart Grid solutions provides organizations, which have successfully achieved high maturity levels, with valuable information for automation. Also, it enables power flow management to reduce losses and maximize generation at reduced costs, and gaining higher levels of automation and larger view of the entire system. On the other hand, response times improve in communications and control avoiding cascading failures throughout the system. The benefits of such capabilities allow for larger grid improvements to reach the goal of serving customers with high-quality power, diversified generation, maximize asset utilization, and efficient operations.

The Work and Asset Management (WAM) domain represents the organization capabilities in managing assets and personnel. Maturity in the domain reflects improvements in predictive and reactive maintenance resulting in improving reliability, safety, and efficient operations. Advances in this domain represent an incremental ability to utilize information obtained from Smart Grid implementations to reduce causeless maintenance and out of service time, identify failure origin and indicate corrective actions, identify failures beforehand, minimize problem identification and solution time, and improve personnel resources and planning results.

The Technology (TECH) domain focuses on utilizing IT infrastructure that serves as a base to develop and support services that open new markets. It reflects compliance with relevant standards and integrates a strategic technology to connect and support various data sources and users (applications, systems, and persons).

Advances in the technology domain represent compliance with industry and government standards and integration of applications in Smart Grid with overwhelming data. The utilization of the IT structures promotes new business solutions and promising new markets.

The transformation to Smart Grid provides benefits to the organization that require efforts beyond the acquisition of new Smart Grid technologies which can provide a positive or negative support to the organization's efforts to materialize their Smart Grid plans. Smart Grid involves several technologies like wide-area monitoring, two-way digital communications, and advanced control. This comprehensive technological platform developed through advanced engineering and organization business initiatives requires processing large data sources and control systems that comprehend current Smart Grids and future applications.

The Customer (CUST) domain represents the organizational capabilities and characteristics that enable customer participation toward achieving the benefits of the Smart Grid transformation. The participation could be passive (the utility manages customer loads and select the power source) or active (provides tools to customer allowing them to manage the usage, sources, and energy cost based on cost and available options in the market).

Achieving important levels of maturity in the CUST domain represents big benefits for customer, allowing them to decide and manage power usage selecting from different sources and energy cost, maintaining network security, and protecting customer privacy. Additionally, exemplifying high levels of maturity in the domain represents organizational ability to fulfill utility, regional, and network wide goals, with the use of modern Smart Grids, with regards to grid stability, energy efficiency, peak load reduction, and green energy integration employing distributed generation while minimizing the usage of foreign resources.

The Value Chain Integration (VCI) domain exhibits the utility potential to advance in reaching the goals set to successfully capitalize the Smart Grid initiatives by integrating the different utility departments with the production and delivery of energy demands. Smart Grid automations go beyond traditional boundaries (sub-station automation) offering innovation in load management, distributed generation, and market operations. Smart Grid serves as a platform for planning, implementation, and energy management from generation to final user consumption.

As a result of environmental concerns and the need for increased efficiencies, market forces and regulatory bodies will again force the industry to change, providing new opportunities for organizations with Smart Grid competence and causing new value chains to emerge. Automation will extend beyond traditional boundaries and across the entire value chain to provide opportunities for innovation and efficiencies in load management, distributed generation, and market structure. As a utility matures, the cooperative planning, implementation, and management of electricity from the sources of production to end-use consumption will optimize profitability and improve performance of the utility's value chain. Networked information technology and data sharing, aligned with value chain business units' requirements, are critical for success.

Finally, the Societal and Environmental (SE) domain represents the organizational capabilities and characteristics that allow contributing to meet social goals associated with reliability and safety of the electric network infrastructure, type of energy sources employed, and the environmental impact and quality of life.

Proper application of these social initiatives benefits the organization and solidifies the connection with users and regulators. Efficient operation enabled by enhanced Smart Grid solutions represents higher profits while reducing environmental impact. Continued prevention and risk mitigation of network security events are an important part of Smart Grid implementations.

12.6   Results and analysis using SGMM

Once the results of the maturity model are obtained (the actual state and the desired future), a gap analysis is needed to determine what actions to take. Figure 12.2 helps to illustrate an example of results obtained by applying the SGMM.

It is important to mention that establishing a high future condition is not precluded by a low present rating. On the other hand, not aiming for a level of 5 in a particular domain does not mean that the organization is not focused on the Smart Grid technology. It is possible for the business model not to involve all aspects in the domain or for the organization not to consider it profitable to improve their level in that domain. Similar to the software industry, where ideas are born from maturity models, some organizations consider an optimization condition acceptable in the processes in a domain.

12.7   Example case

The methodology presented was applied to the evaluation of a real utility around 500,000 users. The utility has a vertical integration as it handles generation, transmission, distribution, and energy marketing. The results presented here refer to some of the questions of the SEI-SGMM model which correspond to the Grid Operations (GO) domain and the answers given by the utility.

GO-1.3
Do you have proof-of-concept projects and/or component testing for grid monitoring and control underway?
(A)   *No*
(B)   *To some extent, not directly for Smart Grid*
(C)   *To some extent, for Smart Grid*
(D)   *To a great extent (i.e., numerous evaluations underway or completed)*

GO-2.3
Aside from SCADA, are you piloting remote asset monitoring of key grid assets to support manual decision making?
(A)   *No*
(B)   *In documented plan including committed schedule and budget*
(C)   *Piloting*
(D)   *Pilots complete or technology being deployed*

GO-3.4
Have smart meters become important grid management sensors within your network?
(A)   *No*
(B)   *In documented plan including committed schedule and budget*
(C)   *Moderately (< 40% of grid is using meters as management sensors)*
(D)   *To a great extent ($\geq$ 40%)*

It can be observed in the questions above that the degree of difficulty increases with the levels. At the end, a score is obtained indicating the maturity level of the organization. Figure 12.3 shows the results from this organization. The black line indicates the required limits to satisfy the requirements of each level. For the case of the example, the values are established by a certification institution by using particular standards.

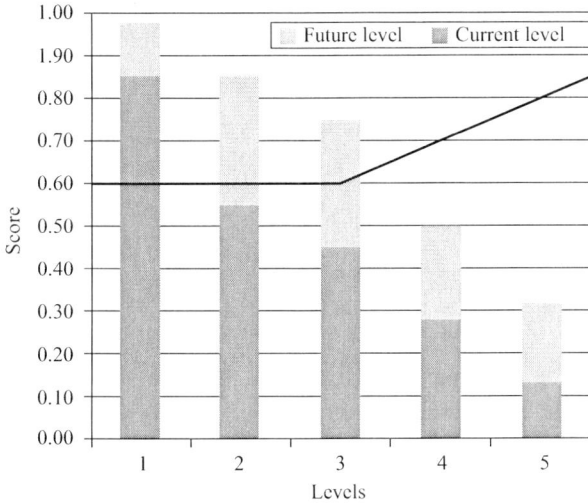

Figure 12.3 Example of results in Grid Operation (GO) domain

It is assumed that the criterion of the certification institution to pass the first three levels establishes a score minimum of 0.6 for each one. After the third level, higher score is required for approval (level 4 specifically requires at least 0.7 and level 5 requires 0.8). Once the score gets below one of the defined limits at a level, the total is then calculated with the summation of the scores obtained up to that level, including the score of the last one.

The GO domain for this organization indicates a maturity around 1.4, which indicates the organization manages higher levels of operation automation and network optimization processes. The example shows that the current condition (dark gray bars) only reach the black line in the first level with a 0.85 score (minimum is 0.6), and in the second level only 0.55 is achieved. Since the second level is the first level that does not approve the objective, the equivalent level will be the score summation up to this point (1.4). According to the results shown in Figure 12.3, this organization can reach at the most for the GO domain, the third level in 5 years. That could be acceptable for that company. In that case the total score would be 3.1, which is calculated from the results of the table (0.98 points for the first level, 0.85 for the second level, 0.75 for the third level, and 0.5 for the fourth level). The organization aims to improve its current condition by 1.7 points in a 5 year term for a total of 3.1.

This demonstrates that the organization currently has instituted Smart Grid initiatives where technologies and equipment have been elected for pilot testing, some of which may have concluded, and implementations are in process. For example, the distribution automation and the interoperable data system employ tools like CIM.

Once current and future state results have been obtained for the company under review, an analysis is conducted to evaluate what efforts must be made to reach such condition. It can also be defined based on the conditions the model expects for the particular level. For the particular example, efforts to reach the desired level must focus on the following:

- The organization must have functional-level approved business cases for improvements in the employee resources and Smart Grid assets in order to determine necessary changes in abilities of management personnel and assets, guaranteeing increased productivity and better prediction of events. They are designed to reduce maintenance due to faults and optimization of life cycle costs, ensuring availability. For the expected level, the business cases do not need to be integrated with the rest of the organization.
- In the asset management area, the organization must have finalized the remote supervision usage evaluation. Remote monitoring offers more than the basic control using SCADA. This implies that a dual real-time communication channel must be established to obtain detailed asset information in contrast with previously used performance reports that provided support to conduct corrective actions.

  At this level, the organization must have completed equipment and field personnel assessment along with focusing assets on the Smart Grid vision of the organization. This evaluation includes the geospatial equipment to connect assets with their geographical location. An example would be the field crew integration with the remote monitoring communication system.
- The organization should invest in technologies to support asset monitoring and field crew performance. A documented plan must be established to explore Smart Grid capabilities in order to create inventories and provide for event record keeping and asset follow-up. The equipment history should include tendencies and profile information based on actual data. Asset supervision must be developed by trial testing, and an integrated view of the GIS, state, and connectivity.
- Additionally, the organization should develop a strategy for field crew operations with the aim of generating a significant impact to the organization. It means the organization has recognized the need to optimize mobile assets with specific performance objectives. Ideally, the field crew strategy must be interconnected with smart networks for higher performance.
- The organization should have started implementing Smart Grid solutions with asset management and field crews to employ the information available.
- The organization must have performance data, tendency data, and event control for at least 25% of individual components; computer systems are SCADA and Remote Terminals (RTU); and physical elements include breakers, transformers, meters, etc., that support generation, transmission, and distribution.
- A maintenance plan must be implemented for a small percentage of equipment (smaller than 25%) to pilot the maintenance and replacement plans which will be developed based on real-time data obtained from an asset supervision program. This allows equipment capacity prediction and prevents damage to the system. It is important to integrate remote asset supervision with asset management while

they are both in the planning stages. Integration at this level might be the ability to automate the order being processed and to relocate the crews based on the requirements in the network. GIS integration with the asset supervision system may be conducted based on location, state, and interconnectivity in order to improve the visual operability in at least one type of asset.

- The organization must make an effort to have a follow-up process with some level of automation for a small percentage (around 1–25%) of assets from when they leave the factory until they reach their destination. The follow-up must include information like asset location, usage state, scheduled maintenance date, and end-of-service date. Bar code and RFID (radio frequency identification) are among the available technologies to accomplish this.

Table 12.1   Example of results after applying the SGMM

Domains	Current maturity level	Future maturity level (5 years)
SMR	2.2	4.7
OS	2.6	3.6
GO	1.4	3.1
WAM	1.3	2.7
TECH	1.5	3.5
CUST	2.2	3.2
VCI	1.6	2.5
SE	1.5	3.5
Mean maturity level	**1.80**	**3.36**

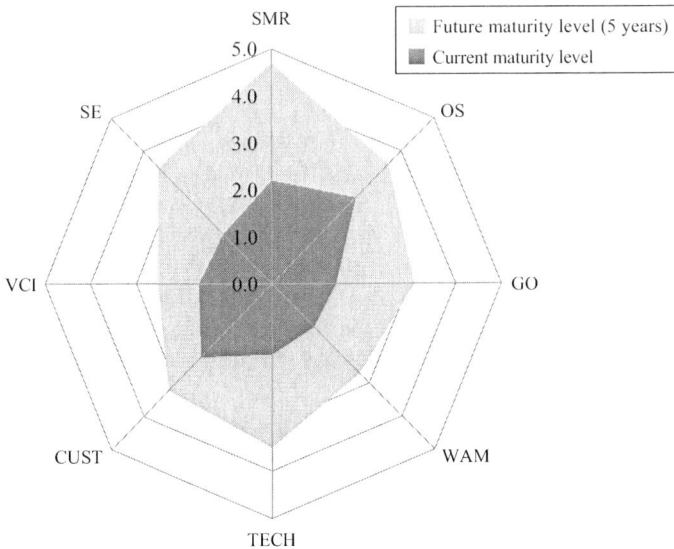

Figure 12.4   Example of results in a radar chart

- The organization must have a documented plan to develop an asset investment model for key components based on Smart Grid data that allow integration and developments in asset control.

Table 12.1 summarizes the maturity level in each of the domains discussed in the model for this example, and Figure 12.4 shows a graphical representation of this score. Based on the results obtained, an analysis will identify the gaps to be improved. The distance from actual to future in the evaluated conditions represents the efforts required to reach the desired future condition. The larger the gap between actual and future conditions (radar chart area differences), the more aspects need to be improved. Based on the estimated time and budget, necessary efforts must be defined and the probability of success evaluated.

Bibliography

ABB: 'Distribution protection', ABB buyer's guide, vols. I, II, and III, 1991–1992

Aguilar, R. and Ariza, J.: 'Testing and configuration of IEC 61850 multivendor protection schemes', *IEEE Transmission and Distribution Conference and Exposition*, New Orleans, LA, 2010

Alstom: *Network Protection & Automation*, 1st edn, Cayfosa, Barcelona, Spain, July 2002

Anderson, P.M.: *Analysis of Faulted Power Systems*, The Iowa State University Press, Ames, IA, 1995

Aravinthan, V., Karimi, B., Namboodiri, V., Jewell, W.: 'Wireless communication for smart grid applications at distribution level – Feasibility and requirements', *Power and Energy Society General Meeting, 2011 IEEE*, pp. 1, 8, 24–29, July 2011

Arrillaga, J. and Arnold, C.P.: *Computer Analysis of Power Systems*, John Wiley & Sons, New Delhi, 1990

Arrillaga, J., Bradley, D.A., and Bodger, P.S.: *Power System Harmonics*, John Wiley & Sons, New Delhi, 1985

Baghzouz, Y.: 'Effects of nonlinear loads on optimal capacitor placement in radial feeders', *IEEE Transactions on Power Delivery,* vol. 6, no. 1, pp. 245–251, January 1991

Baldini and Mansell: *Protective Relays Application Guide*, GEC Measurements, 3rd edn., The General Electric Company, Stafford, England, 1987

Baran, M.E. and Wu, F.F.: 'Network reconfiguration in distribution systems for loss reduction and load balancing', *IEEE Transactions on Power Delivery*, vol. 4, no. 2, pp. 1401–1407, April 1989

Baran, M.E. and Wu, F.F.: 'Optimal capacitor placement on radial distribution systems', *IEEE Transactions on Power Delivery*, vol. 4, no. 1, pp. 725–734, January 1989

Basler Electric: *'Instruction Manual for Overcurrent BE-951,* Publication 9 3289 00 990, Revision F, Basler Electric, Highland, IL, August 2002

Basler Electric: 'Protection and control devices standards, dimensions and accessories', *Product Bulletin*, SDA-5 8-01, Highland, IL, August 2001

Beckwith Electric Co.: *Instruction Book M-7651 Feeder Protection Relay*, Publication 800-7651-SP-00, Beckwith Electric Co., Largo, FL, June 2010

Bennett, C. and Wicker, S.B.: 'Decreased time delay and security enhancement recommendations for AMI smart meter networks', *Innovative Smart Grid Technologies (ISGT), 2010*, pp. 1–6, 19–21, January 2010

Billinton, R. and Jonnavithula, S.: 'Optimal switching device placement in radial distribution systems', *IEEE Transactions on Power Delivery*, vol. 11, no. 3, pp. 1646–1651, July 1996

Blackburn, J.I.: *Protective Relaying Principles and Applications*, 3rd edn, CRC Press, Boca Raton, FL, 2008

Blackburn, J.L.: *Protective Relaying*, Marcel Dekker, New York, NY, 2nd edn, 1998

Borbely, A. and Kreider, J.F.: *Distributed Generation – The Power Paradigm for the New Millennium*, CRC Press, Boca Raton, FL, 2001

Borozan, V., Rajicic, D., and Ackovski, R.: 'Minimum loss reconfiguration of unbalanced distribution networks', *IEEE Transactions on Power Delivery*, vol. 12, no. I, pp. 435–442, January 1997

Broadwater, R.P. and Thompson, J.C.: 'Computer aided protection system design with reconfiguration', *IEEE Transactions on Power Delivery*, vol. 6, no. 1, pp. 260–266, January 1991

Brown, R.E.: *Electric Power Distribution Reliability*, 2nd edn, CRC Press, Boca Raton, FL, 2009

Burnett, J.: 'IDMT relay tripping of main incoming circuit breakers', *Power Engineering Journal*, 1990, vol. 4, pp. 51–56

Cadick, J.: 'Condition Based Maintenance – How to Get Started', *Cadick Corporation*, pp. 1–11, 1999

Cespedes, R.: 'New method for the analysis of distribution networks', *IEEE Transactions on Power Delivery*, vol. 5, no. 1, pp. 391–396, January 1990

Chakrabarti, S., Kyriakides, E., Bi, T., Cai, D., and Terzija, V.: 'Measurements get together', *Power and Energy Magazine, IEEE*, vol. 7, no. 1, pp. 41–49, January/February 2009

Chandrasekaran, B., Josephson, J.R., Benjamins, V.R.: 'What are ontologies, and why do we need them?', *Intelligent Systems and their Applications, IEEE*, vol. 14, no. 1, pp. 20,26, January/February 1999

Chang, N.E.: 'Determination of primary feeder losses', *IEEE Transactions on Power Apparatus and Systems*, vol. PAS 87, no. 12, pp. 1991–1994, 1968

Chen, A.C.M.: 'Power/energy: Automated power distribution: Increasingly diverse and complex power operation and distribution systems will mean a larger role for microprocessor and communications technologies', *Spectrum, IEEE*, vol. 19, no. 4, pp. 55–60, April 1982

Chen, C.S. and Cho, M.Y.: 'Energy loss reduction by critical switches', *IEEE Transactions on Power Delivery*, vol. 8, no. 3, pp. 1246–1253, July 1993

Cheng, D.K.: *Analysis of Linear Systems*, Addison-Wesley, USA, 1972

Cho, M.Y. and Chen, Y.W.: 'Fixed-switched type shunt capacitor planning of distribution systems by considering customer load patterns and simplified feeder model', *IEE Proceedings on Generation, Transmission and Distribution*, vol. 144, no. 6, pp. 533–540, November 1997

Choi, D., Kim, H., Won, D., and Kim, S.: 'Advanced key-management architecture for secure SCADA communications', *IEEE Transactions on Power Delivery*, vol. 24, no. 3, pp. 1154–1163, July 2009

CIP-002-1: *Cyber Security — Critical Cyber Asset Identification*, NERC Standard, 2006

CIP-009-2: *Cyber Security — Recovery Plans for Critical Cyber Assets*, NERC Standard, 2006

Civanlar, S., Grainger, H., Yin, H., and Lee, S.: 'Distribution feeder reconfiguration for loss reduction', *IEEE Transactions on Power Delivery*, vol. 3, pp. 1217–1223, 1988

Closson, J. and Young, M.: 'Commissioning numerical relays', *XIV IEEE Summer Meeting*, Acapulco, Mexico, July 2001

Cook, R.F.: 'Analysis of capacitor application as affected by load cycle', Power apparatus and systems, part III. *Transactions of the American Institute of Electrical Engineers*, vol. 78, no. 3, pp. 950–956, April 1959

Cook, R.F.: 'Calculating loss reduction afforded by shunt capacitor application', *IEEE Transactions on Power Apparatus and Systems*, vol. 83, no. 12, pp. 1227–1230, December 1964

Cook, R.F.: 'Optimizing the application of shunt capacitors for reactive-volt-ampere control and loss reduction', Power apparatus and systems, part III. *Transactions of the American Institute of Electrical Engineers*, vol. 80, no. 3, pp. 430–441, April 1961

Cooper Power Systems: *Electrical Distribution System Protection Manual*, 3rd edn, Cooper Power Systems, Pittsburgh, PA, 1990

Cooper Power Systems: *How Step-voltage Regulators Operate*, Bulletin 77006, Supersedes 10/91. USA, February 1993

Cooper Power Systems: *Voltage Regulator*, Bulletin B225-97020, Supersedes 4/99. USA, July 2005

Davies, T.: *Protection of Industrial Power Systems*, Pergamon Press, Cleveland, OH, 1983

De Vos, A., Widergren, S.E., and Zhu, J.: 'XML for CIM model exchange', Power Industry Computer Applications, 2001. PICA 2001. *22nd IEEE Power Engineering Society International Conference on Innovative Computing for Power – Electric Energy Meets the Market*, pp. 31–37, 2001

EI, AEIC, and UTC, 'Smart meters and smart meter systems: A metering industry perspective', 2011

Electrical Power Research Institute: *Guidebook for Cost/Benefit Analysis of Smart Grid Demonstration Projects: Volume 1, Measuring Impacts*, Electrical Power Research Institute, Palo Alto, CA, 2011

Electricity Training Association: 'Power system protection edited by The Electricity Training Association', *Power Engineering Review, IEEE*, vol. 15, no. 9, pp. 33, September 1995, doi: 10.1109/MPER.1995.410272

Elmore, W.A.: *Protective Relaying Theory and Applications*, ABB, New York, NY, 1994

European Commission: *Advanced Architectures and Control Concepts for More Microgrids*, ICCS Company, Athens, Greece, 2009

European Commission: *Annex C: Some Relevant Standards*, Expert Group 1 (EG1) of the EC Task Force for Smart Grids (TF), pp. 55–69, 2010

European Commission: *Europe and Wind Energy at the Dawn of the 21st Century*, European Union, Belgium, May 2011

European Commission: *First Interim Evaluation of the Fuel Cell & Hydrogen Joint Undertaking, Expert Group Report*, European Union, Luxembourg, May 2011

EU Commission Task Force for Smart Grids Expert Group 1: *Functionalities of Smart Grids and Smart Meters, Final Deliverable*, European Union, Expert Group 1 (EG1) of the EC Task Force for Smart Grids (TF), December 2010

European Commission: *International Initiatives Related to Smart Grid Standardisation: State of the Art*, EU Commission Task Force for Smart, Expert Group 1 (EG1) of the EC Task Force for Smart Grids (TF), pp. 25–38, 2010

European Environment Agency: *Europe's Onshore and Offshore Wind Energy Potential, an Assessment of Environmental and Economic Constraints – Technical Report*, European Environment Agency, Copenhagen, 2009

European Technology Platform Smartgrids: *Strategic Research Agenda for Europe's Electricity Networks of the Future*, European Commission, Community Research, Brussels, EUR 22580, 2007

Feenan, J.: 'The versatility of high-voltage fuses in the protection of distribution systems', *Power Engineering Journal*, vol. 1, pp. 109–115, 1987

Gaonkar, D.N.: *Distributed Generation*, In-Teh, India, February 2010

GEC Alsthom: *Protective Relays Application Guide*, 3rd edn, Balding & Mansell, London, September 1990

Gelling, C.: *The Smart Grid: Enabling Energy Efficiency and Demand Response*, The Fairmont Press, Estados Unidos, 2009

Gers, J.M.: 'Enhancing numerical relaying performance with logic customization', *XV IEEE Summer Meeting*, Acapulco, Mexico, July 2002

Gers, J.M. and Holmes, E.J.: Protection of Electricity Distribution Networks, 3rd edn, The Institution of Engineering and Technology, Hertfordshire, UK 2011

Glover, J.D. and Sarma, M.S.: *Power System Analysis and Design*, 3rd edn., Brooks/Cole, Pacific Grove, CA, 2002

Gonen, T.: *Electric Power Distribution System Engineering*, 2nd edn., CRC Press, Boca Raton, FL, 2008

Goraj, M. and Herrmann, J.: 'Experiences in IEC 61850 and possible improvements of SCL languages', *Praxis Profiline*, p. 60, April 2007

Goswami, S.K.: 'Distribution system planning using branch exchange technique', *IEEE Transactions on Power Systems*, vol. 12, no. 2, pp. 718–723, May 1997

Goswami, S.K. and Basu, S.K.: 'A new algorithm for the reconfiguration of distribution feeders for loss minimization', *IEEE Transactions on Power Delivery*, vol. 7, no. 3, pp. 1484–1491, 1992

Grainger, J.J. and Civanlar, S.: 'Volt-var control on distribution systems with lateral branches using shunt capacitors and voltage regulators, part I – The overall problem', *IEEE Transactions on Power Apparatus and Systems*, vol. PAS-104, no. 11, pp. 3278–3283, November 1985

Grainger, J.J. and Civanlar, S.: 'Volt-Var Control on Distribution Systems with lateral branches using shunt capacitors and voltage regulators, part II – The solution method', *IEEE Transactions on Power Apparatus and Systems*, vol. PAS-104, no. 11, 3284–3290, November 1985

Grainger, J.J. and Civanlar, S.: 'Volt-var control on distribution systems with lateral branches using shunt capacitors and voltage regulators, part III – The numerical results', *IEEE Transactions on Power Apparatus and Systems*, vol. PAS-104, no. 11, pp. 3291–3297, November 1985

Gregerson, S. and Toporek, D.: 'The Doctor Is In', *Power and Energy Magazine, IEEE*, vol. 8, no. 6, pp. 45, 47, November–December 2010

Gross, C.A.: *Power System Analysis*, 2nd edn, John Wiley & Sons, New York, NY, 1986

Harker, K.: *Power System Commissioning and Maintenance Practice*, Peter Peregrinus, Hertfordshire, 1998

Headley, A., Burdis, E.P., and Kelsey, T.: 'Application of protective devices to radial overhead line networks', *IEE Proceedings on Generation, Transmission and Distribution*, vol. 133, pp. 437–440, 1986

Holbach, J. and Dufaure, T.: 'Comparison of IEC 61850 GOOSE messages and control wiring between protection relays'. *62nd Annual Georgia Tech Protective Relaying Conference*, Atlanta, GA, 2008

IEC 60870: *Communication Networks and Systems in Substations*, 6 Parts, 1988–2000

IEC 61850 edn 1: *Communication Networks and Systems in Substations*, 14 Parts, 2003–2005

IEC 61850 edn 2: *Communication Networks and Systems for Power Utility Automation Scheduled for 2010*

IEC 61968: *Common Information Model (CIM)/Distribution Management*

IEC 61970: *Common Information Model (CIM)/Energy Management*

IEC 62056: *Electricity Metering – Data Exchange for Meter Reading*, tariff and load control, 7 Parts, 2002–2007

IEC/PAS 62559 edn 1: *IntelliGrid Methodology for Developing Requirements for Energy Systems*

IEE Conference Publication: 'Developments in power system protection', *7th International Conference*, vol. 479, 2001

IEEE: 'Microprocessor relays and protection systems', Tutorial Course, 88EH0269-1-PWR, 1987

IEEE: *Guides and Standards for Protective Relaying Systems*, IEEE, New York, NY, 1995

IEEE Std 18-1992: *IEEE Standard for Shunt Power Capacitors*

IEEE Std 141-1993: *IEEE Recommended Practice for Electric Power Distribution for Industrial Plants*

IEEE Std 242-2001 (Revision of IEEE Std 242-1986): *IEEE Recommended Practice for Protection and Coordination of Industrial and Commercial Power Systems (IEEE Buff Book)*

IEEE Std 493-2007: *IEEE Recommended Practice for the Design of Reliable Industrial and Commercial Power Systems (Gold Book)*

IEEE Std 519-1992: *IEEE Recommended Practices and Requirements for Harmonic Control in Electrical Power Systems*

IEEE Std 551-2006: *IEEE Recommended Practice for Calculating AC Short-Circuit Currents in Industrial and Commercial Power Systems (Violet Book)*

IEEE Std 1159-2009 (Revision of IEEE Std 1159-1995): *IEEE Recommended Practice for Monitoring Electric Power Quality*

IEEE Std 1366-2012 (Revision of IEEE Std 1366-2003): *IEEE Guide for Electric Power Distribution Reliability Indices*

IEEE Std 2030: *IEEE Guide for Smart Grid Interoperability of Energy Technology and Information Technology Operation with the Electric Power System (EPS), End-Use Applications, and Loads*

IEEE Std C37.1-2008: *IEEE Standard for SCADA and Automation Systems*

IEEE Std C57.13-2008 (Revision of IEEE Std C57.13-1993): *Standard Requirements for Instrument Transformers*

IEEE Std C37.118-2005 (Revision of IEEE Std 1344-1995): *Standard for Synchrophasors for Power Systems*

IEEE Std C37.118.1-2011 (Revision of IEEE Std C37.118-2005): *Standard for Synchrophasor Measurements for Power Systems*

IEEE 802.15.4j: 'IEEE Std for local and metropolitan area networks - Part 15.4: Low-rate wireless personal area networks (LR-WPANs) amendment 4: Alternative physical layer extension to support medical body area network (MBAN) services operating in the 2360 MHz 2400 MHz Band', *IEEE Std 802.15.4j-2013 (Amendment to IEEE Std 802.15.4-2011 as amended by IEEE Std 802.15.4e-2012, IEEE Std 802.15.4f-2012, and IEEE Std 802.15.4g-2012)*, pp. 1, 24, February 2013

IEEE 1815: 'IEEE Std for electric power systems communications-distributed network protocol (DNP3)', *IEEE Std 1815–2012 (Revision of IEEE Std 1815–2010)*, vol., no., pp. 1, 821, October 2012

Ipakchi, A. and Albuyeh, F.: 'Grid of the future', *Power and Energy Magazine, IEEE*, vol. 7, no. 2, pp. 52–62, March–April 2009

ISO/IEC 8802-3 edn 6: *Information Technology – Telecommunications and Information Exchange between Systems – Local and Metropolitan Area Networks – Specific Requirements*, 2000

ISO 27002: *Information Technology – Security Techniques – Code of Practice for Information Security Management*, 2005

ISO 15408: *Information Technology – Security Techniques – Evaluation Criteria for IT Security*, 3 Parts. 2008–2009

Jonnavithula, S. and Billinton, R.: 'Minimum cost analysis of feeder routing in distribution system planning', *IEEE Transactions on Power Delivery*, vol. 11, no. 4, pp. 1935–1940, October 1996

Kasztenny, B., Whatley, J., and Udren, E.A.: 'IEC 61850: A practical application primer for protection engineer', *60th Annual Georgia Tech Protective Relaying Conference*, Atlanta, GA, 2006

Kennedy, B.W.: *Power Quality Primer*. Primer Series. McGraw-Hill, USA, 1976

Kezunovic, M.: 'Smart fault location for smart grids', *IEEE Transactions on Smart Grid*, vol. 2, no. 1, pp. 11, 22, March 2011

Lakervi, E. and Holmes, E.J.: *Electricity Distribution Network Design*, 2nd edn, Peter Peregrinus, The Institution of Engineering and Technology, Hertfordshire, UK, 1995; revised 2003

Laycock, W.J.: 'Management of protection', *Power Engineering Journal*, vol. 5, pp. 201–207, 1991

Lee, R.E. and Brooks, C.L.: 'A method and its application to evaluate automated distribution control', *IEEE Transactions on Power Delivery*, vol. 3, no. 3, July 1988

Lin, C., Chen, C., Ku, T., Tsai, C., and Ho, C.: 'A multiagent-based distribution automation system for service restoration of fault contingencies', *European Transactions on Electrical Power*, vol. 21, pp. 239–253, June 2010

Lo, K.L. and Gers, J.M.: 'Feeder reconfiguration for losses reduction in distribution systems', *Proceedings of the UPEC 94 Conference*, University College Galway, Ireland, September 1994

Lo, K.L. and Nashid, L.: 'Expert systems and their applications to power systems, Part 3 Examples of application', *Power Engineering Journal*, vol. 7, no. 5, pp. 209–213, 1993

Lo, K.L., McDonald, J.R., and Young, D.J.: 'Expert systems applied to alarm processing in distribution control centres', *Conference proceedings*, UPEC 89, Belfast, 1989

Long, W., Cotcher, D., Ruiu, D., Adam, P., Lee, S., and Adapa, R.: 'EMTP – A powerful tool for analysing power system transients', *Computer Applications in Power, IEEE*, vol. 3, no. 3, pp. 36, 41, July 1990

Mavrommati, I. and Darzentas, J.: 'An overview of AMI from a user centered design perspective', *2nd IET International Conference on Intelligent Environments, 2006* (IE 06), vol. 2, pp. 81–88, July 5–6, 2006

Maxwell, M.: 'The economic application of capacitors to distribution feeders', *Power Apparatus and Systems, Part III. Transactions of the American Institute of Electrical Engineers*, vol. 79, no. 3, pp. 353–358, April 1960

McGranaghan, M.F., Dugan, R.C., King, J.A., and Jewell, W.T.: 'Distribution feeder harmonic study methodology', *IEEE Transaction on Power Apparatus and Systems*, vol. PAS-103, no. 12, pp. 3663–3671, December 1984

McGranaghan, M.F., Dugan, R.C., and Sponsler, W.L.: 'Digital simulation of distribution system frequency-response characteristics', *IEEE Transactions on Power Apparatus and Systems*, vol. PAS-100, no. 3, pp. 1362–1369, March 1981

McMorran, A.: *An Introduction to IEC 61970-301 & 61968-11: The Common Information Model*, Institute for Energy and Environment, University of Strathclyde. Glasgow, Scotland, January 2007

Meliopoulos, A.P.S., Cokkinides, G.J., Huang, R., Farantatos, E., Choi, S., Lee, Y., and Yu, X.: 'Smart grid technologies for autonomous operation and control', *IEEE Transactions on Smart Grid*, vol. 2, no. 1, pp. 1, 10, March 2011

Merlin, A. and Back, G.: 'Search for a minimal-loss operating spanning tree configuration in an urban power distribution system', *Proceedings of the Fifth Power System Conference (PSCC)*, Cambridge, pp. 1–18, 1975

Mohd, A., Ortjohann, E., Schmelter, A., Hamsic, N., and Morton, D: 'Challenges in integrating distributed energy storage systems into future smart grid', *IEEE International Symposium on Industrial Electronics*, (ISIE 2008), pp. 1627–1632, June 30–July 2, 2008

Momoh, J.: *Electric Power Distribution, Automation, Protection, and Control*, 1st edn, CRC Press, Boca Raton, FL, September 2008

Momoh, J.: Smart Grid: Fundamentals of Design and Analysis, *Wiley-IEEE* Press, Hoboken, NJ, 2012

Munasinghe, M.: 'Economic principles and policy electricity loss reduction', *Latin American Seminar on Electrical Losses Control*, Bogota DE, 1988

National Energy Technology Laboratory: *A Compendium of Modern Grid Technologies*, U.S. Department of Energy, NETL Modern Grid Initiative, USA, 2007

National Institute of Standards and Technology: *NIST Smart Grid Standards Roadmap*, National Coordinator for Smart Grid Interoperability, USA, 2009

National Institute of Standards and Technology: *IEC 61850 Objects/DNP3 Mapping*, NIST, USA, 20090730, 2009–2010

Neagle, N.M. and Samson, D.R.: 'Loss reduction from capacitor installed on primary feeders', *Power Apparatus and Systems, Part III. Transactions of the American Institute of Electrical Engineers*, vol. 75, no. 3, pp. 950–959, January 1956

NEPLAN, *NEPLAN User Guide for Version 5.5.2*, 2013

Northcote-Green, J. and Wilson, R.: *Control and Automation of Electrical Power Distribution Systems*, Taylor & Francis, Boca Raton, FL, 2007

Office of the National Coordinator for Smart Grid Interoperability: *NIST Framework and Roadmap for Smart Grid Interoperability Standards, Release 1.0*, NIST Special Publication 1108, USA, 2010

Ortmeyer, T.H., Hammam, M.S.A.A., Hiyama, T., and Webb, D.B.: 'Measurement of the harmonic characteristics of radial distribution systems', *Power Engineering Journal*, vol. 2, no. 3, pp. 163–172, May 1988

Parikh, P.P., Kanabar, M.G., and Sidhu, T.S.: 'Opportunities and challenges of wireless communication technologies for smart grid applications', *Power and Energy Society General Meeting, 2010 IEEE*, pp. 1, 7, 25–29, July 2010

Peponis, G.J., Papadopoulos, M.P., and Hatziargyriou, N.D.: 'Distribution network reconfiguration to minimize resistive line losses', *IEEE Transactions on Power Delivery*, vol. 10, no. 3, pp. 1338–1342, July 1995

Pereira, R.A.F., Da Silva, L.G.W., Kezunovic, M., and Mantovani, J.R.S.: 'Improved fault location on distribution feeders based on matching during-fault voltage sags', *IEEE Transactions on Power Delivery*, vol. 24, no. 2, pp. 852, 862, April 2009

Phadke, A.G. and Horowitz, S.H.: 'Adaptive relaying', *IEEE Computer Applications in Power*, July 1990

Power System Relaying Committee Working Group H6: *Application Consideration of IEC 61850/UCA2 for Substation Ethernet Local Area Network Communication for Protection and Control*, 2005

Prabhakara, F.S., Smith, R.L., Stratford, R.P.: *Industrial and Commercial Power System Handbook*, 1st edn, McGraw-Hill Professional, New York, NY, October 1995

Pradeep, Y., Seshuraju, P., Khaparde, S.A., and Joshi, R.K.: 'CIM-based connectivity model for bus-branch topology extraction and exchange', *IEEE Transactions on Smart Grid*, vol. 2, no. 2, pp. 244, 253, June 2011

Press, W.M., Teukolsky, S.A., Vetterling, W.T., Flannery, B.P., and Metcalf, M.: *Numerical Recipes in FORTRAN*, Cambridge University Press, England, 1986

Rajagopalan, S.: 'A new computational algorithm for load flow study of radial distribution systems', Computer & Electrical Engineering, vol. 5, pp. 225–231, Pergamon Press, Oxford, 1978

Rizy, D.T., Gunther, E.W., and McGranaghan, M.F.: 'Transient and harmonic voltages associated with automated capacitor switching on distribution systems', *Power Engineering Review, IEEE*, vol. PER-7, no. 8, pp. 49, 50, August 1987

Sabin, D., McGranaghan, M., and Sundaram, A.: 'A systems approach to power quality monitoring for performance assessment', *Proceedings of the 7th Annual Canadian Electricity Forum on Power Quality and Power Harmonics*, Toronto, Canada, May 1997

Sallam, A.A. and Malik, O.P.: *Electric Distribution Systems*, John Wiley & Sons, Hoboken, NJ, 2011

Sarfi, R.J., Salama, M.M.A., and Chikhani, A.Y.: 'Distribution system reconfiguration for loss reduction – An algorithm based on network partitioning theory', *IEEE Transactions on Power Systems*, vol. 11, no. 1, pp. 504–510, February 1996

Schmill, J.V.: 'Optimum size and location of shunt capacitors on distribution feeders', *IEEE Transactions on Power Apparatus and Systems*, vol. 84, no. 9, pp. 825–832, September 1965

Shaibon, H., Mohd-Zin, A.A., Lim, Y.S., and Lo, K.L.: 'Loss minimisation using islanding technique for district of Klang', *IEE Proceedings on Generation, Transmission and Distribution*, vol. 142, no. 5, pp. 523–526, September 1995

Shirmohammady, D. and Hong, H.W.: 'Reconfiguration of electric distribution networks for resistive line losses reduction', *IEEE Transactions on Power Apparatus and Systems*, vol. 4, pp. 1492–1498, 1989

Short, T.A., *Electric Power Distribution Handbook*, CRS Press, Boca Raton, FL, 2004

Shreyas, A.: Analysis of Communication Protocols for Neighborhood Area Network for Smart Grid, California State University, Sacramento, CA, 2010

Siemens: *Switching, Protection and Distribution in Low Voltage Networks*, 2nd edn, Publicist MCD Verlag, Germany, 1994

SMB Smart Grid Strategic Group: *IEC Smart Grid Standardization Roadmap*, pp. 14–32, 2010

Software Engineering Institute: *SGMM Compass Assessment Survey*, Version 1.2. CERT® Program Research, Technology, and System Solutions Program, Software Engineering Process Management Program, Carnegie Mellon University, 2011

Software Engineering Institute: *SGMM Model Definition. A Framework for Smart Grid Transformation*, Version 1.2. Technical Report CMU/SEI-2011-TR-025, ESC-TR-2011-025. CERT® Program Research, Technology, and System Solutions Program, Software Engineering Process Management Program, Carnegie Mellon University, USA, September 2011

Solanki, J.M., Khushalani, S., and Schulz, N.N.: 'A multi-agent solution to distribution systems restoration', *IEEE Transactions on Power Systems*, vol. 22, no. 3, pp. 1026,1034, August 2007

Song, Y.H., Wang, G.S., Johns, A.T., and Wang, P.Y.: 'Distribution network reconfiguration for loss reduction using fuzzy controlled evolutionary programming', *IEE Proceedings on Generation, Transmission and Distribution*, vol. 144, no. 4, pp. 345–350, July 1997

Sortomme, E., Mapes, G.J., Foster, B.A., and Venkata, S.S.: 'Fault analysis and protection of a microgrid'. *40th North American Power Symposium, 2008 (NAPS '08)*, Calgary, Alberta, pp. 1–6, 28–30, September 2008

Stott, B.: 'Decoupled Newton load flow', *IEEE Transactions on Power Apparatus and Systems*, vol. PAS-91, no. 5, pp. 1955, 1959, September 1972

Stott, B.: 'Review of load-flow calculation methods', *Proceedings of the IEEE*, vol. 62, no. 7, pp. 916, 929, July 1974

Stott, B., Alsac, O.: 'Fast decoupled load flow', *IEEE Transactions on Power Apparatus and Systems*, vol. PAS-93, no. 3, pp. 859, 869, May 1974

Sui, H., Wang, H., Lu, M.-S., and Lee, W.-J.: 'An AMI system for the deregulated electricity markets', *IEEE Transactions on Industry Applications*, vol. 45, no. 6, pp. 2104–2108, November–December 2009

Taylor, T. and Kazemzadeh, H.: 'Integrated SCADA/DMS/OMS: Increasing distribution operations efficiency', *Electric Energy T&D*, no. I, pp. 31–34, March–April 2009

Teo, C.Y. and Chan, T.W.: 'Development of computer-aided assessment for distribution protection', *Power Engineering Journal*, vol. 4, pp. 21–27, 1990

Tosic, V. and Djordjevic-Kajan, S.: 'The common information model (CIM) standard – An analysis of features and open issues', *Proceedings of TELSIKS '99 – 4th International Conference on Telecommunications in Modern Satellite, Cable, and Broadcasting Services*, Nis, Yugoslavia, October 1999, vol. 2, pp. 677–680, 1999

Uluski, R.W.: 'Economic justification of DA: The benefit side', *Power Engineering Society General Meeting, 2007 IEEE*, Tampa, FL, pp. 1, 7, 24–28, June 2007

Uluski, R.W.: 'The role of advanced distribution automation in the smart grid', *Power and Energy Society General Meeting, 2010 IEEE*, Minneapolis, MN, pp. 1, 5, 25–29, July 2010

Uluski, R.W.: 'Using distribution automation for a self-healing grid', *Transmission and Distribution Conference and Exposition (T&D), 2012 IEEE PES*, Orlando, FL, pp. 1, 5, 7–10, May 2012

Uluski, R.W.: 'VVC in the smart grid era', *Power and Energy Society General Meeting, 2010 IEEE*, Minneapolis, MN, pp. 1, 7, 25–29, July 2010

Urdaneta, A.J., Restrepo, H., Marquez, J., and Sanchez, J.: 'Co-ordination of directional overcurrent relay timing using linear programming', *IEEE PAS Winter Meeting*, New York, NY, February 1995

U.S. Department of Energy: *Home Area Networks and the Smart Grid*, Edited by Battelle in Richland, Washington, USA, 2011

U.S. Department of Energy: *Smart Grid System Report*, Edited by U.S. Department of Energy, USA, July 2009

U.S. Department of Energy: *Smart Grid System Report – Annex A and B*, Edited by U.S. Department of Energy, USA, July 2009

U.S. Department of Energy: *The Smart Grid Stakeholder Roundtable Group Perspectives*, Edited by AllianceOne, USA, September 2009

Venkata, S.S.: *Distribution Automation – Course Notes*, Minneapolis, MN, September 2007

Wagner, T.P. and Chikhani, A.Y.: 'Feeder reconfiguration for loss reduction: An application of distribution automation', *IEEE Transactions on Power Delivery*, vol. 6, no. 4, pp. 1922–1931, 1991

Westinghouse/ABB Power T&D Co.: *Protective Relaying Theory and Application,* Marcel Dekker, Inc., copyright, USA, 1994

Wright, A.: 'Application of fuses to power networks', *Power Engineering Journal*, vol. 4, pp. 293–296, 1991, ibid. vol. 5, pp. 129–134, 1991

Wright, A. and Newberry, P.G.: *Electric Fuses*, 2nd edn, Peter Peregrinus, Hertfordshire, UK, 1994

Zhou, Q., Shirmohammadi, D., and Edwin-Liu, W.H.: 'Distribution feeder reconfiguration for operation cost reduction', *IEEE Transactions on Power Systems*, vol. 12, no. 2, pp. 730, 735, May 1997

Zhou, Q., Shirmohammadi, D., and Edwin-Liu, W.H.: 'Distribution feeder reconfiguration for service restoration and load balancing', *IEEE Transactions on Power Systems*, vol. 12, no. 2, pp. 724, 729, May 1997

Index